Current Research on Probiotics and Fermented Products

Current Research on Probiotics and Fermented Products

Editors

Xiqing Yue
Mohan Li

MDPI

Basel • Beijing • Wuhan • Barcelona • Belgrade • Novi Sad • Cluj • Manchester

Editors
Xiqing Yue
Shenyang Agricultural
University
Shenyang
China

Mohan Li
Shenyang Agricultural
University
Shenyang
China

Editorial Office
MDPI AG
Grosspeteranlage 5
4052 Basel, Switzerland

This is a reprint of articles from the Special Issue published online in the open access journal *Foods* (ISSN 2304-8158) (available at: https://www.mdpi.com/journal/foods/special_issues/Probiotics_Fermented_Products).

For citation purposes, cite each article independently as indicated on the article page online and as indicated below:

Lastname, A.A.; Lastname, B.B. Article Title. *Journal Name* **Year**, *Volume Number*, Page Range.

ISBN 978-3-7258-1573-9 (Hbk)
ISBN 978-3-7258-1574-6 (PDF)
doi.org/10.3390/books978-3-7258-1574-6

Cover image courtesy of Mohan Li

Contents

About the Editors

Xiqing Yue

Xiqing Yue, Ph.D. is a Full Professor at the College of Food Science, Shenyang Agricultural University, China, where he earned his degree in Food Science and Engineering. Xiqing Yue's research interests include the processing and utilization of animal-derived foods and the product development of probiotics, mainly including the identification, screening, characterization and functional research of nutrients in dairy products, as well as the delivery system and targeted screening of probiotics. Xiqing Yue serves as the Scientific Editor of the publication *The Food Science of Animal Products* and holds editorial roles in prestigious journals such as *The Frontiers in Nutrition, Food Science*, and *The Journal of Shenyang Agricultural University*. To date, Xiqing Yue has contributed to 617 scientific publications, 206 of which have featured in international conferences, books, and theses, and 411 of which have featured in international journals with referees.

Mohan Li

Mohan Li, Ph.D., is an Associate Professor at College of Food Science, Shenyang Agricultural University, China, where he earned his degree in Food Science and Engineering. During his Ph.D. studies, Dr Li conducted his research between Shenyang Agricultural University and the National University of Singapore. His research interests include the application of multi-omic techniques in dairy products; screening, extraction, and functional validation of characteristic nutritional components from different milk resources; the interaction of components in milk and their impact on dairy processing; and the development of functional dairy products. Mohan Li serves as the Associate Editor for *The International Journal of Dairy Technology* and holds editorial roles in prestigious journals such as *The Current Opinion in Food Science, The Journal of Food Quality*, and *The Journal of Dairy Research*. To date, Mohan Li has contributed to 103 scientific publications, 47 of which have featured in international conferences, books, and theses, and 56 of which have featured in international journals with referees.

foods

MDPI

Editorial

Current Research on Probiotics and Fermented Products

Yushi Dong, Mohan Li * and Xiqing Yue *

College of Food Science, Shenyang Agricultural University, Shenyang 110866, China; 2022200035@stu.syau.edu.cn
* Correspondence: limohan@syau.edu.cn (M.L.); yuexq@syau.edu.cn (X.Y.)

The history of probiotics and fermented products has evolved over millennia [1,2]. Traditional food fermentation benefited from people's appreciation of its unique flavor and texture, while modern fermentation technology places more emphasis on the production of bioactive metabolites during fermentation [3–5]. These metabolites can enhance the bioavailability of functional components in food and offer numerous health advantages [6–8]. Modern technology relies on the selection of specific starter cultures to achieve the optimally expected target, such as improving the flavor, texture, bioavailability, or preventing specific diseases [9–13]. In recent years, an increasing number of studies have revealed the beneficial roles of probiotics and fermented products in human wellness [14–16]. Therefore, their applications in food, medicine, and other functional products has garnered growing attention. However, there are still many aspects of these probiotics and fermented products that have not been fully elucidated [17]. These include the identification and screening of strains in fermented products, optimizing of the fermentation process, understanding the succession law of flora in fermented products, exploring the correlation between microbiota and flavor in fermented products, understanding their roles in the preparation of various fermented foods, their varying tolerances to temperature, pH, and other environmental factors, improvement strategies, and the detailed molecular mechanisms and key active ingredients of these probiotics and fermented products in health regulation functions, among others [18–20].

In an era where the quest for optimal health intersects with the science of food and nutrition, the exploration of probiotics and fermented products emerges as a pivotal domain. This Special Issue of *Foods*, titled "Current Research on Probiotics and Fermented Products", offers a comprehensive dive into the current state and future trajectory of this vibrant field. With three review papers and seven research articles, this collection illuminates the multifaceted roles of microorganisms in foods, health, and diseases, underscoring the ancient art of fermentation as a cornerstone of modern food and nutrition science.

Lactic acid bacteria (LABs) are beneficial microbes known for their health-promoting properties. LABs are well known for their ability to produce substantial amounts of bioactive compounds during fermentation. Peptides, exopolysaccharides (EPS), bacteriocins, some amylases, proteases, lipase enzymes, and lactic acid are the most important bioactive compounds generated by LAB activity during fermentation. Furthermore, the product produced by LAB is dependent on the type of fermentation used. LABs derived from the genera *Lactobacillus* and *Enterococcus* are the most popular probiotics at present. Consuming fermented foods has previously been associated with several health-promoting benefits, such as antibacterial activity and modulation of the immune system. Furthermore, functional food implementations lead to the application of LABs in therapeutic nutrition, such as prebiotic, immunomodulatory, antioxidant, antitumor, and blood-glucose-lowering actions. Understanding the characteristics of LABs from various sources, and its potential as a functional food, is crucial for therapeutic applications. The review by Hakim et al. presents an overview of functional food knowledge regarding interactions between LABs isolated from dairy products (dairy LABs) and fermented foods, as well as the prospect of functioning LABs in human health. Finally, the health advantages of bioactive compounds from LABs are emphasized.

Citation: Dong, Y.; Li, M.; Yue, X. Current Research on Probiotics and Fermented Products. *Foods* **2024**, *13*, 1406. https://doi.org/10.3390/foods13091406

Received: 11 April 2024
Accepted: 17 April 2024
Published: 3 May 2024

Plant-based drinks have garnered significant attention as viable substitutes for traditional dairy milk, providing options for people who are lactose intolerant or allergic to dairy proteins and those who adhere to vegan or vegetarian diets. In recent years, the demand for plant-based drinks has expanded rapidly. Each variety has unique characteristics in terms of flavor, texture, and nutritional composition, offering consumers a wide range of choices tailored to meet individual preferences and nutritional needs. In the review by Xie et al., they aimed to provide a comprehensive overview of the various types of plant-based beverages and explore potential considerations, including their nutritional compositions, health benefits, and processing technologies, as well as the challenges facing the plant-based beverage processing industry. This review explores the scientific evidence supporting the consumption of plant-based beverages, discusses their potential role in meeting dietary requirements, and addresses current limitations and concerns about their use. In conclusion, this review illuminates the growing importance of plant-based drinks as sustainable and nutritious alternatives to dairy milk, and assists people in making informed choices about their dietary habits, expanding the potential applications for plant-based drinks, and providing the necessary theoretical and technical support for the development of a plant-based drink processing industry.

Functional dyspepsia (FD) is a common functional gastrointestinal disorder. Pathophysiology remains poorly understood; however, alterations in the small intestinal microbiome have been observed. Current treatments for FD with drugs are limited, and there are certain safety problems. A class of active probiotic bacteria can control gastrointestinal homeostasis, nutritional digestion and absorption, and energy balance, when taken in certain doses. Probiotics play many roles in maintaining the intestinal microecological balance, improving the intestinal barrier function, and regulating the immune response. The presence and composition of intestinal microorganisms play a vital role in the onset and progression of FD, and serve as a critical factor for both regulation and potential intervention regarding the treatment of this condition. Thus, there are potential advantages to alleviating FD by regulating the intestinal flora using probiotics that target intestinal microorganisms. The review by Shen et al. summarizes the progress of probiotic research in improving FD by regulating intestinal flora and provides a reference basis for probiotics to improve FD.

Fermented dairy foods, such as yogurt, exhibit some beneficial effects on consumers, including relieving symptoms of hypertension. The study by Yu et al. aims to obtain fermented dairy products from a co-starter that has a great flavor and the auxiliary function of reducing blood pressure after long time consumption. Commercial starter cultures composed of *Lactobacillus delbrueckii* subsp. *bulgaricus* CICC 6047 and *Streptococcus thermophilus* CICC 6038 were combined with *Lactobacillus plantarum* strains Y44, Y12 and Y16, respectively, as a combined starter culture to ferment a mix of skim milk and soybean milk. The fermented milk produced using the combined starter culture mixed with *L. plantarum* Y44 showed an inhibitory activity of the angiotensin converting enzyme (ACE) ($53.56 \pm 0.69\%$). Some peptides that regulate blood pressure were released in fermented milk, such as AMKPWIQPK, GPVRGPFPII, LNVPGEIVE, NIPPLTQTPV, and YQEPVL. In spontaneously hypertensive rat (SHR) oral administration experiments compared to the gavage unfermented milk group, gavage feeding of SHR with fermented milk produced using the combined starter culture mixed with *L. plantarum* Y44 significantly reduced SHR blood pressure after long-term intragastric administration, shown with the systolic blood pressure (SBP) and diastolic blood pressure (DBP) decreasing by 23.67 ± 2.49 mmHg and 15.22 ± 2.62 mmHg, respectively. Furthermore, the abundance of short-chain fatty acids (SCFA), bacterial diversity in the gut microbiota, and SCFA levels, including acetic acid, propionic acid, and butyric acid in the feces of SHRs, were increased by oral administration of fermented milk produced using the combined starter culture containing *L. plantarum* Y44. Furthermore, the ACE-angiotensin II (Ang II)-angiotensin type 1 (AT 1) axis was negatively regulated, the angiotensin-converting enzyme 2 (ACE 2)-angiotensin(1-7) (Ang1-7)-Mas receptor axis of SHRs was positively regulated, and then the RAS signal was rebalanced.

Fermented milk obtained from the combined starter culture shows the potential to be a functional food with antihypertensive properties.

The creation of functional foods through the enrichment of new foods with probiotic organisms or spores is a promising approach. The study by Suwanangul et al. investigates the use of encapsulating agents to establish a synbiotic relationship with *Bacillus coagulans* (TISTR 1447). Various proportions of wall materials, such as skim milk powder, maltodextrin, and cellulose acetate phthalate (represented as SMC1, SMC3, SMC5 and SMC7), were examined. In all formulations, 5% inulin was included as a prebiotic. The research evaluated their impact on cell viability and bioactive properties during both the spray-drying process and in vitro gastrointestinal digestion. The results demonstrate that these encapsulating agents efficiently protect *B. coagulans* spores during the spray drying process, resulting in spore viability exceeding 6 log CFU/g. In particular, SMC5 and SMC7 displayed the highest spore viability values. Furthermore, SMC5 showed the most notable antioxidant activity, encompassing DPPH, hydroxy radical, and superoxide radical scavenging, as well as significant antidiabetic effects through the inhibition of α-amylase and α-glucosidase. Furthermore, during simulated gastrointestinal digestion, both SMC5 and SMC7 exhibited a slight reduction in spore viability during the 6 h simulation. Consequently, SMC5 was identified as the optimal condition for synbiotic production, providing protection to *B. coagulans* spores during microencapsulation and gastrointestinal digestion while maintaining bioactive properties after encapsulation. Synbiotic microcapsules containing SMC5 showed a remarkable positive impact, suggesting its potential as an advanced food delivery system and a functional ingredient for various food products.

Wild artisanal cultures, such as a symbiotic culture of bacteria and yeast (SCOBY) and water kefir grains (WKG), represent a complex microorganism consortium composed of yeasts and lactic and acetic acid bacteria, with large strains of diversity and abundance. Fermented products (FP) obtained by the contribution of the microbiome can be included in functional products, due to their metabiotics (pre, pro, post and paraprobiotics) resulting from complex and synergistic associations, as well as due to metabolic functionality. In the study performed by Pihurov et al., consortia of both SCOBY and WKG were involved in the cofermentation of a newly formulated substrate, which was further analyzed with the aim of increasing the postbiotic composition of the FPs. Plackett–Burman (PBD) and Response Surface Methodology (RSM) techniques were used for the experimental designs to select and optimize several parameters that influence the lyophilized starter cultures of SCOBY and WKG activity as a multiple inoculum. Tea concentration (1–3%), sugar concentration (5–10%), raisins concentration (3–6%), SCOBY lyophilized culture concentration (0.2–0.5%), WKG lyophilized culture concentration (0.2–0.5%), and fermentation time (5–7 days) were considered the independent variables for mathematical analysis and optimization of fermentation conditions. Antimicrobial activity against *Bacillus subtilis* MIUG B1, *Staphylococcus aureus* ATCC 25923, *Escherichia coli* ATCC 25922, and *Aspergillus niger* MIUG M5, antioxidant capacity (DPPH), pH and total acidity (TA) were evaluated as responses. The rich postbiotic bioactive composition of the FP obtained under optimized biotechnological conditions highlighted the usefulness of the artisanal cocultures through their symbiotic metabolic interactions for the improvement of bioactive potential.

To develop a new or better combination of probiotics and/or probiotic-prebiotics for a healthy functional food ingredient or a remedial agent to treat or prevent obesity, the potential antiobesity efficacy of *Lactobacillus rhamnosus* BST-L.601 and its fermented product (named SPY) with mashed sweet potato paste were investigated using 3T3-L1 preadipocytes and obese mice induced by the high fat diet (HD) by Kang et al. SPY (0–0.5 mg/mL) dose dependently and significantly reduced lipid accumulation and TG content and the expression of adipogenic markers (C/EBPα, PPAR-γ, and aP2) and fatty acid synthetic pathway proteins (ACC and FAS) in adipocytes 3T3-L1 adipocytes, demonstrating that SPY suppresses adipocyte differentiation and lipogenesis. Oral administration of SPY (4×107 CFU/kg body weight) to HD-induced obese mice for 12 weeks significantly reduced body and liver weight, adipocyte size, and epididymal, visceral, and subcutaneous fat tissues weight. SPY was more effective in decreasing body weight gain in HD mice than in treatment with BST-L.601 alone. The administration of SPY or BST-L.601 also reduced the serum level of total cholesterol and LDL cholesterol and leptin secretion to a similar level. These results revealed that both SPY and BST-L.601 effectively suppress HD-induced adipogenesis and lipogenesis, suggesting that these materials would be useful in the functional foods industry to improve and/or prevent obesity.

Lacticaseibacillus paracasei (formerly Lactobacillus paracasei) is a nomadic LAB that inhabits a wide variety of ecological niches, from fermented foodstuffs to host-associated microenvironments. Many isolated strains of *L. paracasei* have been used as single-strain probiotics or as part of a symbiotic consortium within formulations. The study proposed by Moiseenko et al. contributes to the exploration of different strains of *L. paracasei* derived from non-conventional isolation sources, the traditional South African fermented drink mahewu (strains MA2 and MA3) and kefir grains (strains KF1 and ABK). The microbiological, biochemical, and genomic comparative analyses performed on the studied strains demonstrated a correlation between the properties of the strains and their isolation source, suggesting the presence of at least partial adaptation of the strain to the isolation environments. Furthermore, for the studied strains, antagonistic activities against common pathogens and against each other were observed, and the ability to release bioactive peptides with antioxidant and angiotensin I-converting enzyme inhibitory properties (ACE-I) during milk fermentation was investigated. The results obtained may be useful for a deeper understanding of the nomadic lifestyle of *L. paracasei*, and for the development of new starter cultures and probiotic preparations based on this LAB in the future.

Greek yogurt is a strained yogurt with a high protein content that provides nutritional benefits. To enhance the functional benefits of Greek yogurt, Yang et al. formulated Greek yogurt with various combinations of probiotic LAB (*Streptococcus thermophilus, Lactobacillus bulgaricus, Lactobacillus gasseri* BNR17, and *Lactobacillus plantarum* HY7714) by Yang et al. Effects of probiotic LAB were compared on the quality, sensory and microbiological characteristics of Greek yogurt. Among the samples, Greek yogurt fermented with *S. thermophilus* and *L. bulgaricus* showed the highest changes in pH and titratable acidity during 21 days of storage at 4 °C. Greek yogurt fermented with *L. plantarum* HY7714 had a higher viscosity than other samples. Greek yogurt fermented with S. thermophilus, *L. bulgaricus, L. gasseri* BNR17 and *L. plantarum* HY7714 showed superior physicochemical properties, and received the highest preference score from sensory evaluation among samples. In general, the enterohaemorrhagic *Escherichia coli* (EHEC) population was more effectively reduced in fermented Greek yogurt with probiotic LAB than in commercial Greek yogurt during storage at 4, 10, and 25 °C. Therefore, the addition of *L. gasseri BNR17* and *L. plantarum* HY7714 as starter cultures could improve the microbial safety of Greek yogurt and the sensory acceptance of consumers.

Taking into account the need for functional foods and the use of by-products of the food industry, Mendonça et al. developed a potentially functional ice cream using soy extract, soy kefir, and dehydrated jaboticaba peel. Five ice creams were made using soy kefir (K) and soy extract (S): (1) GS-100% S; (2) GK1-75% S/25% K; (3) GK2-50% S/50% K; (4) GK3-25% S/75% K; and (5) GK-100% K. The products were evaluated by physicochemical, microbiological, and sensory analyses. The addition of kefir was found to increase the acidity of the products. The concentrations of total phenolic compounds in the kefir formulations were approximately ten times higher than those of the GS formulation. All products presented concentrations of thermotolerant coliforms < 3 NMP/g and absence of *Salmonella* ssp. The viability of *Lactobacillus* ssp., *Streptococcus* spp., and *Bifidobacterium* ssp. was greater than 10 log CFU/g during the storage period. GS and GK1 formulations had the lowest scores, while GK ice cream was preferred. The formulations showed distinct sensory profiles in the CATA and ice cream with 100% kefir was associated with desirable attributes. The ice creams exhibited microbiological and sensory characteristics that meet the expectations of the target audience of the product.

The investigation into probiotics and fermented products represents an interdisciplinary endeavor, integrating scientific inquiry, food science innovation, and societal impact. Ongoing research efforts are uncovering new advancements and potential in probiotics and fermented foods, offering insights that go beyond traditional fermentation practices. As we delve further into this realm of complexity, the opportunities for creativity, health promotion, and sustainability in probiotics and fermented products continue to expand and remain ready for exploration.

Conflicts of Interest: The authors declare no conflicts of interest.

List of Contributions

1. Abdul Hakim, B.N.; Xuan, N.J.; Oslan, S.N.H. A comprehensive review of bioactive compounds from lactic acid bacteria: Potential functions as functional food in dietetics and the food industry. *Foods* **2023**, *12*, 2850.
2. Xie, A.; Dong, Y.; Liu, Z.; Li, Z.; Shao, J.; Li, M.; Yue, X. A Review of Plant-Based Drinks Addressing Nutrients, Flavor, and Processing Technologies. *Foods* **2023**, *12*, 3952.
3. Shen, X.; Xie, A.; Li, Z.; Jiang, C.; Wu, J.; Li, M.; Yue, X. Research Progress for Probiotics Regulating Intestinal Flora to Improve Functional Dyspepsia: A Review. *Foods* **2024**, *13*, 151.
4. Yu, J.; Sun, M.; Jiang, S.; Jiang, C.; Mu, G.; Tuo, Y. Oral Administration of Fermented Milk from Co-Starter Containing Lactobacillus plantarum Y44 Shows an Ameliorating Effect on Hypertension in Spontaneously Hypertensive Rats. *Foods* **2024**, *13*, 641.
5. Suwanangul, S.; Jaichakan, P.; Narkprasom, N.; Kraithong, S.; Narkprasom, K.; Sangsawad, P. Innovative Insights for Establishing a Synbiotic Relationship with Bacillus coagulans: Viability, Bioactivity, and In Vitro-Simulated Gastrointestinal Digestion. *Foods* **2023**, *12*, 3692.
6. Pihurov, M.; Păcularu-Burada, B.; Cotârleț, M.; Grigore-Gurgu, L.; Borda, D.; Stănciuc, N.; Kluz, M.; Bahrim, G.E. Kombucha and water kefir grains microbiomes' symbiotic contribution to postbiotics enhancement. *Foods* **2023**, *12*, 2581.
7. Kang, T.; Ree, J.; Park, J.-W.; Choe, H.; Park, Y.I. Anti-obesity effects of SPY fermented with lactobacillus rhamnosus BST-L. 601 via suppression of adipogenesis and lipogenesis in high-fat diet-induced obese mice. *Foods* **2023**, *12*, 2202.
8. Moiseenko, K.V.; Begunova, A.V.; Savinova, O.S.; Glazunova, O.A.; Rozhkova, I.V.; Fedorova, T.V. Biochemical and genomic characterization of two new strains of Lacticaseibacillus paracasei isolated from the traditional corn-based beverage of South Africa, mahewu, and their comparison with strains isolated from kefir grains. *Foods* **2023**, *12*, 223.
9. Yang, S.-Y.; Yoon, K.-S. Effect of probiotic lactic acid bacteria (LAB) on the quality and safety of Greek yogurt. *Foods* **2022**, *11*, 3799.
10. Mendonça, G.M.; Oliveira, E.M.; Rios, A.O.; Pagno, C.H.; Cavallini, D.C. Vegan ice cream made from soy extract, soy kefir and jaboticaba peel: Antioxidant capacity and sensory profile. *Foods* **2022**, *11*, 3148.

References

1. Yue, X.; Li, M.; Liu, Y.; Zhang, X.; Zheng, Y. Microbial diversity and function of soybean paste in East Asia: What we know and what we don't. *Curr. Opin. Food Sci.* **2021**, *37*, 145–152. [CrossRef]
2. Ray, R.C.; Joshi, V. Fermented foods: Past, present and future. In *Microorganisms and Fermentation of Traditional Foods*; CRC Press: Boca Raton, FL, USA, 2014; pp. 1–36. Available online: https://www.taylorfrancis.com/chapters/edit/10.1201/b17307-4/fermented-foods-past-present-future-ramesh-ray-joshi (accessed on 10 April 2024).
3. Ding, R.; Li, M.; Zou, Y.; Wang, Y.; Yan, C.; Zhang, H.; Wu, R.; Wu, J. Effect of normal and strict anaerobic fermentation on physicochemical quality and metabolomics of yogurt. *Food Biosci.* **2022**, *46*, 101368. [CrossRef]
4. Hu, Y.; Zhang, L.; Wen, R.; Chen, Q.; Kong, B. Role of lactic acid bacteria in flavor development in traditional Chinese fermented foods: A review. *Crit. Rev. Food Sci. Nutr.* **2022**, *62*, 2741–2755. [CrossRef] [PubMed]
5. Purutoğlu, K.; İspirli, H.; Yüzer, M.O.; Serencam, H.; Dertli, E. Diversity and functional characteristics of lactic acid bacteria from traditional kefir grains. *Int. J. Dairy Technol.* **2020**, *73*, 57–66. [CrossRef]
6. García-Burgos, M.; Moreno-Fernández, J.; Alférez, M.J.; Díaz-Castro, J.; López-Aliaga, I. New perspectives in fermented dairy products and their health relevance. *J. Funct. Foods* **2020**, *72*, 104059. [CrossRef]
7. Qi, F.; Zhu, J.; Li, M.; Ren, J.; Hu, Y.; Sun, Q. Preparation of W/O/W Lactiplantibacillus plantarum L3 microcapsules using modified low methoxy pectin as wall materials and their application in simulated yoghurt fermentation systems. *Int. J. Dairy Technol.* **2024**, *77*, 114–131. [CrossRef]

8. Mituniewicz–Małek, A.; Zielińska, D.; Ziarno, M. Probiotic monocultures in fermented goat milk beverages–sensory quality of final product. *Int. J. Dairy Technol.* **2019**, *72*, 240–247. [CrossRef]
9. Xie, A.; Zhao, S.; Liu, Z.; Yue, X.; Shao, J.; Li, M.; Li, Z. Polysaccharides, proteins, and their complex as microencapsulation carriers for delivery of probiotics: A review on carrier types and encapsulation techniques. *Int. J. Biol. Macromol.* **2023**, *242*, 124784. [CrossRef] [PubMed]
10. Ozcan, T.; Eroglu, E. Effect of stevia and inulin interactions on fermentation profile and short-chain fatty acid production of Lactobacillus acidophilus in milk and in vitro systems. *Int. J. Dairy Technol.* **2022**, *75*, 171–181. [CrossRef]
11. Dimov, S.G.; Posheva, V.; Georgieva-Miteva, D.; Peykov, S.; Kitanova, M.; Ilieva, R.; Dimitrov, T.; Iliev, M.; Gotcheva, V.; Strateva, T. Artisanal cheeses relying on spontaneous fermentation as sources of unusual microbiota–The example of the Bulgarian 'mehovo sirene'skin bag cheese. *Int. J. Dairy Technol.* **2023**, *76*, 1019–1024. [CrossRef]
12. Bandara, T.A.; Munasinghe-Arachchige, S.P.; Gamlath, C.J. Fermented whey beverages: A review of process fundamentals, recent developments and nutritional potential. *Int. J. Dairy Technol.* **2023**, *76*, 737–757. [CrossRef]
13. Sharma, R.; Garg, P.; Kumar, P.; Bhatia, S.K.; Kulshrestha, S. Microbial fermentation and its role in quality improvement of fermented foods. *Fermentation* **2020**, *6*, 106. [CrossRef]
14. Kerry, R.G.; Patra, J.K.; Gouda, S.; Park, Y.; Shin, H.-S.; Das, G. Benefaction of probiotics for human health: A review. *J. Food Drug Anal.* **2018**, *26*, 927–939. [CrossRef] [PubMed]
15. Maldonado Galdeano, C.; Cazorla, S.I.; Lemme Dumit, J.M.; Vélez, E.; Perdigón, G. Beneficial effects of probiotic consumption on the immune system. *Ann. Nutr. Metab.* **2019**, *74*, 115–124. [CrossRef] [PubMed]
16. Şanlier, N.; Gökcen, B.B.; Sezgin, A.C. Health benefits of fermented foods. *Crit. Rev. Food Sci. Nutr.* **2019**, *59*, 506–527. [CrossRef] [PubMed]
17. Mora-Villalobos, J.A.; Montero-Zamora, J.; Barboza, N.; Rojas-Garbanzo, C.; Usaga, J.; Redondo-Solano, M.; Schroedter, L.; Olszewska-Widdrat, A.; López-Gómez, J.P. Multi-product lactic acid bacteria fermentations: A review. *Fermentation* **2020**, *6*, 23. [CrossRef]
18. Chai, K.F.; Voo, A.Y.H.; Chen, W.N. Bioactive peptides from food fermentation: A comprehensive review of their sources, bioactivities, applications, and future development. *Compr. Rev. Food Sci. Food Saf.* **2020**, *19*, 3825–3885. [CrossRef] [PubMed]
19. Wang, S.; Naumovski, N.; Ajlouni, S.; Ayyash, M.; Silva, R.; Balthazar, C.F.; Esmerino, E.A.; Freitas, M.Q.; da Silva, M.C.; Sant'Ana, A.S. Nonbovine milk and its products as sources of probiotics delivery: An overview of its viability, functionality and product quality characteristics. *Int. J. Dairy Technol.* **2023**, *76*, 482–511. [CrossRef]
20. Barat, A.; Ozcan, T. Growth of probiotic bacteria and characteristics of fermented milk containing fruit matrices. *Int. J. Dairy Technol.* **2018**, *71*, 120–129. [CrossRef]

foods

MDPI

Article

Oral Administration of Fermented Milk from Co-Starter Containing *Lactobacillus plantarum* Y44 Shows an Ameliorating Effect on Hypertension in Spontaneously Hypertensive Rats

Jiang Yu [1,2], Mengying Sun [1,2], Shilong Jiang [3], Chuqi Jiang [3], Guangqing Mu [1,2] and Yanfeng Tuo [1,2,*]

1 School of Food Science and Technology, Dalian Polytechnic University, Dalian 116034, China; yujiang0324@163.com (J.Y.); smy493006403@163.com (M.S.); guangqingmu@163.com (G.M.)
2 Dalian Probiotics Function Research Key Laboratory, Dalian Polytechnic University, Dalian 116034, China
3 Heilongjiang Feihe Dairy Co., Ltd., Beijing 100016, China; jiangshilong@feihe.com (S.J.); jiangchuqi1224@163.com (C.J.)
* Correspondence: tuoyf@dlpu.edu.cn; Tel./Fax: +86-0411-86322121

Abstract: Fermented dairy foods such as yogurt exhibit some beneficial effects on consumers, including relieving the symptoms of hypertension. This study aims to obtain fermented dairy products from a co-starter that have a great flavor and the auxiliary function of reducing blood pressure after longtime consumption. Commercial starter cultures composed of *Lactobacillus delbrueckii* subsp. *bulgaricus* CICC 6047 and *Streptococcus thermophilus* CICC 6038 were combined with *Lactobacillus plantarum* strains Y44, Y12, and Y16, respectively, as a combined starter culture to ferment the mixed milk of skim milk and soybean milk. The fermented milk produced using the combined starter culture mixed with *L. plantarum* Y44 showed an angiotensin-converting-enzyme (ACE) inhibitory activity (53.56 ± 0.69%). Some peptides that regulate blood pressure were released in the fermented milk, such as AMKPWIQPK, GPVRGPFPII, LNVPGEIVE, NIPPLTQTPV, and YQEPVL. In spontaneously hypertensive rat (SHR) oral-administration experiments compared with the gavage unfermented milk group, the gavage feeding of SHRs with the fermented milk produced using the combined starter culture mixed with *L. plantarum* Y44 significantly reduced the blood pressure of the SHRs after long-term intragastric administration, shown with the systolic blood pressure (SBP) and diastolic blood pressure (DBP) decreasing by 23.67 ± 2.49 mmHg and 15.22 ± 2.62 mmHg, respectively. Moreover, the abundance of short-chain fatty acids (SCFA), bacterial diversity in the gut microbiota, and SCFA levels including acetic acid, propionic acid, and butyric acid in the feces of the SHRs were increased via oral administration of the fermented milk produced using the combined starter culture containing *L. plantarum* Y44. Furthermore, the ACE-angiotensin II (Ang II)-angiotensin type 1 (AT 1) axis was downregulated, the angiotensin-converting-enzyme 2 (ACE 2)-angiotensin(1-7) (Ang1-7)-Mas receptor axis of the SHRs was upregulated, and then the RAS signal was rebalanced. The fermented milk obtained from the combined starter culture shows the potential to be a functional food with antihypertension properties.

Keywords: hypertension; ACE inhibitory peptide; fermented milk; gut microbiota; SCFAs

Citation: Yu, J.; Sun, M.; Jiang, S.; Jiang, C.; Mu, G.; Tuo, Y. Oral Administration of Fermented Milk from Co-Starter Containing *Lactobacillus plantarum* Y44 Shows an Ameliorating Effect on Hypertension in Spontaneously Hypertensive Rats. *Foods* **2024**, *13*, 641. https://doi.org/10.3390/foods13050641

Academic Editors: Xiqing Yue and Mohan Li

Received: 16 January 2024
Revised: 7 February 2024
Accepted: 12 February 2024
Published: 20 February 2024

1. Introduction

Hypertension is a prevalent chronic disease that stands as a leading contributor to the global disease burden. It serves as a clear risk factor for conditions such as coronary heart disease, stroke, and atherosclerosis, and causes damage to both the kidneys and cerebral blood vessels [1]. The induction of hypertension is attributed to a complex interplay of various factors, including genetics, environmental influences, vascular remodeling, excessive endothelin release, and the upregulation of the renin–angiotensin-system (RAS), among others. The RAS is a major factor controlling blood pressure changes and is mediated by the important stress hormone angiotensin II (Ang II), leading to powerful vasoconstriction effects. The heart of this system is the angiotensin-converting-enzyme (ACE), which is responsible for catalyzing the conversion of angiotensin I (Ang I) to Ang II, and it is considered to be the core enzyme of the RAS [2]. Consequently, ACE inhibitors have emerged as the primary pharmacological agents for hypertension treatment. Extracting more efficient and safer ACE inhibitors from natural food ingredients has promising prospects for the treatment of hypertension.

A growing line of the literature is based on research of the health benefits of probiotics and their fermented dairy products, such as maintaining an intestinal microecological balance, improving the intestinal barrier function, regulating the immune response, and lowering blood pressure [3–5]. As reported, a daily intake of some probiotic products of more than 10^9 CFU would significantly reduce blood pressure [6]. A number of mechanisms of oxidative stress and gut microbiota can induce blood pressure regulation [7,8]. The gut microbiome is made up of several types of bacteria, fungi, archaea, viruses, and protozoa, and they are critical for maintaining host physiological homeostasis [9]. In recent years, mounds of data have shown that the gut microbiota has a major impact on host blood pressure and plays a key role in hypertension development [10]. In SHRs, microbial richness, diversity, and evenness were drastically decreased compared with normal rats [11]. Short-chain fatty acids (SCFAs), produced by gut bacteria during fermentation, have been proposed as modulators of the gut microbiota–host interaction [12]. Pluznick [13] showed in numerous experiments that new SCFA receptors are involved in blood pressure control, supporting the hypothesis that SCFAs are key mediators of the microbiota and host blood pressure regulation. These studies show that the gut microbiota could offer benefits for patients with hypertension. Notably, some lactic acid bacteria can hydrolyze the proteins in dairy products to produce short peptides with blood pressure-lowering effects [14]. For example, the short peptides Ile–Pro–Pro and Val–Pro–Pro have the same effect as antihypertensive drugs in the treatment of hypertension and play a role in lowering blood pressure by inhibiting the activities of ACEs [15]. Plant-derived beverages have many nutritional benefits and can provide essential nutrients, including vitamins, minerals, healthy fats, and fiber [16]. In addition, ACE inhibitors can also be derived from the fermentation or enzymatic breakdown of soybeans, peas, pigeon peas, chickpeas, or adzuki beans [17]. For example, the fermented mixed milk of skim milk and soy milk by lactic acid bacteria has shown improved immunomodulatory properties and the potential to prevent gastric mucosal damage [18]. Therefore, we used a combined starter culture to ferment the mixed milk of skim milk and soybean milk.

In our previous study, *L. plantarum* Y44, *L. plantarum* Y12, and *L. plantarum* Y16 exhibited high extracellular protease activities [19,20], which may lead to the release of specific bioactive peptides during milk fermentation. The intention of this study was to analyze the difference in the antihypertensive effect of fermented milk consisting of skim milk and soy milk using the commercial starter combined with *L. plantarum* strains Y44, Y12, and Y16, respectively, on spontaneously hypertensive rats (SHRs). Therefore, this study would obtain fermented milk products with a great flavor and the adjuvant effect of reducing blood pressure.

2. Materials and Methods

2.1. Materials

De Man, Rogosa, and Sharpe (MRS) broth medium was obtained from Baisi Biotechnology Co., Ltd. (Hangzhou, China). Skim milk powder was obtained from San Yuan Foods Co., Ltd. (Beijing, China). Soybeans were obtained from Ruinonghe Trading Co., Ltd. (Harbin, China). Captopril, N-benzoyl-Gly-His-Leu (HHL) and ACEs were obtained from Sigma-Aldrich Co., Ltd. (Saint Louis, MO, USA).

2.2. Preparation of Combined Starter Culture

L. plantarum Y44, *L. plantarum* Y12, and *L. plantarum* Y16 were isolated from wild turbot intestines in Dalian and stored in the Dalian Probiotics Function Research Key Laboratory with glycerin protectant at −80 °C. The commercial yogurt starter culture, consisting of *Streptococcus thermophilus* CICC 6038 and *Lactobacillus delbrueckii* subsp. *bulgaricus* CICC 6047, were obtained from the China Center of Industrial Culture Collection. The strains were subcultured three times in MRS culture medium at 37 °C for 18 h, washed with 0.85% normal saline three times, and resuspended in sterilized saline, respectively. The living cell concentrations of Y12, Y16, and Y44 strains were nearly 3×10^8 CFU/mL, obtained using the plate counting method [21]. Y12, Y16, and Y44 were mixed with 6047 and 6038 bacterial suspensions at a ratio of 1:1:1 (volume ratio), respectively. The mixed bacterial suspensions were used as combined starter cultures, namely, 6047 + 6038 + Y12, 6047 + 6038 + Y16, and 6047 + 6038 + Y44.

2.3. Fermentation of the Mixture of Soybean Milk and Skim Milk

Soybeans were washed with ultrapure water 2–3 times to remove dust and then soaked in ultrapure water overnight. Eight times the weight of water was added to the soaked soybeans and put into the soybean milk machine for grinding for 10 min to obtain soybean milk. After that, the soybean milk was filtered with a double-filter cloth, and the residue was removed before cooling. The skim milk was prepared according to our previous study [22]. The mixed milk was made by mixing skim milk and soybean milk in a ratio of 8:2 (v/v), which was then continuously homogenized and pasteurized (95 °C, 5 min). The combined starter culture was inoculated into the pasteurized mixed milk at a ratio of 4% (v/v). The inoculated mixed milk was cultured in a constant temperature incubator at 42 °C until curding; the curdling time was about 5 h.

2.4. Determination of Fermented Mixed Milk Characteristics

2.4.1. Titratable Acidity

The determination of titratable acidity [TA(°T)] of the fermented mixed milk samples was carried out according to the National Standards of the People's Republic of China (GB5009.239-2016) [23].

2.4.2. Viable Counts

The viable counts of lactic acid bacteria in the fermented mixed milk at the end of fermentation were determined using the pour plate method on MRS agar under an anaerobic incubation at 37 °C for 72 h, recorded as CFU/mL [21].

2.4.3. Free Amino Acids

Free amino acids in the fermented milk at the end of fermentation were quantified using the *o-phthalaldehyde* (OPA) method, which was similar to our previous studies [21,24].

2.4.4. Sensory Evaluation

At the end of fermentation, the sensory quality of the fermented mixed milk was evaluated according to the requirements specified in the National Standard of the People's Republic of China (GB 19302-2010) [25]. Ten volunteers were informed consent and participated in the sensory evaluation of the fermented mixed milk. The volunteers ranged in age from 20 to 45, with five women and five men. Their daily diet include consumption yogurt 4-6 times per week. Before sensory evaluation, all fermented mixed milk samples were transferred from refrigeration at 4 °C. About 30 g of each fermented milk sample was placed in a 50 mL unscented transparent plastic cup, coded with a 3-digit random number, and equilibrated at room temperature for 30 minutes. During sensory evaluation, prepare palate-cleaning water at intervals between samples to prevent cross-over effects between samples. The criteria for the sensory evaluation of fermented milk are shown in Table A1. This experiment did not require Ethics Committee approval, because there were no risks associated for panelists who tasted samples, and this experiment met the national standards of the People's Republic of China (GB 19302-2010) [25].

2.4.5. Determination of Angiotensin-Converting-Enzyme Inhibitory (ACE I) Activity

The ACE inhibitory activity of the fermented milk samples at the end of fermentation were determined according to the method of Wu et al., with some modifications [26]. The principle is that ACE catalyzes the decomposition of the analogue of angiotensin I, hippuryl histaminoylleucine (HHL), producing hippuric acid under the conditions of 37 °C and pH 8.3, which showed a characteristic absorption peak at 228 nm. When ACEI was added, the catalytic decomposition of HHL by ACE was inhibited, and the production of hippuric acid was reduced. The analysis employed HPLC, with a Zorbax C18 column (4.6 × 250 mm, particle size 5 μm; Agilent Technologies, Santa Clara, CA, USA) at a column temperature of 30 °C and a 1 mL/min flow rate. The detection wavelength was set at 228 nm. The mobile phase consisted of acetonitrile (A) and 0.5% TFA (B) with the following elution conditions: 0–11 min, A: 20%, B: 80%; 11–12 min, A: 35%, B: 65%; 12–15 min, A: 35%, B: 65%; 16–17 min, A: 20%, B: 80%; and 18–30 min, A: 20%, B: 80%.

2.4.6. Peptide Sequence Analysis

The free peptides of the fermented milk samples were determined using the method described by Mao et al. [27]. BIOPEP-UWM (https://biochemia.uwm.edu.pl/biopep-uwm, accessed on 15 April 2023) was used to screen out identified ACEIPs while combining features of ACEIP sequences to find potential ACEIPs.

2.5. Assay to Determine the Effects of the Fermented Milk Oral Administration on the Spontaneously Hypertensive Rats (SHRs)

Ten-week-old female SHR rats (n = 30) and Wistar Kyoto (WKY) rats (n = 6) were purchased from Vital River Experimental Technology Co., Ltd. (SCXK 2021-0006, Beijing, China). All animal experiments were approved by the Animal Ethics Committee of Dalian Polytechnic University (SYXK2017-0005). The rats were raised according to the method previously described in our lab [20]. After the adaptation period (7 d), thirty SHR rats were randomly divided into five groups (n = 6). After the rats were stable, they were orally administered milk at a fixed time once a day for 4 weeks. The methods of gavage are detailed in Table 1. The systolic blood pressure (SBP) and diastolic blood pressure (DBP) were monitored using the tail sleeve method (BP-2010A, Softron Beijing Biotechnology Co., Ltd. Beijing, China).

Table 1. Group information and administration of animal experiments.

Animals	Groups	Feeding
SHR	Positive control group (GY)	Captopril
SHR	Negative control group (KB)	Pasteurized milk
WKY	Blank group (WKY)	Pasteurized milk
SHR	Y12 group (Y12)	6038 + 6047 + Y12 fermented milk
SHR	Y44 group (Y44)	6038 + 6047 + Y44 fermented milk
SHR	Y16 group (Y16)	6038 + 6047 + Y16 fermented milk
The gavage dose was consistent among the groups, with a dosage of 10 mg/kg/d.		

GY: Gavage feeding of captopril to SHRs; KB: gavage feeding of pasteurized milk to SHRs; WKY: gavage feeding of pasteurized milk to WKYs; Y12: gavage feeding of the fermented milk produced using the combined starter culture mixed with *L. plantarum* Y12 to SHRs; Y44: gavage feeding of the fermented milk produced using the combined starter culture mixed with *L. plantarum* Y44 to SHRs; Y16: gavage feeding of the fermented milk produced using the combined starter culture mixed with *L. plantarum* Y16 to SHRs.

2.5.1. Histological Analyses and Organ Index

After 4 weeks of gavage, all rats were weighed and euthanized. Blood samples were collected, and the serum was separated by centrifugation at 4000× g at 4 °C for 10 min, and frozen at −80 °C for further analysis. The kidney, heart, and liver were isolated and weighed; we soaked and fixed the entire heart in 4% paraformaldehyde fixative for the subsequent sectioning and pathological staining analysis [28]. The ImageJ2×(2.1.4.7, Rawak Software Inc.; Stuttgart, Germany) was used to analyze images. The organ index was calculated as follows: organ index = organ weight (g)/rat weight (g).

2.5.2. Determination of Biochemical Indexes in Rat Serum

The levels of alanine transaminase (ALT), aspartate transaminase (AST), total cholesterol (TC), triglycerides (TG), high-density lipoprotein cholesterol (HDL-C), low-density lipoprotein cholesterol (LDL-C), superoxide dismutase (SOD), glutathione (GSH), catalase (CAT), nitric oxide (NO), and malondialdehyde (MDA) in the sera of rats from each group were determined following the instructions provided with the kits (Nanjing Jiancheng Bioengineering Institute, China). Angiotensin-converting-enzyme (ACE), angiotensin-converting-enzyme 2 (ACE 2), angiotensin 1-7 (Ang 1–7), and angiotensin II (Ang II) levels were measured according to the ELISA kit instructions (Jingmei Biotech Co., Ltd., Shenzhen, China).

2.5.3. RT-PCR Analysis

We took 10 mg of kidney tissue samples from rats in each group. We crushed the samples by adding nine times the volume of 0.85% normal saline, and then centrifuged them to obtain the supernatant. The total RNA was transcribed into cDNA using the PrimeScript™ RT Master Mix kits. The mRNA expressions of ACE, ACE 2, MAS, and β-actin (with the β-actin gene serving as an endogenous control for the assay) were quantified using a StepOne Plus real-time PCR system (Applied Biosystems, Waltham, MA, USA) following the instructions provided with the ChamQ SYBR qPCR Master Mix kit [29]. The RT-reaction was as follows: the first stage was 30 s at 95 °C, the second stage was 40 cycles of 10 s at 95 °C and 30 s at 60 °C, and the third stage was 15 s at 95 °C, 60 s at 60 °C and 15 s at 90 °C.

2.5.4. Fecal SCFA Analysis

We used a previously published method to prepare the fecal sample (50 mg) from the tested animals and analyzed the SCFA levels using a gas chromatography–mass spectrometry system [30].

2.5.5. Rat Fecal Microbiota

All rat feces samples were forwarded to Maiwei Technology Co., Ltd. (Wuhan, China) to extract the total DNA from rat feces. The extracted DNA was analyzed using 1% agarose gel electrophoresis to ensure the appropriate amount was presented in the centrifuge tubes, and the DNA was diluted with sterile water to a concentration of 1 ng/μL. PCR was performed using the Phusion® High-Fidelity PCR Master Mix with GC buffer (New England Biolabs, MA, USA), and primer targeting was in the 16S V4 region (515F GTGCCAGCMGCCGCGGTAA and 806R GGACTACHVGGGTWTCTAAT). The PCR products were analyzed using 2% agarose gel electrophoresis. The TruSeq® DNA PCR-Free Sample Preparation Kit was employed to construct the library. Then, the library was quantified using Qubit and qPCR methods, followed by sequencing on the NovaSeq6000 platform. The results from each sample were processed and filtered using FLASH (V1.2.7) based on the barcode and PCR amplification primer sequences. For clustering, and the effective grouping of tags, the Uparse algorithm (Uparse v7.0.1001) was utilized with a default 97% identity threshold to group sequences into OTUs (pperational taxonomic units). The OTU sequences were annotated using the Mothur method and searching the SILVA132 SSUrRNA database (http://www.arb-silva.de/). The diversity and richness, Chao1, and Shannon calculations of samples were performed using Qiime software (Version 1.9.1).

2.6. Statistical Analysis

The data were expressed with the "mean ± standard deviation" of n replicates of samples. SPSS version 20.0 (SPSS, Inc., Chicago, IL, USA) software was used for mean comparisons to determine significant differences ($p < 0.05$) among the groups. The results were obtained using a one-way ANOVA. The LSD test and Dunnett's T3 were used to analyze the significant differences when equal variances were assumed and when they were not assumed, respectively.

3. Results

3.1. Physicochemical Properties of the Fermented Mixed Milk from Combined Starter Cultures

As shown in Table 2, the mixed milk fermented with starter cultures combined with different *L. plantarum* strains exhibited different physicochemical properties. Compared with the fermented milk only produced using a commercial starter culture, the combination of the commercial starter culture with *L. plantarum* strains endowed the fermented mixed milk with different properties, particularly in terms of the ACE inhibitory activity and free amino acid contents. The ACE inhibitory activity of the fermented milk was significantly increased ($p < 0.05$) from the commercial starter mixed with *L. plantarum*, and the ACE inhibitory activity of the fermented milk produced using the combined starter culture mixed with *L. plantarum* Y44 was 55.36 ± 0.69%, significantly higher than that of the commercial starter culture fermented milk (25.98 ± 1.19%). The viable lactic acid bacteria count of fermented milk remains above 10^9 CFU/mL and meets the national yogurt standard (GB19302-2010) [25]. Compared with the commercial starter culture fermented milk, the viable lactic acid bacteria count in the combined starter culture fermented milk was significantly higher, which may lead to the higher free amino acids, ACE inhibition activity, and TA of the fermented milk. Compared with commercial starter culture fermented milk, the combined starter culture fermented milk showed a better taste, flavor, ratio of sour to sweet, and sensory score. These results indicated that *L. plantarum* could improve the flavor of the fermented milk.

Table 2. Chemical characteristics of the fermented mixed milk with compound starter at 4 °C.

Type of Fermented Milk / Chemical Characteristics	6038 + 6047	6038 + 6047 + Y12	6038 + 6047 + Y44	6038 + 6047 + Y16
TA (°T)	90.09 ± 3.09 ^{b}	98.96 ± 2.27 ^{a}	97.75 ± 5.91 ^{a}	95.80 ± 1.77 ^{a}
Viable count (CFU mL^{-1})	3.88 ± 0.52 × 10^9 ^{b}	4.35 ±0.47 × 10^9 ^{a}	4.59 ± 0.54 × 10^9 ^{a}	4.54 ± 0.36 × 10^9 ^{a}
Free amino acids (mg/mL)	0.303 ± 0.097 ^{c}	0.390 ± 1.06 ^{b}	0.417 ± 0.93 ^{b}	0.493 ± 0.94 ^{a}
Sensory score (points)	85.25 ± 3.39 ^{c}	91.25 ± 4.11 ^{a}	88.67 ± 3.17 ^{b}	92.65 ± 2.47 ^{a}
ACE I activity (%)	25.98 ± 1.19 ^{c}	46.80 ± 2.10 ^{ab}	53.56 ± 0.69 ^{a}	42.98 ± 1.19 ^{b}

Data represent the mean ± SD $^{(a–c)}$; mean values with different letters are significantly different ($p < 0.05$) according to the Duncan multiple range test.

3.2. Analysis of Differential Peptides of the Fermented Mixed Milk from Combined Starter Cultures

The Venn diagram of the detected peptide in the fermented milk is shown in Figure 1. A total of 2561, 2384, 2404, and 2371 peptide segments were identified in samples 6038 + 6047, Y12, Y44, and Y16, respectively. Among them, 861 peptide segments were detected in all four samples, while 599, 636, 1034, and 731 peptide segments were only detected in the corresponding samples 6038 + 6047, Y12, Y44, and Y16, respectively. Among the fermented milk samples, the fermented milk from the combined starter culture with *L. plantarum* Y44 not only identified the most peptide segments, but also had the highest number of unique peptides. Out of the 2404 peptide segments, 1034 were not identified in the three other fermented milk samples, and their numbers were significantly higher than those of the three other fermented milk samples. The specific numbers of differentially expressed peptides are shown in Table 3. Whether the expression was upregulated or downregulated, there were significant differences in the numbers of peptides produced in the fermented milk using the commercial starter mixed with *L. plantarum* Y12 and *L. plantarum* Y44, compared with the commercial fermented milk. The numbers of common peptides in each of the two fermented milks were different, and the existence of these differential peptides indicated that the milk proteins in different fermented milks were decomposed by bacteria at different sites. As shown in Table 4, known ACEI peptides were screened, including AMKPWIQPK, GPVRGPFPII, LNVPGEIVE, NIPPLTQTPV, and YQEPVL, according to the BIOPEP database. We found that more types and contents of peptides were extracted from the fermented milk produced using the combined starter culture mixed with *L. plantarum* Y44, such as GPVRGPFPII (relative content of 68.8615), AMKPWIQPK (relative content of 57.9351), and NIPPLTQTPV (relative content of 76.8847). So, we inferred that the increase in the ACE inhibitory activity of the fermented milk containing *L. plantarum* 44 might be related to the type and content of ACE inhibitory peptides.

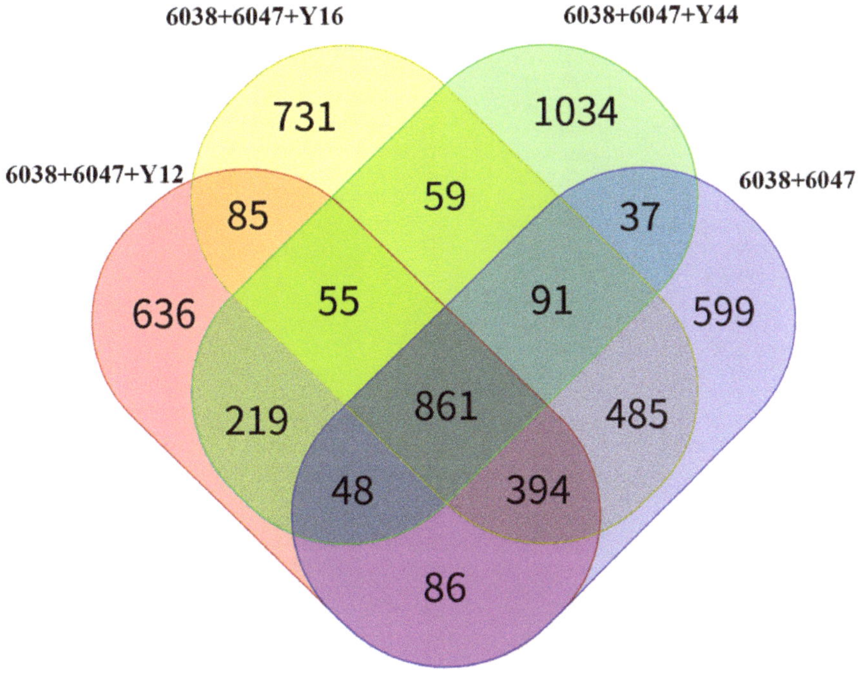

Figure 1. Venn diagram of differential peptides between different fermented milk groups.

Table 3. Peptide differential analysis between different fermented mixed milks.

Comparisons	All	Up	Down
Y vs. Y12	750	331	419
Y vs. Y44	982	374	608
Y vs. Y16	96	37	59
Y12 vs. Y44	887	511	376
Y12 vs. Y16	673	346	337
Y44 vs. Y16	886	371	443

Note: Y: 6038 + 6047; Y12: 6038 + 6047 + Y12; Y44: 6038 + 6047 + Y44; Y16: 6038 + 6047 + Y16.

Table 4. Peptides predicted to have ACE inhibitory ability.

Polypeptide Sequence	Mass	Length	*m*/*z*	Relative Content			
				Y	Y12	Y44	Y16
DAYPSGAW	1141.5079	8	571.7617	0.0429	0	0	0
EMPFPK	1007.4786	6	504.745	0.2865	0.1977	0	0
FVAPFPEVFG	1108.5593	10	555.2863	0.3365	0.4932	0.1881	0.6551
HLPLP	575.3431	5	288.6788	0.3818	0.2455	0	0
AMKPWIQPK	1097.6056	9	549.8098	9.1141	56.9335	57.9351	18.3578

Table 4. *Cont.*

Polypeptide Sequence	Mass	Length	m/z	Relative Content			
				Y	Y12	Y44	Y16
DKIHPF	755.3966	6	378.7053	1	4.1244	2.9753	4.7795
EIVPNSAEERLH	1392.6997	12	465.2398	0	0	0.0714	0
FALPQY	737.3748	6	369.6942	2.4255	0	0	0.4849
FALPQYLK	978.5538	8	490.2833	0	0	0	0.1396
FPEVFGK	822.4276	7	412.2204	0.0138	0.0573	0	0
FVAPFPEVFG	1108.5593	10	555.2863	0.3365	0.3817	0.1881	0.6551
GPVRGPFPII	1051.6178	10	526.8159	1.9852	18.2349	68.8615	3.4358
IPPLTQTPV	964.5593	9	483.287	0	1.0812	6.5293	0
KAAAAP	726.4388	6	364.2252	0	0	0.0717	0
KYIPIQ	760.4483	6	381.2311	0	0.4559	0.8266	0
LDAQSAPLR	969.5243	9	485.77	0.7855	0	0	0.2509
LGPVRGPFP	938.5338	9	470.2735	0.4908	0.7913	0	0.2365
LHLPLPL	801.5112	7	401.7628	9.2085	1.8828	3.6409	1.8227
LNVPGEIVE	968.5178	9	485.2659	0.0144	0.6265	16.0615	0.8688
LTQTPVVVPPF	1196.6804	11	599.3472	6.9091	0	0	1.2390
LVYPFPGPIH	1138.6174	10	570.316	0	2.2781	0	0.1767
LVYPFPGPIPNSLPQN	1751.9246	16	876.968	0	0.0204	0.0696	0.1572
LVYPFPGPIPNSLPQNIPP	2059.1143	19	1030.5667	0.7600	0	0	0.6918
MKPWIQPK	1026.5685	8	514.2914	1.2883	8.5013	2.3931	1.8913
NIPPLTQTPV	1078.6023	10	540.308	5.0910	56.5832	76.8847	15.6471
NLHLPLP	802.4701	7	402.2415	0	4.5065	1.0877	0
PFPEVFGK	919.4803	8	460.7472	7.9391	6.6073	1.0813	13.6954
QEPVLGPVRGPFP	1391.7561	13	696.8857	37.6619	95.2655	1.9809	80.9390
TTMPLW	747.3625	6	374.6879	0	0.6130	0.1559	0
VPSERYL	862.4548	7	432.2333	0.0000	0.0983	0.5551	0.2403
VRGPFP	671.3755	6	336.6948	1.6800	0	0	0
VRGPFPIIV	996.612	9	499.313	1.7564	2.6911	0.6419	1.0535
VVVPPF	656.3897	6	329.2016	0.2691	0	0.2359	0
YAKPA	548.2958	5	275.1553	0	0	1	0
YIPIQY	795.4167	6	398.715	0	0.3886	0.5081	0
YPQRDMPIQ	1146.5492	9	574.2802	0	0	1	0
YQEPVL	747.3802	6	374.6969	0.2386	6.2676	5.2048	0.6514

Note: Y: 6038 + 6047; Y12: 6038 + 6047 + Y12; Y44: 6038 + 6047 + Y44; Y16: 6038 + 6047 + Y16; the content of the polypeptide is normalized to the median; that is, each column of data is divided by the median of the column of data.

3.3. Effects of Fermented Milk Oral Administration on Blood Pressures of SHRs and WKY Rats

The fermented milk showed a high ACE inhibitory activity in vitro and the presence of some peptides that could regulate blood pressure. Therefore, we investigated the antihypertensive effects of the fermented milk in the SHRs and WKY rats by monitoring the blood pressure changes after 4 weeks of oral administration of the fermented milk. As shown in Figure 2A,B, the SBP and DBP of the KB group continuously increased. Before the end of the experiment, compared with the KB group, the SBP and DBP of SHRs in the treated GY, Y44, Y12, and Y16 groups were significantly reduced ($p < 0.05$). All fermented milk by the combined starter containing *L. plantarum* alleviated elevated blood pressure, with no significant difference among the three groups. The blood pressure of the SHRs in the Y44 group showed decreased values, with an SBP and DBP of 23.67 ± 2.49 mmHg ($p < 0.05$) and 15.22 ± 2.62 mmHg ($p < 0.05$), respectively. Therefore, the long-term oral administration of the fermented milk produced using the combined starter culture with *L. plantarum* Y44 showed significant antihypertensive effects in vivo.

3.4. Effect of Fermented Milk Oral Administration on Renal RAS in SHRs and WKY Rats

As shown in Figure 3, compared with the WKY group, the contents of ACE and Ang II in the sera of the SHRs in the KB group were significantly increased ($p < 0.001$,

$p < 0.001$), while the contents of ACE 2 and Ang 1-7 were significantly decreased ($p < 0.05$, $p < 0.01$), which was in agreement with the results reported by Sun et al., who found that hypertension can affect the contents of substances related to ACE regulation in the rat sera [30]. After the long-term oral administration of the fermented milk, the contents of ACE, Ang 1-7, ACE2, and Ang II in the sera of the SHRs in the Y44 group were significantly different from those in the KB group ($p < 0.05$, $p < 0.05$, $p < 0.05$, and $p < 0.01$, respectively), while the contents of ACE, Ang 1-7, ACE 2, and Ang II in the sera of the SHRs in the GY group were significantly different from those in the KB group ($p < 0.01$, $p < 0.05$, $p < 0.001$, and $p < 0.01$, respectively). The results suggest that the fermented milk produced using the combined starter culture mixed with *L. plantarum* Y44 could rebalance RAS signaling to regulate the blood pressure by downregulating the ACE-AngII-AT1 axis and upregulating the ACE 2-Ang (1-7)-Mas axis. However, the fermented milk produced using the combined starter culture mixed with *L. plantarum* Y44 showed a lesser effectiveness than captopril in the overall regulation of the RAS.

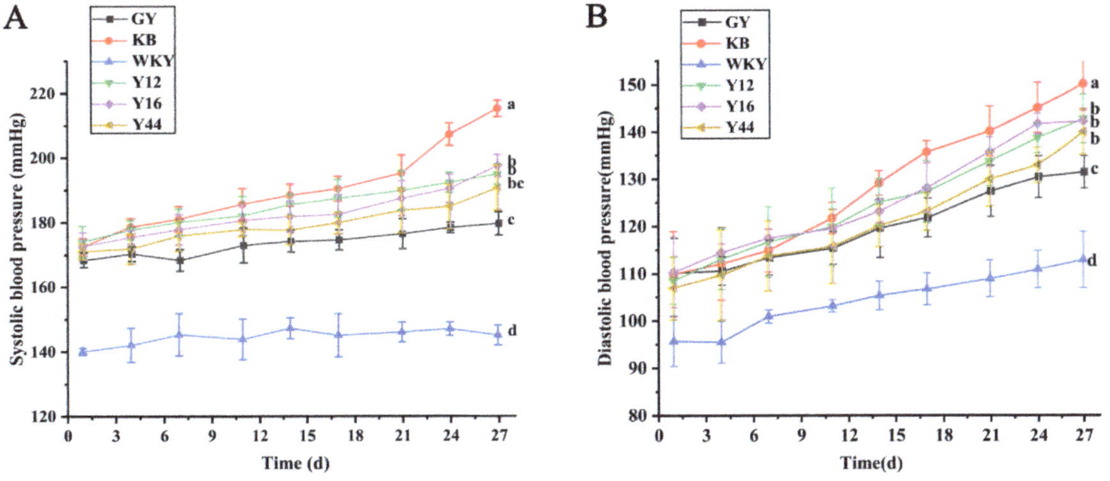

Figure 2. Effects of fermented milk oral administration on systolic blood pressure (SBP) (**A**) and diastolic blood pressure (DBP) (**B**) in spontaneously hypertensive rats (SHRs). Data represent the mean ± SD ($n = 6$, a–d); mean values with different letters over the bars are significantly different ($p < 0.05$) according to Duncan's multiple range test. (GY: gavage feeding of captopril to SHRs; KB: gavage feeding of pasteurized milk to SHRs; WKY: gavage feeding of pasteurized milk to WKYs; Y12: gavage feeding of the fermented milk produced using the combined starter culture mixed with *L. plantarum* Y12 to SHRs; Y44: gavage feeding of the fermented milk produced using the combined starter culture mixed with *L. plantarum* Y44 to SHRs; Y16: gavage feeding of the fermented milk produced using the combined starter culture mixed with *L. plantarum* Y16 to SHRs).

3.5. Effects of Fermented Milk Oral Administration on Blood Lipid Indexes and Oxidation Indexes in the Sera of SHRs and WKY Rats

Hypertension could cause an abnormal blood lipid metabolism. As shown in Figure 4, compared with the WKY group, the contents of TG, TC, and LDL-C in the sera of the SHRs rats in the KB group were significantly increased, while the HDL-C content was significantly decreased ($p < 0.01$), which was in agreement with the results reported by Kazemi et al. [31], who found that hypertension could cause an abnormal blood lipid metabolism. Compared to the KB group, the Y44 and GY groups showed varying degrees of decrease in the TG, TC, and LDL-C contents ($p < 0.01$), but they were still higher than those in the WKY group ($p < 0.05$).

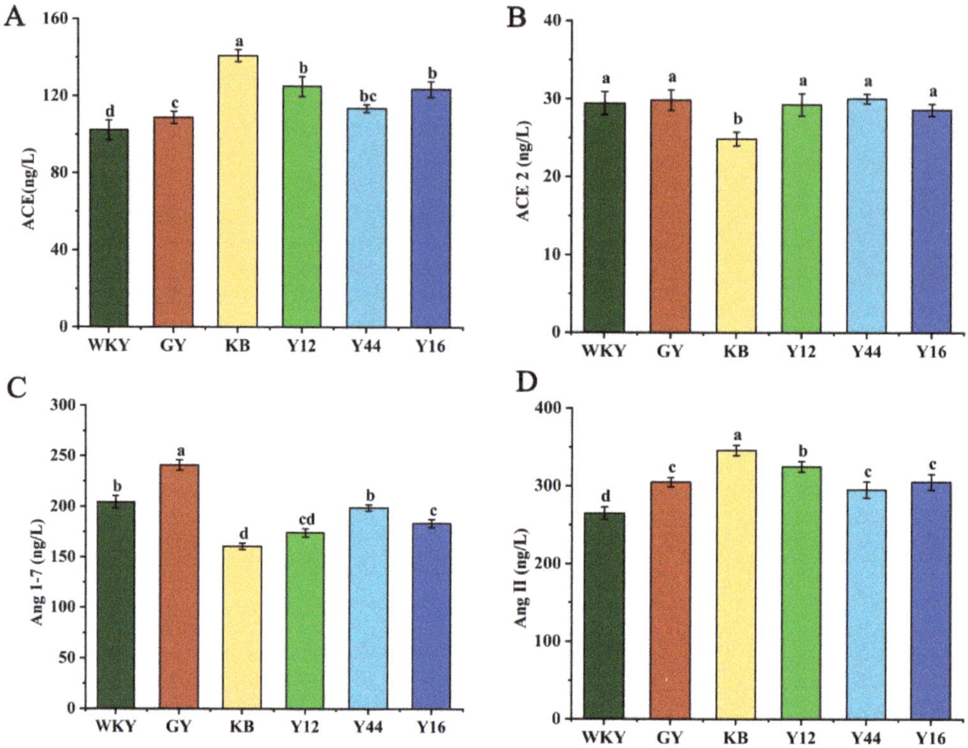

Figure 3. Effects of fermented milk oral administration on ACE (**A**), ACE 2 (**B**), Ang1-7 (**C**), and Ang II (**D**) in spontaneously hypertensive rats (SHRs). Data represent the mean ± SD (n = 6, a–d); mean values with different letters over the bars are significantly different ($p < 0.05$) according to Duncan's multiple range test. (GY: gavage feeding of captopril to SHRs; KB: gavage feeding of pasteurized milk to SHRs; WKY: gavage feeding of pasteurized milk to WKYs; Y12: gavage feeding of the fermented milk produced using the combined starter culture mixed with *L. plantarum* Y12 to SHRs; Y44: gavage feeding of the fermented milk produced using the combined starter culture mixed with *L. plantarum* Y44 to SHRs; Y16: gavage feeding of the fermented milk produced using the combined starter culture mixed with *L. plantarum* Y16 to SHRs).

Oxidative stress is considered a cause, consequence, or enhancing factor in the development of hypertension [32]. As shown in Figure 5, compared with the WKY group, the KB group showed significant decreases in the sera of the SHRs in terms of the antioxidant indicators (SOD, GSH, NO, and CAT, $p < 0.05$), and the content of MDA was significantly increased ($p < 0.05$). After the oral administration of the fermented milk, the MDA content was significantly decreased in the sera of the SHRs ($p < 0.05$), while the CAT activity, GSH activity, NO content, and SOD activity were significantly increased ($p < 0.05$). Among them, the oral administration of the fermented milk produced using the combined starter culture mixed with *L. plantarum* Y44 significantly decreased the content of MDA, increased the SOD activity and NO content, and relieved oxidative stress. The long-term administration of the fermented milk produced using the combined starter culture mixed with *L. plantarum* Y44 significantly relieved the increased blood pressure, oxidative stress, and hyperlipidemia in the SHRs.

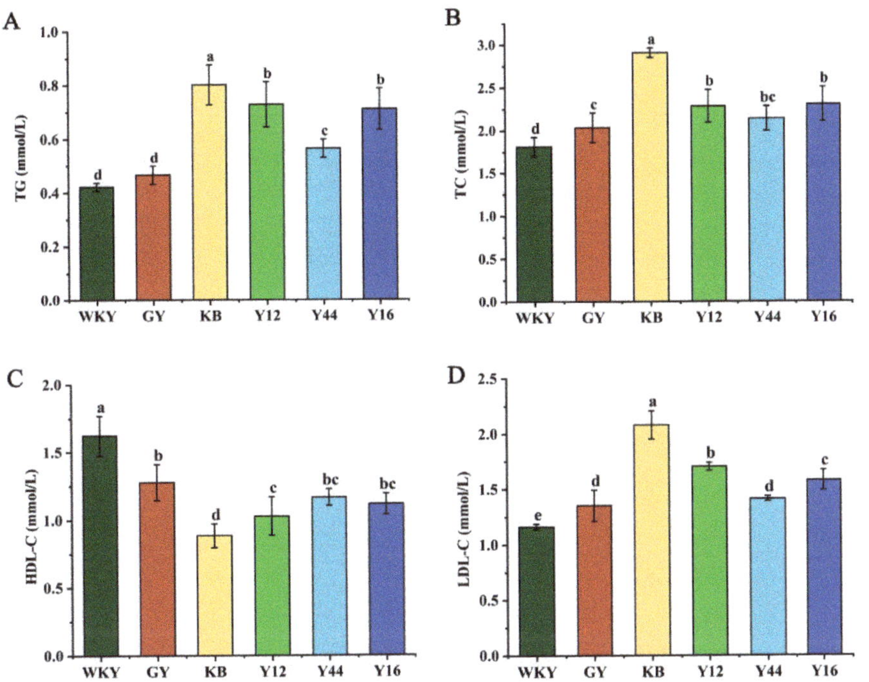

Figure 4. Effects of fermented milk oral administration on serum TG (**A**), TC (**B**), HDL-C (**C**), and LDL-C (**D**) in spontaneously hypertensive rats (SHRs). Data represent the mean ± SD (n = 6, a–d); mean values with different letters over the bars are significantly different ($p < 0.05$) according to Duncan's multiple range test. (GY: gavage feeding of captopril to SHRs; KB: gavage feeding of pasteurized milk to SHRs; WKY: gavage feeding of pasteurized milk to WKYs; Y12: gavage feeding of the fermented milk produced using the combined starter culture mixed with *L. plantarum* Y12 to SHRs; Y44: gavage feeding of the fermented milk produced using the combined starter culture mixed with *L. plantarum* Y44 to SHRs; Y16: gavage feeding of the fermented milk produced using the combined starter culture mixed with *L. plantarum* Y16 to SHRs).

3.6. Ameliorating Effects of Fermented Milk Oral Administration on the Liver, Heart, and Kidney Injury of SHRs and WKY Rats

It is well known that an aminotransferase composed of ALT and AST is the main marker of liver injury [33]. As shown in Figure 6, compared with the WKY group, the contents of AST and ALT in the sera of the SHRs in the GY group exhibited a significant increase ($p < 0.05$), which may be related to the hepatotoxicity of captopril [34]. There was no significant difference in the contents of ATL and AST in the sera of SHRs in the Y44 group, which underwent oral administration of the fermented milk. The results indicate that captopril caused hepatotoxicity; meanwhile, the fermented milk produced by the combined starter culture mixed with *L. plantarum* Y44 was safer.

As shown in Figure 7, compared with the WKY group, the relative weights of the kidneys, livers, and hearts of the SHRs in the KB group were significantly increased ($p < 0.01$). After the gavage treatment, the relative weights of the livers and hearts of the SHRs in the Y44 group significantly decreased ($p < 0.05$), and there was no significant difference among the other groups. This displayed that the fermented milk produced using the combined starter culture mixed with *L. plantarum* Y44 could reduce the organ damage caused by hypertension.

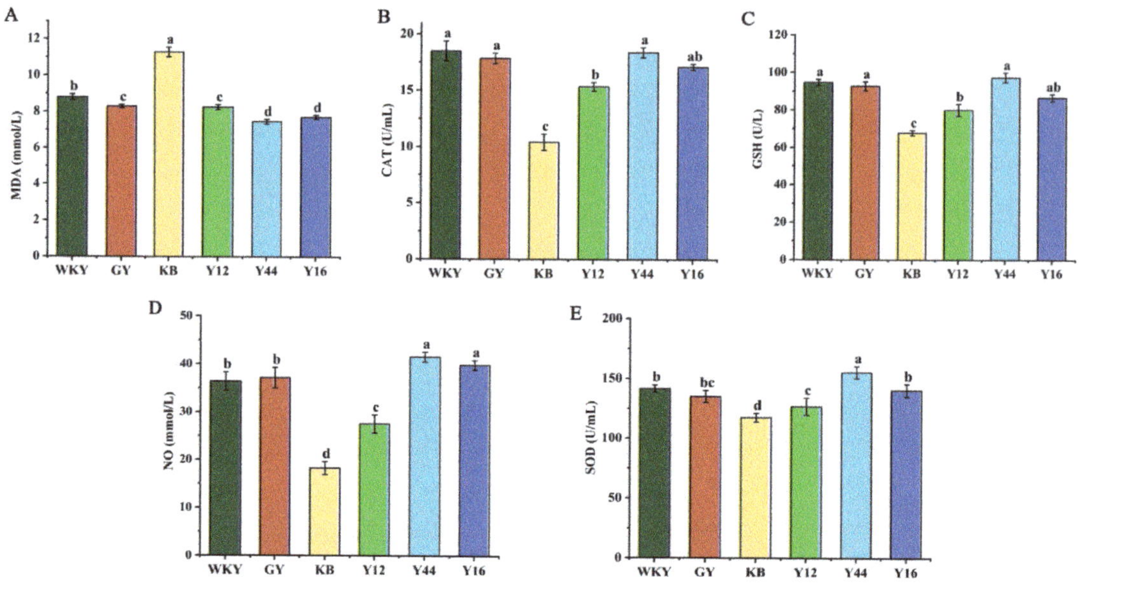

Figure 5. Effects of compound fermented milk on serum antioxidant enzymes in spontaneously hypertensive rats (SHRs), and the effects of MDA (**A**), CAT (**B**), GSH (**C**), NO (**D**), and SOD (**E**). Data represent the mean ± SD (n = 6, a–d); mean values with different letters over the bars are significantly different ($p < 0.05$) according to Duncan's multiple range test. (GY: gavage feeding of captopril to SHRs; KB: gavage feeding of pasteurized milk to SHRs; WKY: gavage feeding of pasteurized milk to WKYs; Y12: gavage feeding of the fermented milk produced using the combined starter culture mixed with *L. plantarum* Y12 to SHRs; Y44: gavage feeding of the fermented milk produced using the combined starter culture mixed with *L. plantarum* Y44 to SHRs; Y16: gavage feeding of the fermented milk produced using the combined starter culture mixed with *L. plantarum* Y16 to SHRs).

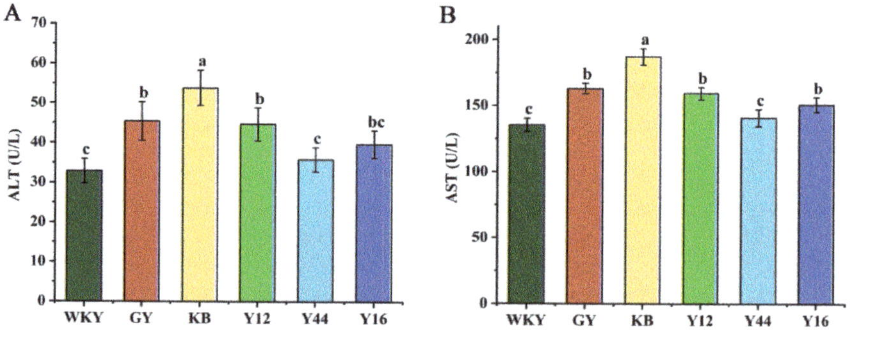

Figure 6. Effects of compound fermented milk on liver damage markers ALT (**A**) and AST (**B**) in sera of spontaneously hypertensive rats (SHRs). Data represent the mean ± SD (n = 6, a–c); mean values with different letters over the bars are significantly different ($p < 0.05$) according to Duncan's multiple range test. (GY: gavage feeding of captopril to SHRs; KB: gavage feeding of pasteurized milk to SHRs; WKY: gavage feeding of pasteurized milk to WKYs; Y12: gavage feeding of the fermented milk produced using the combined starter culture mixed with *L. plantarum* Y12 to SHRs; Y44: gavage feeding of the fermented milk produced using the combined starter culture mixed with *L. plantarum* Y44 to SHRs; Y16: gavage feeding of the fermented milk produced using the combined starter culture mixed with *L. plantarum* Y16 to SHRs).

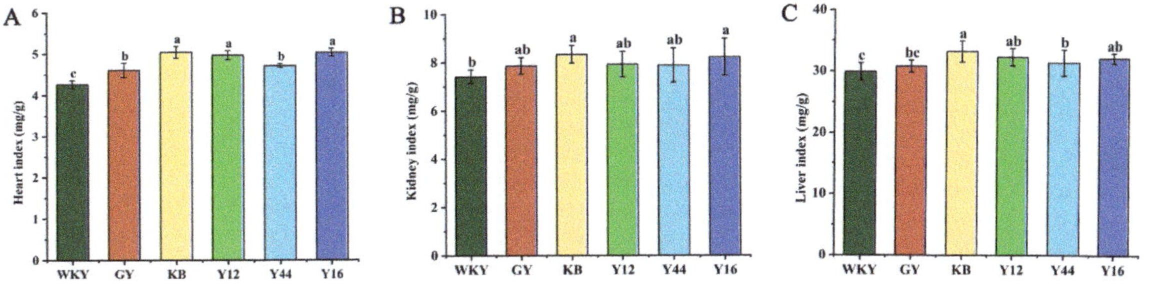

Figure 7. Effects of fermented milk administration on heart index (**A**), kidney index (**B**) and liver index (**C**) in spontaneously hypertensive rats (SHRs). Data represent the mean ± SD (n = 6, a–c); mean values with different letters over the bars are significantly different ($p < 0.05$) according to Duncan's multiple range test. (GY: gavage feeding of captopril to SHRs; KB: gavage feeding of pasteurized milk to SHRs; WKY: gavage feeding of pasteurized milk to WKYs; Y12: gavage feeding of the fermented milk produced using the combined starter culture mixed with *L. plantarum* Y12 to SHRs; Y44: gavage feeding of the fermented milk produced using the combined starter culture mixed with *L. plantarum* Y44 to SHRs; Y16: gavage feeding of the fermented milk produced using the combined starter culture mixed with *L. plantarum* Y16 to SHRs).

As shown in Figure 8, it was observed that hypertension led to an upregulation in the ACE gene expression in the KB group ($p < 0.001$), while ACE 2 and Mas gene expressions were inhibited ($p < 0.001$). However, with the oral administration of the fermented milk produced using the combined starter culture mixed with *L. plantarum* Y44, the expression of ACE 2 and Mas genes was upregulated ($p < 0.01$). In addition, compared with the Y12 and Y16 groups, the gene expression levels of the Y44 group were more significantly upregulated. These results suggest that the fermented milk produced using the combined starter culture mixed with *L. plantarum* Y44 could reverse the imbalance between ACE and ACE 2 by downregulating the ACE-Ang II-AT1 axis and upregulating the ACE 2-Ang (1-7)-Mas axis, increasing the expression of Mas and reversing the RAS, thereby regulated blood pressure.

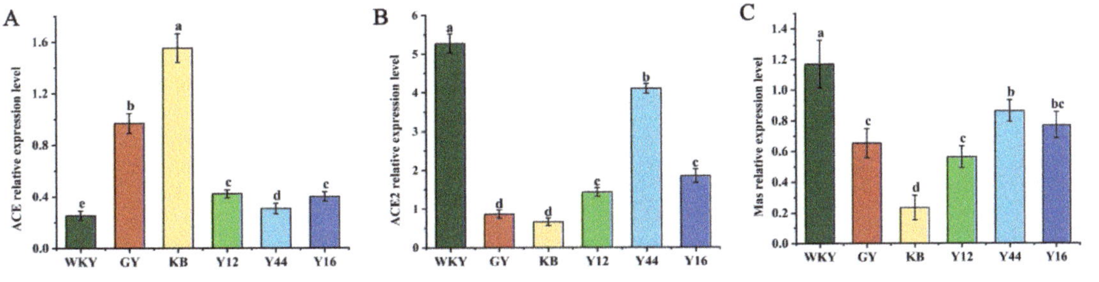

Figure 8. Effect of compound fermented milk on expressions of ACE mRNA (**A**), ACE 2 mRNA (**B**), and MAS mRNA (**C**) in the hearts of spontaneously hypertensive rats (SHRs).. Data represent the mean ± SD (n = 6, a–d); mean values with different letters over the bars are significantly different ($p < 0.05$) according to Duncan's multiple range test. (GY: gavage feeding of captopril to SHRs; KB: gavage feeding of pasteurized milk to SHRs; WKY: gavage feeding of pasteurized milk to WKYs; Y12: gavage feeding of the fermented milk produced using the combined starter culture mixed with *L. plantarum* Y12 to SHRs; Y44: gavage feeding of the fermented milk produced using the combined starter culture mixed with *L. plantarum* Y44 to SHRs; Y16: gavage feeding of the fermented milk produced using the combined starter culture mixed with *L. plantarum* Y16 to SHRs).

Cardiac hypertrophy in rats was primarily characterized by the enlargement of individual cardiomyocytes rather than an increase in their number. Staining techniques could be used to visually observe the morphological changes in the myocardial tissue, providing a clear representation of the damage caused by hypertension to the heart. As shown in Figure 9, the cardiomyocytes in the KB group exhibited signs of swelling and had granular contents. The cell arrangement appeared disordered, and the perinuclear space decreased in size due to myofibril extrusions. Some nuclei also displayed signs of nuclear contraction, and the area of individual cardiomyocytes increased significantly. After the intervention of the fermented milk combined with *L. plantarum* strains and captopril, some cardiomyocytes regained their normal cell shape and organized themselves in a more orderly manner. In all groups, the Y44 group demonstrated a recovery effect that was closer to the normal group (WKY). These results suggest that the fermented milk produced using the combined starter culture mixed with *L. plantarum* Y44 could effectively repair the myocardial fibrosis damage caused by hypertension.

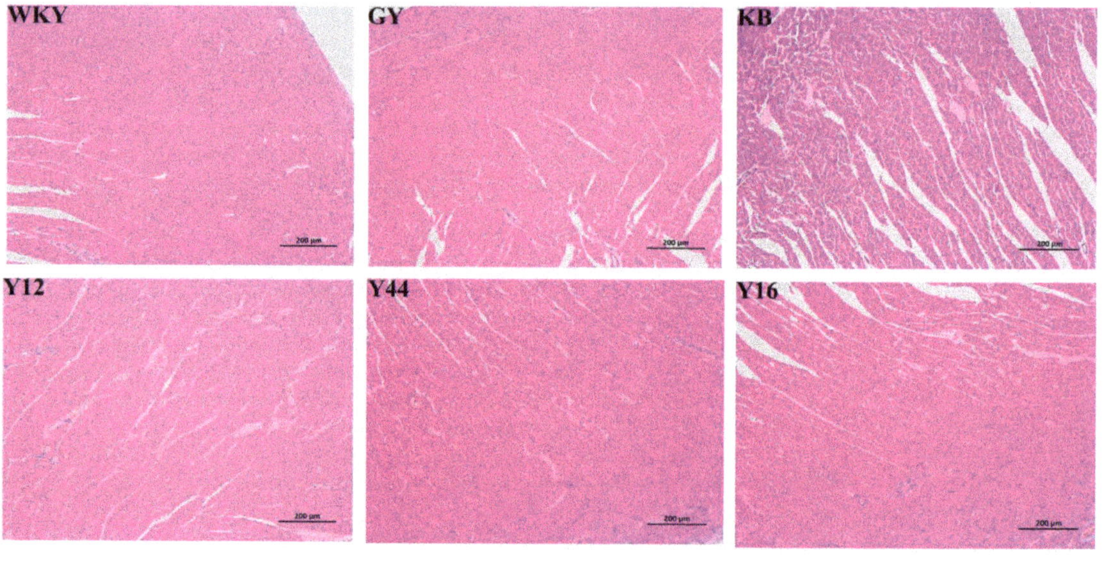

Figure 9. Effects of oral fermented milk on rat heart tissue (50× magnification). GY: gavage feeding of captopril to SHRs; KB: gavage feeding of pasteurized milk to SHRs; WKY: gavage feeding of pasteurized milk to WKYs; Y12: gavage feeding of the fermented milk produced using the combined starter culture mixed with *L. plantarum* Y12 to SHRs; Y44: gavage feeding of the fermented milk produced using the combined starter culture mixed with *L. plantarum* Y44 to SHRs; Y16: gavage feeding of the fermented milk produced using the combined starter culture mixed with *L. plantarum* Y16 to SHRs.

3.7. Effect of Fermented Milk Oral Administration on SCFA Concentration in Rats Feces

SCFAs are end metabolites produced from carbohydrate fermentation by gut microbes, which play an active role in regulating hypertension [35,36]. As shown in Figure 10, it could be observed that the levels of acetic acid, propionic acid, and butyric acid in the feces of rats in the KB group were significantly lower compared to the WKY group ($p < 0.05$). Compared with the KB group, the contents of acetic acid, propionic acid and butyric acid in the feces of the SHRs were significantly increased ($p < 0.05$) after the intervention of fermented milk and captopril. Among the three groups of fermented milk, the SCFA levels (acetic acid, propionic acid, and butyric acid) in the feces showed a higher value in the Y44 group than those in the Y12 and Y16 groups. These results suggest that the fermented milk produced using the combined starter culture mixed with *L. plantarum* Y44 could restore the homeostasis of SCFAs.

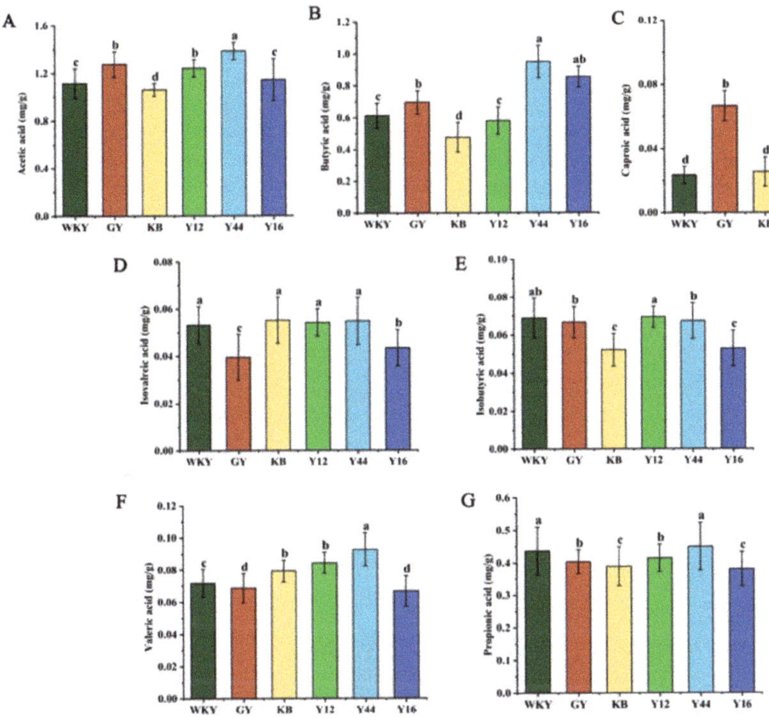

Figure 10. Changes of short-chain fatty acid content in the rat feces: (**A**) acetic acid, (**B**) butyric acid, (**C**) caproic acid, (**D**) isovaleric acid, (**E**) isobutyric acid, (**F**) valerate acid, and (**G**) propionic acid. Data represent the mean ± SD (n = 6, a–d); mean values with different letters over the bars are significantly different ($p < 0.05$) according to Duncan's multiple range test. (GY: gavage feeding of captopril to SHRs; KB: gavage feeding of pasteurized milk to SHRs; WKY: gavage feeding of pasteurized milk to WKY rats; Y12: gavage feeding of the fermented milk produced using the combined starter culture mixed with *L. plantarum* Y12 to SHRs; Y44: gavage feeding of the fermented milk produced using the combined starter culture mixed with *L. plantarum* Y44 to SHRs; Y16: gavage feeding of the fermented milk produced using the combined starter culture mixed with *L. plantarum* Y16 to SHRs).

3.8. Effect of Fermented Milk Oral Administration on the Gut Microbiota of Rats

As shown in Figure 11, compared with the KB group, hypertension caused significant decreases in the Sobs, ACE index, Simpson index, and Shannon index in the KB group, indicating that hypertension was associated with a reduced intestinal flora diversity. The levels of the Sobs, Simpson index, and ACE index were significantly increased ($p < 0.05$) after the oral administration of the fermented milk produced using the combined starter culture mixed with *L. plantarum* Y44. These results indicate that the fermented milk produced using the combined starter culture mixed with *L. plantarum* Y44 treatment had a significant effect on improving the diversity of intestinal flora in hypertensive rats.

As shown in Figure 12A,B, the dominant phyla of intestinal microorganisms were Firmicutes (F) and Bacteroides (B), accounting for 75–80%. Hypertension significantly increases the relative abundance of Firmicutes in the guts of rats, while significantly reducing the relative abundance of Bacteroides. Following the fermented milk intervention, the relative abundance of Bacteroides and Firmicutes of mice in the Y44 and Y16 groups was restored ($p < 0.05$), and the increase in the F/B ratio caused by hypertension was mitigated. As shown in Figure 12C, at the family level, compared with the WKY group, the levels of Lactobacillusceae and Lachnospiraceae were significantly decreased in the KB group, as well as the relative abundance of *Murinobacteriaceae*, *Ruminobacteriaceae*, and *Pseudomonadaceae*.

The relative abundance of *Murinobacteriaceae*, Pseudomonadaceae, and *Ruminobacteriaceae* was downregulated, and the relative abundance of *Lachnospiraceae*, *Prevotellaceae*, and *Lactobacillusceae* in the SHRs was upregulated, as well as the relative abundance of rats orally administrated the fermented milk produced using the combined starter culture with *L. plantarum* Y44. As shown in Figure 12D,E, at the genus level, hypertension led to a decrease in the proportions of Turicibacter, Lactobacillus, Ruminococcus, Bacteroides, Ligilactobacillus, and Oscillibacter in the guts of SHRs in the KB group. Compared with the KB group, the oral administration of the fermented milk produced using the combined starter culture mixed with *L. plantarum* Y44 restored the relative abundance of Romboutsia, Alloprevotella, and Colidextribacter induced by hypertension, and the relative abundance of Turicibacter, Lactobacillus, Ruminococcus, Bacteroides, Ligilactobacillus, and Oscillibacter improved.

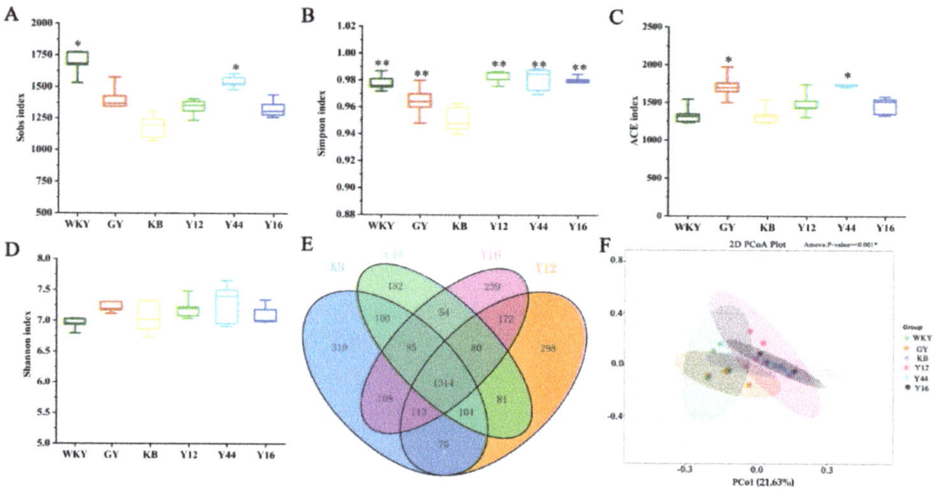

Figure 11. Intestinal flora diversity of rats in different treatment groups: (**A**) Sobs diversity analysis of different groups, (**B**) Simpson index diversity analysis, (**C**) ACE index analysis, (**D**) Shannon index diversity analysis, out Venn diagram outOTU number, (**E**) Venn diagram of OTU number and (**F**) PCoA analysis based on OTU. The data are expressed as the median; * $p < 0.05$, ** $p < 0.01$. (GY: gavage feeding of captopril to SHRs; KB: gavage feeding of pasteurized milk to SHRs; WKY: gavage feeding of pasteurized milk to WKY rats; Y12: gavage feeding of the fermented milk produced using the combined starter culture mixed with *L. plantarum* Y12 to SHRs; Y44: gavage feeding of the fermented milk produced using the combined starter culture mixed with *L. plantarum* Y44 to SHRs; Y16: gavage feeding of the fermented milk produced using the combined starter culture mixed with *L. plantarum* Y16 to SHRs).

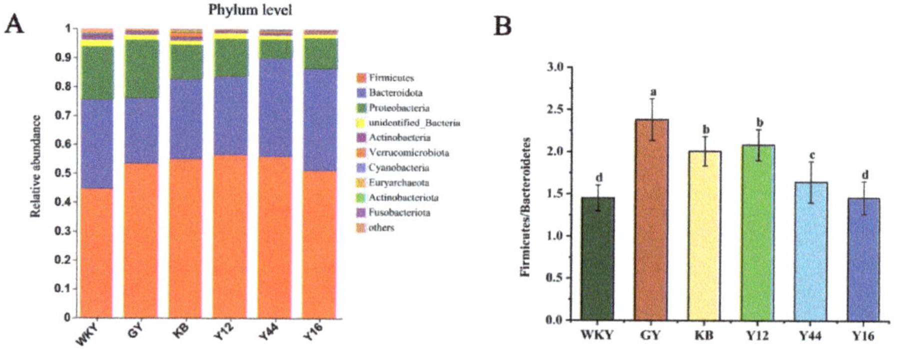

Figure 12. *Cont.*

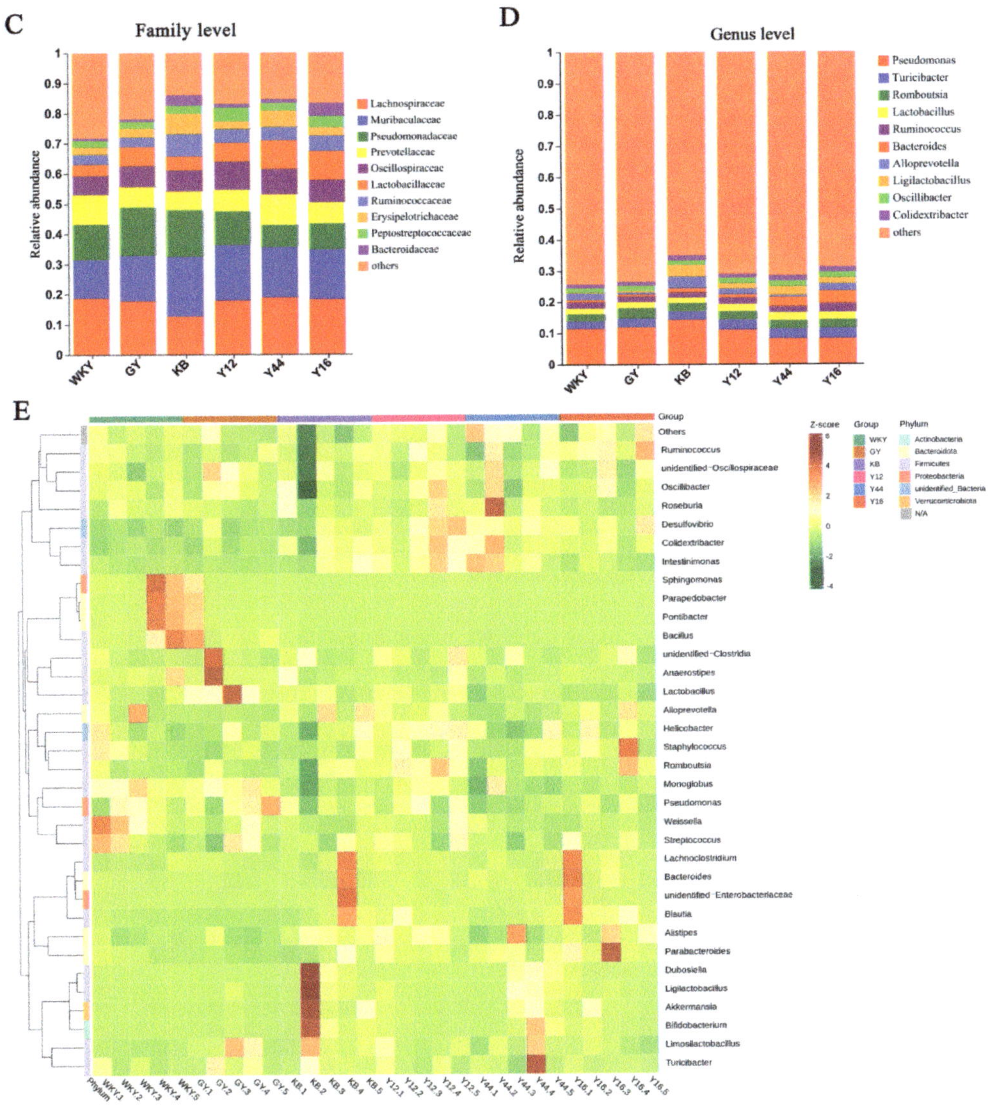

Figure 12. (**A**) Bar graph of gate levels for different groups, (**B**) Firmicutes/Bacteroidetes ratio for each group, (**C**) columnar distribution of intestinal flora in rats at the family level, (**D**) histogram of community structure distribution of rat intestinal flora at the genus level, and (**E**) heat map of the 30 most abundant species of bacteria in rat feces. The data in Figure 12B represent the mean ± SD (n = 6, a–d); mean values with different letters over the bars are significantly different ($p < 0.05$) according to Duncan's multiple range test. (GY: gavage feeding of captopril to SHRs; KB: gavage feeding of pasteurized milk to SHRs; WKY: gavage feeding of pasteurized milk to WKY rats; Y12: gavage feeding of the fermented milk produced using the combined starter culture mixed with *L. plantarum* Y12 to SHRs; Y44: gavage feeding of the fermented milk produced using the combined starter culture mixed with *L. plantarum* Y44 to SHRs; Y16: gavage feeding of the fermented milk produced using the combined starter culture mixed with *L. plantarum* Y16 to SHRs).

4. Discussion

Hypertension is a common and severe chronic disease that poses a serious threat to human health and is also the primary cause of the global burden of diseases [1]. ACEI can effectively treat hypertension in patients with renal hypertension, but the chemical synthesis of ACEI will bring in many adverse effects, so there is a growing interest in extracting more effective and safer ACEIs from dietary sources [37]. Probiotic-rich foods are currently dominating the market, and dairy products are widely regarded as the most effective carriers of probiotics [38]. Fermented milks have gained increasing recognition among the public due to their positive health effects [39]. It was reported that fermented milk showed antihypertensive effects, and the long-term consumption of fermented dairy products may reduce the risk of hypertension [40]. In this study, we evaluate the antihypertension effect of a fermented milk produced using the combined starter culture mixed with *L. plantarum* Y44 in the SHRs. The results indicate that the long-term oral administration of the fermented milk produced using the combined starter culture mixed with *L. plantarum* Y44 reduced the blood pressure, alleviated the hypertension-associated intestinal flora disorder, and increased the intestinal SCFA contents of the SHRs, which were attributed to the ACE-inhibitory peptides, such as AMKPWIQPK, GPVRGPFPII, LNVPGEIVE, NIPPLTQTPV, and YQEPVL, produced in the fermented milk.

In addition, studies had reported that the fermented milk from *Lactiplantibacillus plantarum* SR37-3 and SR61-2 showed an antihypertensive activity in the NOS inhibitor L-NAME-induced hypertensive rats [41]. It was reported that different starter culture strains have different proteolytic activities and proteolytic systems, and produce ACE-inhibitory peptides with different activities in fermented milk [42]. In addition, the heterogeneity between different strains leads to differences in their metabolites, ultimately leading to different antihypertensive potentials. In this study, the mixed milk fermented using the three strains of *L. plantarum* Y44, Y12, and Y16 combined with commercial starter cultures showed different effects on relieving hypertension in SHRs.

Oxidative stress is the main cause of inflammation and plays a crucial role in the development of hypertension and organ damage [43]. Mice lacking mitochondrial superoxide dismutase exhibit elevated levels of oxidative and inflammatory markers, leading to an increase in blood pressure [44]. Studies have shown that the main cause of microalbuminuria in hypertensive patients is oxidative stress, and excessive ROS can cause glomerular cell dysfunction and renal vascular endothelial damage [45]. The oral administration of the fermented milk produced using the combined starter culture mixed with *L. plantarum* Y44 significantly decreased the MDA content and increased the NO content and SOD activity in the SHRs, indicating that the intragastric administration of the fermented milk produced using the combined starter culture mixed with *L. plantarum* Y44 relieved oxidative stress in SHRs.

Research has shown that chronic kidney disease is usually caused by persistent hypertension, and the disorder of the RAS has an undeniable impact on the occurrence and development of hypertension [46,47]. An ACE-AngII-AT1 axis and an ACE2-Ang (1-7)-Mas axis make up the RAS. The excessive activation of the ACE regulatory axis and inhibition of the ACE 2 regulatory axis can lead to increased blood pressure and kidney damage [48]. Previous studies have reported that RAS diseases play an undeniable role in the occurrence and development of hypertension [49]. In this study, the contents of ACE and Ang II in the sera of SHRs were decreased by the fermented milk administration, and the contents of ACE 2 and Ang 1-7 in the sera of SHRs were significantly increased. The RAS was regulated at the gene level by upregulating the expression of ACE 2 and the MAS receptor and downregulating the expression of the ACE receptor, thereby controlling blood pressure. Many physiological effects of Ang (1-7) were mediated by Mas receptors. Xu et al. found that the overexpression of the ACE 2 gene in the brain of a hypertensive mouse model induced by a continuous infusion of Ang II could improve the baroreflex sensitivity and parasympathetic nerve activity and slow down the development of neurogenic hypertension [50]. The results of this study prove that the antihypertensive effect of ACE 2 was mediated by Ang

(1-7)-Mas. In addition, the transfection of the ACE 2 gene could inhibit Ang II-induced oxidative stress, which could be reversed by Mas receptor antagonist A779. These results suggested that ACE-inhibiting peptides produced in the fermented milk containing *L. plantarum* Y44, such as AMKPWIQPK, GPVRGPFPII, LNVPGEIVE, NIPPLTQTPV, and YQEPVL, may be involved in the blood pressure regulation of the ACE 2-Ang (1-7)-Mas axis, which rebalance RAS signaling to regulate blood pressure by downregulating the ACE-Ang II-AT1 axis and upregulating the ACE 2-Ang (1-7)-Mas axis. The contents of ALT and AST could indicate the degree of liver cell damage. The changes in ALT and AST in the sera of the SHRs indicated that the fermented milk produced using the combined starter culture mixed with *L. plantarum* Y44 was safer than captopril.

Studies have shown that hypertension can affect the structure of gut microbiota, reduce the diversity of the gut microbiota, and reduce the number of beneficial bacteria [51]. The difference in the intestinal flora composition between hypertensive and normal rats was mainly due to the increase in the ratio of Firmicutes to Bacteroidetes (F/B) and the decrease in microbial diversity in the SHRs. In this study, the fermented milk containing *L. plantarum* Y44 may regulate the disruption of gut microbiota caused by hypertension at the phylum level. Previous studies have demonstrated a connection between hypertension and an intestinal microbiota imbalance and circulating SCFA levels [52]. An increase in the number of bacteria (Bacteroides, Firmicutes, Lactobacillaceae, Prevotellaceae, Lachnospiraceae, and Alloprevotella) that produce SCFAs (acetic acid, propionic acid, and butyric acid), might be one reason why the fermented milk lowers blood pressure [52]. At the same time, SCFAs could, through the regulation of T cells, promote Ang II to produce an effector response and improve the function of the immune system to achieve the effect of a lower blood pressure [35]. Lal et al. reported that SCFAs, particularly butyrate, could help regulate blood pressure by directly activating the vagus nerve's afferent nerve fibers after being absorbed into the intestine's mucosal lamina propria nerve terminals [53]. Moreover, Goswami et al. showed that SCFAs can significantly activate vagal afferent neurons by increasing phosphorylation, exhibiting the order of butyrate > propionate > acetate [54]. Additionally, butyrate can directly act on the central nervous system to regulate blood pressure by crossing the blood–brain barrier through specific transport proteins. In addition, the expression of butyric acid receptors in the hypothalamus of SHRs was relatively low, leading to a decrease in reactivity. Therefore, the role of butyrate in blood pressure regulation was affected. Some studies have found that the propionate SCFAs have a good antihypertensive effect [36]. Acetate also has a positive effect on regulating hypertension and preventing atherosclerosis, and malonate can improve cardiac function after myocardial infarction [55]. Similarly, Onyszkiewicz et al. demonstrated that butyrate can enter the bloodstream through the intestinal vascular barrier and act on GPR41 and GPR43 to relax the mesenteric arteries, thereby significantly reducing hypertension [56]. Therefore, we speculated that our fermented milk combined with *L. plantarum* Y44 can alleviate hypertension-associated intestinal microbiota dysbiosis by recovering the diversity of the gut microbiota and altering the key floras, which are SCFA producers, and that SCFA further reduces blood pressure and improves cardiovascular effects through G protein-coupled receptors.

5. Conclusions

In this study, we used the commercial starter combined with *L. plantarum* strains to ferment the mixed milk of cow milk and soy milk. Our results showed that the combined starter culture containing *L. plantarum* Y44 can generate ACE-inhibitory peptides in the fermented milk, such as AMKPWIQPK, GPVRGPFPII, LNVPGEIVE, NIPPLTQTPV, and YQEPVL, and its characteristics depend on the strain. And the long-term oral administration of the fermented milk can reduce blood pressure, alleviate the hypertension-associated intestinal flora disorder by restoring the intestinal flora diversity, and increase the intestinal SCFA levels in the SHRs. SCFAs can further reduce the blood pressure and improve the cardiovascular function through G protein-coupled receptors. The results indicate that the fermented milk from the combined starter culture containing *L. plantarum* Y44 can be used as a functional food with antihypertension effects.

Author Contributions: J.Y.: investigation, methodology, formal analysis, software, visualization, writing—original draft; M.S.: validation, investigation, and data curation; S.J.: project administration, resources, validation, and funding acquisition; C.J.: project administration, resources, validation, and funding acquisition; G.M.: investigation, supervision, and methodology; Y.T.: conceptualization, supervision, and writing—review and editing. All authors have read and agreed to the published version of the manuscript.

Funding: This research was financially supported by the Key Technology and Product Development for Functional Dairy Food Project from Heilongjiang Feihe Dairy Co., Ltd. (Funding number: 2020210205000212).

Institutional Review Board Statement: All animal experiments were carried out in accordance with the guidelines of the Experimental Animal Ethics Committee of Dalian Polytechnic University (SYXK2017-0005).

Informed Consent Statement: Informed consent was obtained from all subjects involved in the study.

Data Availability Statement: The original contributions presented in the study are included in the article, further inquiries can be directed to the corresponding author.

Conflicts of Interest: The authors declare that this study received funding from Heilongjiang Feihe Dairy Co., Ltd. The funder had the following involvement with the study: project administration, resources, validation, and funding acquisition.

Appendix A

Table A1. Sensory evaluation standard of compound fermented milk.

Evaluation Index	Evaluation Index	Score
Sour–sweet ratio (15)	Moderate sweet and sour	11–15
	Slightly sour or slightly sweet	6–10
	Bitter, too sour, or too sweet	0–5
Viscosity (15)	Medium viscosity, thick	11–15
	Viscosity is viscous or thin	6–10
	Viscosity is too sticky or too thin	0–5
Consistency (15)	Good viscosity, fine and silky	11–15
	Better viscosity and fineness	6–10
	Poor viscosity and no wire drawing	0–5
Taste (20)	No sense of graininess, smoothness and astringency	16–20
	The taste of the granules is slightly soft or hard and slightly powdery	11–15
	The texture of the granules is soft or hard and slightly powdery	6–10
	The taste of the granules is too soft or too hard, and the powder is very astringent	0–5
Flavor (20)	Fermented milk has an obvious flavor, natural fermentation flavor and smell, and no peculiar smell	16–20
	The flavor of the fermented milk is not enough, and the flavor of natural fermentation is low	11–15
	The flavor of the fermented milk is poor, and the flavor of natural fermentation is poor	6–10
	Fermented milk has no characteristic flavor or produces an unpleasant flavor	0–5
Texture (15)	Uniform color, milky white, good organization, no bubbles, no whey precipitation	11–15
	Light gray or grayish white color, good structure, slight whey precipitation	6–10
	Abnormal color, rough tissue and severe whey precipitation	0–5

References

1. Hall, J.E.; Granger, J.P.; do Carmo, J.M.; da Silva, A.A.; Dubinion, J.; George, E.; Hamza, S.; Speed, J.; Hall, M.E. Hypertension: Physiology and pathophysiology. *Compr. Physiol.* **2012**, *2*, 2393–2442. [CrossRef]

2. Guang, C.; Phillips, R.D.; Jiang, B.; Milani, F. Three key proteases—angiotensin-I-converting enzyme (ACE), ACE2 and renin—within and beyond the renin-angiotensin system. *Arch. Cardiovasc. Dis.* **2012**, *105*, 373–385. [CrossRef]
3. Shen, X.; Xie, A.; Li, Z.; Jiang, C.; Wu, J.; Li, M.; Yue, X. Research Progress for Probiotics Regulating Intestinal Flora to Improve Functional Dyspepsia: A Review. *Foods* **2024**, *13*, 151. [CrossRef] [PubMed]
4. Companys, J.; Pla-Pagà, L.; Calderón-Pérez, L.; Llauradó, E.; Solà, R.; Pedret, A.; Valls, R.M. Fermented Dairy Products, Probiotic Supplementation, and Cardiometabolic Diseases: A Systematic Review and Meta-Analysis. *Adv. Nutr.* **2020**, *11*, 834–863. [CrossRef]
5. Xie, A.; Zhao, S.; Liu, Z.; Yue, X.; Shao, J.; Li, M.; Li, Z. Polysaccharides, proteins, and their complex as microencapsulation carriers for delivery of probiotics: A review on carrier types and encapsulation techniques. *Int. J. Biol. Macromol.* **2023**, *242 Pt 1*, 124784. [CrossRef]
6. Khalesi, S.; Sun, J.; Buys, N.; Jayasinghe, R. Effect of probiotics on blood pressure: A systematic review and meta-analysis of randomized, controlled trials. *Hypertension* **2014**, *64*, 897–903. [CrossRef]
7. Touyz, R.M.; Briones, A.M. Reactive oxygen species and vascular biology: Implications in human hypertension. *Hypertens. Res. Off. J. Jpn. Soc. Hypertens.* **2011**, *34*, 5–14. [CrossRef] [PubMed]
8. Marques, F.Z.; Mackay, C.R.; Kaye, D.M. Beyond gut feelings: How the gut microbiota regulates blood pressure. *Nat. Rev. Cardiol.* **2018**, *15*, 20–32. [CrossRef]
9. Lynch, S.V.; Pedersen, O. The Human Intestinal Microbiome in Health and Disease. *N. Engl. J. Med.* **2016**, *375*, 2369–2379. [CrossRef] [PubMed]
10. Li, J.; Zhao, F.; Wang, Y.; Chen, J.; Tao, J.; Tian, G.; Wu, S.; Liu, W.; Cui, Q.; Geng, B.; et al. Gut microbiota dysbiosis contributes to the development of hypertension. *Microbiome* **2017**, *5*, 14. [CrossRef]
11. Yang, T.; Santisteban, M.M.; Rodriguez, V.; Li, E.; Ahmari, N.; Carvajal, J.M.; Zadeh, M.; Gong, M.; Qi, Y.; Zubcevic, J.; et al. Gut dysbiosis is linked to hypertension. *Hypertension* **2015**, *65*, 1331–1340. [CrossRef]
12. Macfarlane, G.T.; Macfarlane, S. Bacteria, colonic fermentation, and gastrointestinal health. *J. AOAC Int.* **2012**, *95*, 50–60. [CrossRef] [PubMed]
13. Pluznick, J.L.; Protzko, R.J.; Gevorgyan, H.; Peterlin, Z.; Sipos, A.; Han, J. Olfactory receptor responding to gut microbiota-derived signals plays a role in renin secretion and blood pressure regulation. *Proc. Natl. Acad. Sci. USA* **2013**, *110*, 4410–4415. [CrossRef] [PubMed]
14. Li, C.; Kwok, L.Y.; Mi, Z.; Bala, J.; Xue, J.; Yang, J.; Ma, Y.; Zhang, H.; Chen, Y. Characterization of the angiotensin-converting enzyme inhibitory activity of fermented milks produced with *Lactobacillus casei*. *J. Dairy Sci.* **2017**, *100*, 9495–9507. [CrossRef]
15. Song, T.; Lv, M.; Sun, B.; Zheng, L.; Zhao, M. Tripeptides Val-Pro-Pro (VPP) and Ile-Pro-Pro (IPP) Regulate the Proliferation and Migration of Vascular Smooth Muscle Cells by Interfering Ang II-Induced Human Umbilical Vein Endothelial Cells Derived EVs Delivering RNAs to VSMCs in the Co-Culture Model. *J. Agric. Food Chem.* **2020**, *68*, 6628–6637. [CrossRef]
16. Xie, A.; Dong, Y.; Liu, Z.; Li, Z.; Shao, J.; Li, M.; Yue, X. A Review of Plant-Based Drinks Addressing Nutrients, Flavor, and Processing Technologies. *Foods* **2023**, *12*, 3952. [CrossRef]
17. Li, H.; Prairie, N.; Udenigwe, C.C.; Adebiyi, A.P.; Tappia, P.S.; Aukema, H.M.; Jones, P.J.; Aluko, R.E. Blood pressure lowering effect of a pea protein hydrolysate in hypertensive rats and humans. *J. Agric. Food Chem.* **2011**, *59*, 9854–9860. [CrossRef]
18. Liu, C.F.; Hu, C.L.; Chiang, S.S.; Tseng, K.C.; Yu, R.C.; Pan, T.M. Beneficial preventive effects of gastric mucosal lesion for soy-skim milk fermented by lactic acid bacteria. *J. Agric. Food Chem.* **2009**, *57*, 4433–4438. [CrossRef]
19. Liu, Y.; Gao, Y.; Ma, F.; Sun, M.; Mu, G.; Tuo, Y. The ameliorative effect of *Lactobacillus plantarum* Y44 oral administration on inflammation and lipid metabolism in obese mice fed with a high fat diet. *Food Funct.* **2020**, *11*, 5024–5039. [CrossRef] [PubMed]
20. Sun, M.; Liu, Y.; Song, Y.; Gao, Y.; Zhao, F.; Luo, Y.; Qian, F.; Mu, G.; Tuo, Y. The ameliorative effect of *Lactobacillus plantarum*-12 on DSS-induced murine colitis. *Food Funct.* **2020**, *11*, 5205–5222. [CrossRef] [PubMed]
21. Ge, X.; Tang, N.; Huang, Y.; Chen, X.; Dong, M.; Rui, X.; Zhang, Q.; Li, W. Fermentative and physicochemical properties of fermented milk supplemented with sea buckthorn (*Hippophae eleagnaceae* L.). *LWT* **2022**, *153*, 112484. [CrossRef]
22. Yu, J.; Guo, H.; Sun, M.; Jiang, C.; Jiang, S.; Mu, G.; Tuo, Y.; Gao, P. Milk fermented by combined starter cultures comprising three *Lactobacillus* strains exerts an alleviating effect on loperamide-induced constipation in BALB/c mice. *Food Funct.* **2023**, *14*, 5264–5276. [CrossRef]
23. *GB 5009.239-2016*; National Food Safety Standard Determination of Acidity in Food. National Standards of the People's Republic of China: Beijing, China, 2016.
24. Laroque, D.; Inisan, C.; Berger, C.; Vouland, É.; Dufossé, L.; Guérard, F. Kinetic study on the Maillard reaction. Consideration of sugar reactivity. *Food Chem.* **2008**, *111*, 1032–1042. [CrossRef]
25. *GB 19302-2010*; National Food Safety Standard Fermented Milk. National Standards of the People's Republic of China: Beijing, China, 2010.
26. Wu, N.; Xu, W.; Liu, K.; Xia, Y.; Shuang, Q. Angiotensin-converting enzyme inhibitory peptides from *Lactobacillus delbrueckii* QS306 fermented milk. *J. Dairy Sci.* **2019**, *102*, 5913–5921. [CrossRef]
27. Mao, J.; Zhou, Z.; Yang, H. Microbial succession and its effect on the formation of umami peptides during sufu fermentation. *Front. Microbiol.* **2023**, *14*, 1181588. [CrossRef] [PubMed]
28. Wang, A.; Li, P.; Ma, F.; Li, X.; Mu, G.; Tuo, Y. Mixed *Lactiplantibacillus plantarum* strains alleviated DSS-induced intestinal inflammation of Balb/c mice via the 5-HT/5-HT7R/NF-κB signaling pathway. *J. Funct. Foods* **2023**, *102*, 105435. [CrossRef]

29. Wang, J.; Liu, R.; Qi, H.; Wang, Y.; Cui, L.; Wen, Y.; Li, H.; Yin, C. The ACE2-angiotensin-(1-7)-Mas axis protects against pancreatic cell damage in cell culture. *Pancreas* **2015**, *44*, 266–272. [CrossRef] [PubMed]
30. Sun, X.; Wang, M.; Xu, C.; Wang, S.; Li, L.; Zou, S.; Yu, J.; Wei, Y. Positive Effect of a Pea-Clam Two-Peptide Composite on Hypertension and Organ Protection in Spontaneously Hypertensive Rats. *Nutrients* **2022**, *14*, 4069. [CrossRef] [PubMed]
31. Kazemi, M.; McBreairty, L.E.; Chizen, D.R.; Pierson, R.A.; Chilibeck, P.D.; Zello, G.A. A Comparison of a Pulse-Based Diet and the Therapeutic Lifestyle Changes Diet in Combination with Exercise and Health Counselling on the Cardio-Metabolic Risk Profile in Women with Polycystic Ovary Syndrome: A Randomized Controlled Trial. *Nutrients* **2018**, *10*, 1387. [CrossRef] [PubMed]
32. Franco, C.; Sciatti, E.; Favero, G.; Bonomini, F.; Vizzardi, E.; Rezzani, R. Essential Hypertension and Oxidative Stress: Novel Future Perspectives. *Int. J. Mol. Sci.* **2022**, *23*, 14489. [CrossRef] [PubMed]
33. McGill, M.R. The past and present of serum aminotransferases and the future of liver injury biomarkers. *EXCLI J.* **2016**, *15*, 817–828. [CrossRef]
34. Kelleni, M.T.; Ibrahim, S.A.; Abdelrahman, A.M. Effect of captopril and telmisartan on methotrexate-induced hepatotoxicity in rats: Impact of oxidative stress, inflammation and apoptosis. *Toxicol. Mech. Methods* **2016**, *26*, 371–377. [CrossRef]
35. Yang, F.; Chen, H.; Gao, Y.; An, N.; Li, X.; Pan, X.; Yang, X.; Tian, L.; Sun, J.; Xiong, X.; et al. Gut microbiota-derived short-chain fatty acids and hypertension: Mechanism and treatment. *Biomed. Pharmacother.* **2020**, *130*, 110503. [CrossRef] [PubMed]
36. Cui, J.; Lian, Y.; Zhao, C.; Du, H.; Han, Y.; Gao, W.; Xiao, H.; Zheng, J. Dietary Fibers from Fruits and Vegetables and Their Health Benefits via Modulation of Gut Microbiota. *Compr. Rev. Food Sci. Food Saf.* **2019**, *18*, 1514–1532. [CrossRef]
37. Marco, M.L.; Heeney, D.; Binda, S.; Cifelli, C.J.; Cotter, P.D.; Foligné, B.; Gänzle, M.; Kort, R.; Pasin, G.; Pihlanto, A.; et al. Health benefits of fermented foods: Microbiota and beyond. *Curr. Opin. Biotechnol.* **2017**, *44*, 94–102. [CrossRef] [PubMed]
38. Lin, P.P.; Hsieh, Y.M.; Kuo, W.W.; Lin, C.C.; Tsai, F.J.; Tsai, C.H.; Huang, C.Y.; Tsai, C.C. Inhibition of cardiac hypertrophy by probiotic-fermented purple sweet potato yogurt in spontaneously hypertensive rat hearts. *Int. J. Mol. Med.* **2012**, *30*, 1365–1375. [CrossRef]
39. Ahrén, I.L.; Xu, J.; Önning, G.; Olsson, C.; Ahrné, S.; Molin, G. Antihypertensive activity of blueberries fermented by *Lactobacillus plantarum* DSM 15313 and effects on the gut microbiota in healthy rats. *Clin. Nutr.* **2015**, *34*, 719–726. [CrossRef]
40. Beltrán-Barrientos, L.M.; Hernández-Mendoza, A.; Torres-Llanez, M.J.; González-Córdova, A.F.; Vallejo-Córdoba, B. Invited review: Fermented milk as antihypertensive functional food. *J. Dairy Sci.* **2016**, *99*, 4099–4110. [CrossRef] [PubMed]
41. Yuan, L.; Li, Y.; Chen, M.; Xue, L.; Wang, J.; Ding, Y.; Zhang, J.; Wu, S.; Ye, Q.; Zhang, S.; et al. Antihypertensive Activity of Milk Fermented by *Lactiplantibacillus plantarum* SR37-3 and SR61-2 in L-NAME-Induced Hypertensive Rats. *Foods* **2022**, *11*, 2332. [CrossRef]
42. Chaves-López, C.; Serio, A.; Paparella, A.; Martuscelli, M.; Corsetti, A.; Tofalo, R.; Suzzi, G. Impact of microbial cultures on proteolysis and release of bioactive peptides in fermented milk. *Food Microbiol.* **2014**, *42*, 117–121. [CrossRef]
43. Rubattu, S.; Pagliaro, B.; Pierelli, G.; Santolamazza, C.; Di Castro, S.; Mennuni, S.; Volpe, M. Pathogenesis of target organ damage in hypertension: Role of mitochondrial oxidative stress. *Int. J. Mol. Sci.* **2014**, *16*, 823–839. [CrossRef] [PubMed]
44. Lob, H.E.; Marvar, P.J.; Guzik, T.J.; Sharma, S.; McCann, L.A.; Weyand, C.; Gordon, F.J.; Harrison, D.G. Induction of hypertension and peripheral inflammation by reduction of extracellular superoxide dismutase in the central nervous system. *Hypertension* **2010**, *55*, 277–283. [CrossRef] [PubMed]
45. Miranda-Díaz, A.G.; Pazarín-Villaseñor, L.; Yanowsky-Escatell, F.G.; Andrade-Sierra, J. Oxidative stress in diabetic nephropathy with early chronic kidney disease. *J. Diabetes Res.* **2016**, *2016*, 7047238. [CrossRef]
46. Mennuni, S.; Rubattu, S.; Pierelli, G.; Tocci, G.; Fofi, C.; Volpe, M. Hypertension and kidneys: Unraveling complex molecular mechanisms underlying hypertensive renal damage. *J. Hum. Hypertens.* **2014**, *28*, 74–79. [CrossRef]
47. Feng, Y.; Xia, H.; Cai, Y.; Halabi, C.M.; Becker, L.K.; Santos, R.A.; Speth, R.C.; Sigmund, C.D.; Lazartigues, E. Brain-selective overexpression of human Angiotensin-converting enzyme type 2 attenuates neurogenic hypertension. *Circ. Res.* **2010**, *106*, 373–382. [CrossRef]
48. Lal, S.; Kirkup, A.J.; Brunsden, A.M.; Thompson, D.G.; Grundy, D. Vagal afferent responses to fatty acids of different chain length in the rat. American journal of physiology. *Gastrointest. Liver Physiol.* **2001**, *281*, G907–G915. [CrossRef]
49. Goswami, C.; Iwasaki, Y.; Yada, T. Short-chain fatty acids suppress food intake by activating vagal afferent neurons. *J. Nutr. Biochem.* **2018**, *57*, 130–135. [CrossRef]
50. Chappell, M.C. Emerging evidence for a functional angiotensin-converting enzyme 2-angiotensin-(1-7)-MAS receptor axis: More than regulation of blood pressure. *Hypertension* **2007**, *50*, 596–599. [CrossRef]
51. Duprez, D.A. Role of the renin–angiotensin–aldosterone system in vascular remodeling and inflammation: A clinical review. *J. Hypertens.* **2006**, *24*, 983–991. [CrossRef] [PubMed]
52. Xu, P.; Sriramula, S.; Lazartigues, E. ACE2/ANG-(1–7)/Mas pathway in the brain: The axis of good. *Am. J. Physiol. Regul. Integr. Comp. Physiol.* **2011**, *300*, R804–R817. [CrossRef] [PubMed]
53. Yang, Z.; Wang, Q.; Liu, Y.; Wang, L.; Ge, Z.; Li, Z.; Feng, S.; Wu, C. Gut microbiota and hypertension: Association, mechanisms and treatment. *Clin. Exp. Hypertens.* **2023**, *45*, 2195135. [CrossRef] [PubMed]
54. Kong, C.Y.; Li, Z.M.; Mao, Y.Q.; Chen, H.L.; Hu, W.; Han, B.; Wang, L.S. Probiotic yogurt blunts the increase of blood pressure in spontaneously hypertensive rats *via* remodeling of the gut microbiota. *Food Funct.* **2021**, *12*, 9773–9783. [CrossRef]

55. Park, J.; Kim, M.; Kang, S.G.; Jannasch, A.H.; Cooper, B.; Patterson, J.; Kim, C.H. Short-chain fatty acids induce both effector and regulatory T cells by suppression of histone deacetylases and regulation of the mTOR-S6K pathway. *Mucosal Immunol.* **2015**, *8*, 80–93. [CrossRef] [PubMed]
56. Onyszkiewicz, M.; Gawrys-Kopczynska, M.; Konopelski, P.; Aleksandrowicz, M.; Sawicka, A.; Koźniewska, E.; Samborowska, E.; Ufnal, M. Butyric acid, a gut bacteria metabolite, lowers arterial blood pressure via colon-vagus nerve signaling and GPR41/43 receptors. *Pflug. Arch. Eur. J. Physiol.* **2019**, *471*, 1441–1453. [CrossRef] [PubMed]

foods

Article

Innovative Insights for Establishing a Synbiotic Relationship with *Bacillus coagulans*: Viability, Bioactivity, and *In Vitro*-Simulated Gastrointestinal Digestion

Saranya Suwanangul [1], Pannapapol Jaichakan [2], Nukrob Narkprasom [3], Supaluck Kraithong [4], Kanjana Narkprasom [3,*] and Papungkorn Sangsawad [5,*]

[1] Program in Food Science and Technology, Faculty of Engineering and Agro-Industry, Maejo University, Chiang Mai 50290, Thailand; saranya_sw@mju.ac.th
[2] Faculty of Business Administration, Chitralada Technology Institute, Bangkok 10300, Thailand; pannapapol.jai@cdti.ac.th
[3] Program in Food Engineering, Faculty of Engineering and Agro-Industry, Maejo University, Chiang Mai 50290, Thailand; narkprasom@gmail.com
[4] Guangdong Provincial Key Laboratory of Food Quality and Safety, College of Food Science, South China Agricultural University, Guangzhou 510642, China; supaluck@scau.edu.cn
[5] School of Animal Technology and Innovation, Institute of Agricultural Technology, Suranaree University of Technology, Nakhon Ratchasima 30000, Thailand
* Correspondence: kanjana_n@mju.ac.th (K.N.); papungkorn@sut.ac.th (P.S.)

Abstract: This study investigates the use of encapsulating agents for establishing a synbiotic relationship with *Bacillus coagulans* (TISTR 1447). Various ratios of wall materials, such as skim milk powder, maltodextrin, and cellulose acetate phthalate (represented as SMC1, SMC3, SMC5, and SMC7), were examined. In all formulations, 5% inulin was included as a prebiotic. The research assessed their impact on cell viability and bioactive properties during both the spray-drying process and in vitro gastrointestinal digestion. The results demonstrate that these encapsulating agents efficiently protect *B. coagulans* spores during the spray-drying process, resulting in spore viability exceeding 6 log CFU/g. Notably, SMC5 and SMC7 displayed the highest spore viability values. Moreover, SMC5 showcased the most notable antioxidant activity, encompassing DPPH, hydroxy radical, and superoxide radical scavenging, as well as significant antidiabetic effects via the inhibition of α-amylase and α-glucosidase. Furthermore, during the simulated gastrointestinal digestion, both SMC5 and SMC7 exhibited a slight reduction in spore viability over the 6 h simulation. Consequently, SMC5 was identified as the optimal condition for synbiotic production, offering protection to *B. coagulans* spores during microencapsulation and gastrointestinal digestion while maintaining bioactive properties post-encapsulation. Synbiotic microcapsules containing SMC5 showcased a remarkable positive impact, suggesting its potential as an advanced food delivery system and a functional ingredient for various food products.

Keywords: antidiabetic; antioxidative activity; *Bacillus coagulans* spores; microencapsulation; viability

Citation: Suwanangul, S.; Jaichakan, P.; Narkprasom, N.; Kraithong, S.; Narkprasom, K.; Sangsawad, P. Innovative Insights for Establishing a Synbiotic Relationship with *Bacillus coagulans*: Viability, Bioactivity, and *In Vitro*-Simulated Gastrointestinal Digestion. *Foods* **2023**, *12*, 3692. https://doi.org/10.3390/foods12193692

Academic Editors: Xiqing Yue and Mohan Li

Received: 20 September 2023
Revised: 4 October 2023
Accepted: 6 October 2023
Published: 8 October 2023

1. Introduction

Creating functional foods via the enrichment of novel foods with probiotic organisms or spores is a promising approach. Processed foods with improved health benefits compared to traditional nutritional options can now be produced using the advancements in current food processing technology [1]. In contemporary society, consumers are growing increasingly health-conscious, giving greater significance to the nutritional value of their food choices. Consequently, the successful acceptance of novel foods depends on the concept of food quality and additional value-added food functionalities, particularly those related to probiotics [2].

Probiotics are academically defined as living microorganisms that, when administered in sufficient quantities, bestow advantageous health effects upon the host organism [3]. Products must contain a minimum of 10^6 CFU/g of probiotic microorganisms and ensure the preservation of their viability throughout their designated shelf life to attain probiotic status and access these benefits [4]. *Bacillus coagulans* (*B. coagulans*) find wide application in food and feed additives due to their effective inhibition of harmful bacterial growth. Nonetheless, a significant proportion of these microorganisms lose their activity during the manufacturing, storage, and rigorous processing associated with functional foods, primarily owing to their susceptibility to stomach acid and elevated temperatures [5]. Altun and Erginkaya [6] provide evidence that *B. coagulans* can thrive within a pH range of 5.5–6.2 and release spores at 37 °C, leading to its designation as Generally Recognized as Safe (GRAS). In the small intestine, this bacterium's spores assume a crucial role in the digestion of carbohydrates within the human gastrointestinal tract [7]. Probiotics fulfill a crucial role in maintaining the balance of the gastrointestinal system and energy levels via the production and regulation of essential metabolites, encompassing enzymes, vitamins, peptides, essential amino acids, and antioxidants [8–10]. These metabolites are vital for promoting human well-being. Probiotics offer an array of significant benefits, including the prevention of conditions such as hypertension [11] and diabetes [12], a reduction in cholesterol levels [13], displaying anti-carcinogenic properties [14], and the alleviation of conditions like anxiety, depression [15], and allergies. Furthermore, probiotics actively modulate the immune system [16] and provide protection against food allergies and atopic diseases [17], making them an indispensable component of comprehensive health management. Microencapsulation is a technique that provides physical protection to bacterial spores and bioactive compounds, effectively shielding them from chemical degradation and preserving their functionality, especially in the context of the food industry. For instance, bioactive peptides produced from bacteria such as *Phaseolus lunatus* can be safeguarded from the harsh gastrointestinal environment by microencapsulation using materials like maltodextrin/gum Arabic [18]. This evolving method holds promise for the preservation of microbial isolates. Furthermore, encapsulation holds the potential to enable controlled release and optimize delivery to specific target sites, thereby enhancing the overall efficacy of probiotics. Nonetheless, ensuring the viability of probiotic-containing products remains a substantial challenge. Probiotics must withstand the rigors of industrial processing, various storage conditions, and the expedition through the gastrointestinal tract. Notably, microencapsulation technology has proven effective in safeguarding probiotic cells and spores, significantly bolstering their viability within food products and throughout the gastrointestinal tract [19]. Encapsulation serves as a valuable means for separating probiotics from food matrices. However, its success is contingent upon several critical factors, including the choice of coating materials, the specific encapsulation technology applied, storage conditions, capsule quality, and the sensitivity of the probiotic strain. In essence, encapsulation has the potential to efficiently isolate probiotics from food, but its actual effectiveness hinges on the considerations outlined above [20].

The extensive application of microencapsulation technology in combination with thermal protection agents has been underscored in a study by Misra et al. [21]. Among various methods, spray drying stands out as a preferred choice for probiotics due to its cost-effectiveness and high-yield efficiency. However, challenges arise, as seen in the work of Yao et al. [22] where *Lactobacillus plantarum* (*L. plantarum*) 550 lost viability during this process. To enhance cell survival, proteins and polysaccharides are often added to bacteria to create probiotic microcapsules. Skim milk powder, due to its low viscosity similar to calcium alginate, is a valuable encapsulating agent. It effectively enhances the survival of encapsulated materials in the harsh lower gastric environment compared with free cells [23]. Proteins are commonly employed as wall materials in the spray-drying process to form protective layers around cells swiftly, safeguarding probiotic bacteria from thermal stress [24]. Maltodextrin is favored as a carrier in spray-dried hygroscopic powders, facilitating interactions with liquids and solids, making it a versatile choice for incorporating various substances [25]. Additionally, cellulose acetate phthalate functions as an effective polymer for pH-controlled release, dissolving in mildly acidic or neutral intestinal conditions while withstanding exposure to highly acidic gastric fluids [26]. Researchers have explored multiple wall matrix systems, including ternary wall matrices, to enhance probiotics' viability and physicochemical characteristics [27,28]. Nonetheless, there is a notable gap in the literature regarding specific combinations of encapsulation materials, such as skim milk powder, maltodextrin, and cellulose acetate phthalate, for safeguarding *B. coagulans* spores during the intricate spray-drying process.

As far as current knowledge extends, there is a notable lack of research on the combination of a synbiotic formulation involving *B. coagulans* spores with specific encapsulation materials like skim milk powder, maltodextrin, and cellulose acetate phthalate while also incorporating the prebiotic inulin. Therefore, the primary aim of this study is to investigate how different encapsulation materials affect the viability of probiotic *B. coagulans* spores. Furthermore, the research seeks to assess the multifaceted bioactive properties of these spores, including their antioxidant and antidiabetic activities, which are integral to health promotion, especially in the context of the spray-drying process. Additionally, this study explores the resilience of *B. coagulans* spores during simulated gastrointestinal digestion.

2. Materials and Methods

2.1. Materials

The probiotic culture of *B. coagulans* (TISTR 1447) was sourced from the Thailand Institute of Scientific and Technological Research (TISTR) for our research conducted at Maejo University's Food Safety and Biotechnology Lab. Gastrointestinal fluids were simulated using pepsin porcine (P6887) and pancreatin (P7545) from Sigma-Aldrich Co. in the St. Louis, MI, USA. In this research project, we utilized chemicals of analytical reagent-grade quality.

2.2. Production of Spore Suspensions

The production followed the methodology outlined by Russell et al. [29] with minor adjustments. The first step involved initiating sporulation in *B. coagulans* TISTR 1447 by cultivating cells in Nutrient Broth at 37 °C for 24 h. Subsequently, there was aerobic growth on Nutrient Yeast Extract Salt Medium (NYSM) agar for 24 h at the same temperature. In the second step, culturing was carried out in NYSM broth using a single colony from the agar plate. This culture was shaken at 250 rpm and maintained at 37 °C for five days, resulting in a 90% sporulation rate. Bacterial sediment was successfully obtained by subjecting the bacterial suspension to centrifugation at 4000× *g* for 10 min, followed by a washing step and resuspension in 0.9% Sodium Chloride Sterile Saline, totaling one hundred microliters. To effectively eliminate vegetative bacterial forms, a heat sterilization process was employed (80 °C for 15 min). Ultimately, the original stock solution underwent dilution, decreasing its concentration from 1×10^{12} CFU/g to a final concentration of 1×10^{10} CFU/g. This thinned solution was subsequently stored under refrigeration and prepared for future use, as depicted in Figure 1.

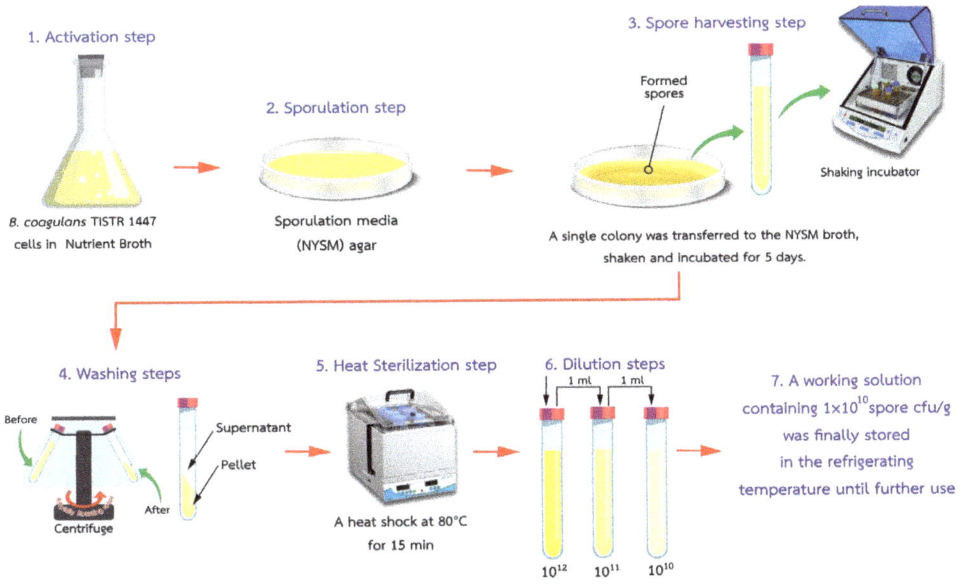

Figure 1. Diagram illustrating *B. coagulans* spore production.

2.3. Production of Synbiotic Microcapsules by Spray Drying

The preparation of microencapsulation beads adhered to the methodology delineated by Arslan-Tontul and Erbas [30], albeit with minor adaptations. In this process, the encapsulation of *B. coagulans* spores within a probiotic culture was executed by dissolving these spores within a 200 mL aqueous solution. This solution was composed of a mixture of skim milk powder, maltodextrin, and cellulose acetate phthalate in different ratios, precisely defined as 1:2:1, 3:2:1, 5:2:1, and 7:2:1. Additionally, it was fortified with 5% (w/v) inulin, a well-established prebiotic. Subsequently, the entire mixture underwent sterilization via a heating regimen at 121 °C for a duration of 15 min, thereby yielding synbiotic microcapsules of distinct formulations, designated as SMC1, SMC3, SMC5, and SMC7. After the solutions were allowed to cool, *B. coagulans* spores were introduced into the mixture in a proportion of 20% v/w relative to the mass of the wall materials. This introduction maintained an initial cell count of 1×10^{10} CFU/g within the microencapsulation process. These inoculated solutions underwent microencapsulation using a laboratory-scale spray dryer, specifically the Buchi-290 from Switzerland. Temperature control was maintained, with an inlet temperature set at 120 °C and an outlet temperature at 50 °C. In a two-fluid nozzle system, compressed air at 0.3 bar is used to aspirate, and the feeding rate of the solutions is approximately 16.5 mL/min. A cyclone separated the synbiotic microcapsules, accumulating them in a designated vessel. The collected microcapsules were then preserved at −18 °C for subsequent analysis (Figure 2). The survival rate (%) in comparison to the initial count prior to spray drying was computed following the approach outlined by Gomez-Mascaraque et al. [31].

$$\text{Survival rate (\%)} = \frac{\text{N(F)}}{\text{N(0)}} \times 100 \quad (1)$$

The equation value was determined by evaluating the probiotic cell count (log CFU/g of dry solid material) both prior to and following the spray-drying process. In this context, N(F) denotes the count of probiotic cells after the spray-drying process, while N(0) signifies the count of probiotic cells before the spray-drying procedure.

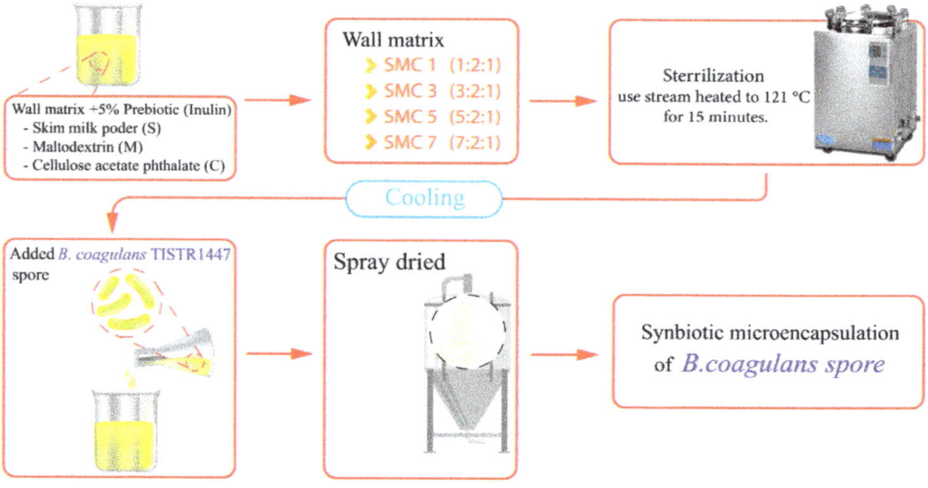

Figure 2. Diagram depicting synbiotic microcapsules produced via a spray-drying process.

2.4. *In Vitro Bioactive Properties of Synbiotic Microcapsules*

We conducted an analysis of the spray-dried samples, comprising free cells, unencapsulated cells, and encapsulated cells, to assess their bioactive properties.

2.4.1. 1,1-Diphenyl-2-Picrylhydrazyl (DPPH) Free-Radical Scavenging

The evaluation of DPPH free-radical-scavenging activity followed a previously established method by Yan et al. [10], with slight modifications to adapt it for the microplate technique. The spray-dried sample was dissolved in 0.1 M sodium phosphate buffer with a pH of 7.0, yielding a concentration range of 0–20 mg solid/mL. In the initial step, one hundred microliters of the sample or buffer (utilized as a positive control) was mixed with an equal volume of an ethanolic DPPH radical solution at a concentration of 0.2 mM or absolute ethanol (used as a blank). Following this, in the second step, the mixture was left to incubate at room temperature in the absence of light for 30 min, after which the absorbance was measured at 517 nm. The DPPH radical-scavenging activity was calculated using the following equation:

$$\text{DPPH radical-scavenging activity } (\%) = 1 - \frac{\text{A sample} - \text{A blank}}{\text{A control}} \times 100 \quad (2)$$

2.4.2. Hydroxyl Radical Scavenging (HRSA)

The assessment of HRSA (Hydroxyl Radical-Scavenging Activity) was carried out according to a previously established protocol by Suwanangul et al. [32], with minor modifications to adapt the method for microplate analysis. The spray-dried sample was dissolved in 0.1 M sodium phosphate buffer at pH 7.4, achieving a concentration range of 0–20 mg solid/mL. In the first step, the reaction mixture was prepared that included a 3 mM solution of 1,10-phenanthroline in water (50 µL), 0.1 M sodium phosphate buffer at pH 7.4 (50 µL), the samples (50 µL), and 3 mM $FeSO_4$ (50 µL) in a 96-well microplate. Subsequently, in the second step, a 20 mM hydrogen peroxide solution (50 µL) was added to initiate the reaction. This was followed by incubation at 37 °C for 90 min. Afterward, the absorbance at 517 nm was measured, and the determination of hydroxyl radical-scavenging ability was calculated according to the procedure outlined below:

$$\text{HRSA } (\%) = \frac{(\text{A samples} - \text{A blank})}{(\text{A control} - \text{A blank})} \times 100 \quad (3)$$

2.4.3. Superoxide Radical Scavenging

The determination of superoxide radical-scavenging ability (SRSA) followed the method outlined by Yan, Li, Yue, Wang, Zhao, Evivie, Li, and Huo [10], with slight adaptations to accommodate a 96-well clear flat-bottom plate. Initially, 80 μL of the spray-dried sample was dissolved in 50 mM Tris–HCl solution at pH 8.2, with a concentration range of 0–20 mg solid/mL. Alternatively, deionized water was used for the control group and was combined with 80 μL of 50 mM Tris-HCl solution. Following this, the resulting mixture was incubated at 25 °C for 20 min. An amount of 40 μL of 25 mM pyrogallol in 10 mM HCl was introduced, and the mixture was left to stand at room temperature for 4 min. The determination of superoxide anion radical-scavenging activity was carried out as described below:

$$\text{SRSA } (\%) = 1 - \frac{\text{A samples}}{\text{A control}} \times 100 \quad (4)$$

The results have been presented as EC_{50} values, which indicate the sample concentrations needed to scavenge 50% of DPPH, HRSA, and SRSA. These EC_{50} values were determined by utilizing a linear regression curve derived from a range of the spray-dried sample concentrations from 0 to 20 mg/mL.

2.4.4. In Vitro Activities of α-Glucosidase

The determination of α-glucosidase-inhibitory activity followed a method previously outlined by Zhang et al. [33], with slight modifications. The spray-dried sample was dissolved in 50 mM Tris–HCl solution at pH 8.2, with a concentration range of 0–20 mg solid/mL. Alternatively, it was solubilized in 0.02 M sodium phosphate buffer at pH 6.9, with a concentration range of 0–40 mg solid/mL. In summary, a mixture of 50 μL of our samples and 50 μL of a 10 mM PNP-glycoside solution (dissolved in 0.1 M sodium phosphate buffer at pH 6.9) was prepared within a 96-well microplate. Subsequently, 50 μL of 0.2 U/mL α-glucosidase enzyme solution was added to initiate the reaction. The microplate was then placed in an incubator set at 37 °C for 30 min. The release of p-nitrophenol from PNP-glycoside was quantified by measuring the absorbance at 405 nm using a microplate reader (Thermo Scientific™, Waltham, MA, USA). Acarbose was evaluated using the same procedure and served as our positive control. The extent of α-glucosidase-inhibitory activity was determined as follows:

$$\alpha\text{-glucosidase inhibition } (\%) = \frac{(\text{A control} - \text{A control blank}) - (\text{A sample} - \text{A sample blank})}{(\text{A control} - \text{A control blank})} \times 100 \quad (5)$$

2.4.5. In Vitro Activities of α-Amylase

The determination of α-amylase-inhibitory activity was conducted following a method previously documented by Mudgil et al. [34], with some modifications. In this adapted approach, p-nitrophenyl α-D maltohexaoside (pNPM; 5 mmol/L) served as the substrate. The procedure was outlined by preparing a mixture consisting of 50 μL of pNPM, 50 μL of our samples (which were dissolved in 0.02 M sodium phosphate buffer at pH 6.9 with a concentration range of 0–40 mg solid/mL), and 50 μL of the PPA enzyme solution (50 mg/mL) in sodium phosphate buffer (0.02 M, pH 6.9). This mixture was then incubated for 30 min at 37 °C. Subsequently, the release of p-nitrophenol at 405 nm was monitored. The extent of α-amylase-inhibitory activity was calculated as follows:

$$\alpha\text{-amylase inhibition } (\%) = 1 - \frac{(\text{A1} - \text{A2})}{(\text{A3} - \text{A4})} \times 100 \quad (6)$$

In this equation, A1 signifies the absorbance measured in the reaction wells containing both the enzyme and the test sample, whereas A2 stands for the absorbance obtained in the reaction blank containing only the test sample without the enzyme. A3 represents the absorbance recorded in the control wells with the enzyme but without the test sample, and finally, A4 indicates the absorbance measured in the control blank, which lacks both the enzyme and the test sample.

2.4.6. Dipeptidyl Peptidase IV (DPP-IV) Inhibition

The assessment of DPP-IV-inhibitory activity followed a method described by Sangsawad et al. [35], with minor adjustments. The spray-dried sample was dissolved in 50 mM Tris–HCl solution at pH 8.2, achieving a concentration range of 0–20 mg solid/mL. Alternatively, it was solubilized in 100 mM Tris-HCl buffer at pH 8.0, with a concentration range of 0–40 mg solid/mL. Specifically, a twenty-five microliter aliquot of the diluted sample (diluted in 100 mM Tris-HCl buffer at pH 8.0) was mixed with twenty-five microliters of the prepared 1.6 mM Gly-Pro-p-nitroanilide substrate. The sample–substrate mixture was pre-incubated for 10 min at 37 °C. Following this, fifty microliters of DPP-IV (0.01 U/mL) were added, and the reaction proceeded for 60 min at 37 °C. To halt the reaction, researchers introduced one hundred microliters of 1.0 M sodium acetate buffer at pH 4.0 and measured the absorbance at 405 nm. The absorbance values were then normalized against sample blanks where DPP-IV was replaced with Tris-HCl buffer. Negative controls (representing no DPP-IV activity) and positive controls (indicating DPP-IV activity without an inhibitor) were prepared by substituting the sample and DPP-IV solution with Tris-HCl buffer, respectively. The DPP-IV inhibition rate for each sample was calculated as follows:

$$\text{DPP-IV inhibition } (\%) = 1 - \frac{(\text{A sample} - \text{A sample blank})}{(\text{A positive control} - \text{A negative control})} \times 100 \quad (7)$$

The IC_{50} values for each sample were determined via curve interpolation, indicating the synbiotic microcapsule concentration needed to reach 50% inhibition of the initial enzyme reaction rate.

2.5. *In Vitro Simulation of Gastrointestinal Digestion*

The in vitro simulated digestion method by Gomez-Mascaraque, Morfin, Pérez-Masiá, Sanchez, and Lopez-Rubio [31] involved several steps: Simulated Gastric Fluid (SGF) was prepared by mixing pepsin and sodium chloride in deionized water and adjusting its pH to 2.0 using hydrochloric acid. Simulated Intestinal Fluid (SIF) was created by adding pancreatin and porcine bile salt to a phosphate-buffered saline (PBS) solution at pH 7.5. Both SGF and SIF were preheated to 37 °C, filtered, and used for digestion. A powdered sample containing free and encapsulated cells was mixed with SGF for gastric digestion at 37 °C, with samples withdrawn at 30, 60, 90, and 120 min. After gastric digestion, SIF was added, and samples were retrieved at 1, 2, 3, and 4 h for simulated intestinal digestion, followed by cell counting.

$$\text{Viability} = \frac{\log \text{cfu N1}}{\log \text{cfu N0}} \quad (8)$$

$$\text{Reduction in cell numbers} = \text{Survival rate (initial time)} - \text{Survival rate (SGF and SIF)} \quad (9)$$

where N1 represents the count of probiotic cells at the time of digestion in simulated gastric fluid (SGF) and simulated intestinal fluid (SIF), and N0 represents the count of probiotic cells at the initial time (0 h).

2.6. *Statistical Analysis*

Data collected in triplicate underwent analysis using a one-way analysis of variance (ANOVA). The detection of significant differences among means ($p \leq 0.05$) was conducted using the Duncan procedure, employing SPSS 16.0 for Windows (SPSS Inc., Chicago, IL, USA).

3. Results and Discussion

3.1. The Effects of Microencapsulation and Spray Drying on the Survival Rate

Table 1 presents the survival rates of *B. coagulans* cells after encapsulation using various wall-matrix ratios and subsequent spray drying. Initially, the probiotic count was consistent at approximately 10.35 log CFU/g across all samples. Our investigation revealed a significant decrease in the viability of free cells after the spray-drying process, showing a reduction of about 4.67 log CFU/g. This decline in viability during spray drying can be attributed to factors such as heat-induced stress, desiccation, and exposure to ambient oxygen, all of which collectively impact the metabolic activity of microbial cells, leading to decreased viability.

Table 1. The survival rates of free cells and cells encapsulated in various wall-matrix ratios before and after the spray-drying process.

Samples	Viable Cells (log CFU/g)		Survival Rate (%)
	Before Spray Drying	After Spray Drying	
Free Cell	10.35 ± 0.15 [a]	4.67 ± 0.26 [d]	45.12 ± 0.25 [d]
SMC1 (1:2:1)	10.91 ± 0.29 [a]	7.28 ± 0.37 [c]	66.73 ± 0.73 [c]
SMC3 (3:2:1)	10.61 ± 0.93 [a]	8.35 ± 0.20 [b]	78.70 ± 0.18 [b]
SMC5 (5:2:1)	10.75 ± 0.06 [a]	9.93 ± 0.53 [a]	92.37 ± 0.36 [a]
SMC7 (7:2:1)	10.58 ± 0.11 [a]	9.77 ± 0.94 [a]	92.34 ± 0.27 [a]

Significant differences within the column ($p \leq 0.05$) are denoted by superscript letters (a–d).

Interestingly, an increase in the concentration of skim milk powder within the encapsulant agents corresponded to improved viability of the encapsulated probiotics. Specifically, SMC1, containing 23.75% skim milk powder, exhibited a viability rate of 66.73%. Similarly, SMC3 (with 47.50% skim milk powder), SMC5 (containing 59.38% skim milk powder), and SMC7 (with 66.50% skim milk powder) displayed viability rates of 78.70%, 92.37%, and 92.34%, respectively. Consequently, post-spray drying, different wall matrices, designated as SMC1-7, showcased a spectrum of probiotic survival rates.

These findings unmistakably delineate a robust association between the concentration of skim milk powder and the viability rate of the probiotics enclosed within. It is noteworthy that when skim milk powder is employed in limited proportions, it may prove inadequate in constructing the requisite stable and protective encasements around the microbial cells, potentially leading to an encumbered encapsulation process wherein the cells are not fully enveloped or adequately shielded within the capsules [36]. Conversely, elevated concentrations of skim milk powder may furnish a surfeit of both quantity and quality in terms of wall materials, thereby facilitating effective encapsulation, averting cell diffusion, and preserving viability [37].

In addition, proteins can effectively shield probiotic cells from the detrimental effects of high temperatures during the spray-drying process. This thermal protection is crucial for preserving the viability and functionality of probiotics [24]. This study suggests that the choice of encapsulation material can have a significant impact on the viability of probiotic microorganisms. This study emphasizes the impact of encapsulation materials on probiotic viability. *B. coagulans* showed higher survival (>66%) with SMC1-7 encapsulation, surpassing *L. acidophilus* encapsulated with maltodextrin, skim milk powder, and trehalose, which had a 59.2% survival rate, aligning with Soukoulis et al. [38]'s findings. Consequently, this study has brought to light that the conditions characterized by SMC5 and SMC7 emerge as the optimal parameters for microencapsulation, particularly with regard to the augmentation of *B. coagulans* spore viability post-spray-drying operation.

3.2. Bioactivity of the Encapsulated Products

3.2.1. In Vitro Antioxidant Activities

Oxidative stress, marked by increased levels of reactive oxygen species (ROS), is correlated with diverse forms of cellular damage and chronic diseases such as cardiovascular diseases in humans, diabetes, neurodegenerative diseases, and cancer [39]. While synthetic antioxidants have traditionally been employed to protect against oxidative stress, concerns have arisen regarding their safety and long-term effects. Consequently, researchers have focused on natural antioxidant solutions, including probiotics [39]. The research conducted by Kodali and Sen [40] has uncovered multiple beneficial properties of *B. coagulans*, including its antibacterial, antiviral, and antioxidant activities. Furthermore, the study demonstrates that supplementing with *B. coagulans* effectively reduces NH_3 levels in the serum, thereby alleviating oxidative stress [41]. However, numerous chemical analyses employing diverse techniques have been performed to evaluate the antioxidant properties of the compounds of interest. The incorporation of these chemical analyses, which operate via a variety of mechanisms, can be instrumental in shedding light on the primary modes of action of these antioxidant compounds.

In our experimental findings, it was discerned that the encapsulated samples, denoted as SMC1-7, exhibited a notable reduction in the EC_{50} value in comparison to the free cells, as illustrated in Figure 3A–C. These observations underscore the advantageous influence of encapsulation on the antioxidant activity exhibited by *B. coagulans* metabolites and spores, particularly in their proficiency to counteract DPPH radicals, hydroxyl radicals, and superoxide radicals. Intriguingly, both SMC5 and SMC7 manifested the most diminished EC_{50} values for DPPH and hydroxyl radical-scavenging activities when juxtaposed with the other specimens. A lower EC_{50} value signifies a heightened potency in the scavenging activity against hydroxyl radicals, indicative of their pronounced efficacy in the realm of antioxidants, particularly concerning mechanisms involving hydrogen atom transfer (HAT), single electron transfer (SET), and the quenching of hydroxy radicals. Among oxygen radicals, hydroxyl radicals are renowned for their extraordinary reactivity and their proclivity to inflict extensive damage upon adjacent biomolecules. This oxidative damage encompasses processes associated with aging, the genesis of cancer, and the onset of various pathological conditions [42]. Furthermore, SMC5 exhibited the most remarkable superoxide radical-scavenging activity, with the lowest EC_{50} value recorded at 0.82 mg/mL. This reactive oxygen species plays a crucial role in oxidative stress. In our study, we observed that the microencapsulated product of *B. coagulans* TISTR 1447 exhibited higher bioactive properties compared to *B. coagulans* T242, as indicated by the research conducted by Sui et al. [43]. Specifically, within the range of 10^6–10^8 CFU/g, *B. coagulans* TISTR 1447 showed superior DPPH-scavenging activity (35.0%), hydroxyl radical-scavenging activity (39.0%), and superoxide anion radical-scavenging activity (14.8%) in comparison to *B. coagulans* T242. In addition, probiotics, such as *B. coagulans*, are renowned for their capacity to generate antioxidant compounds, including butyrate, glutathione (GSH), and folate [44]. These metabolites play a role in diminishing oxidative stress and improving the assimilation of dietary antioxidants via indirect mechanisms [40,45]. Furthermore, the study conducted by Kodali and Sen [40] sheds light on the role of *B. coagulans* in influencing the composition and activity of the gut microbiota. A balanced gut microbiota is crucial for maintaining an overall antioxidant balance, and *B. coagulans* may enhance antioxidant activity by promoting a healthy balance of gut bacteria.

Nonetheless, our observations revealed a noteworthy distinction in the antioxidant potential between the control sample, consisting of encapsulated materials devoid of *B. coagulans* metabolites and spores, and both the free-cell and encapsulated-probiotic-cell counterparts. The control samples containing skim milk powder exhibited markedly diminished antioxidant potency, as evidenced by the highest recorded EC_{50} value in comparison to the free-cell and encapsulated-probiotic-cell variants. This peculiarity in the control samples' antioxidant activity can be ascribed to the presence of casein compounds within skim milk powder. Skimmed milk can serve as a protective matrix for probiotic bacteria;

the probiotic microorganisms are encapsulated or embedded within the milk matrix. This protective matrix helps shield the probiotics from environmental stresses, such as exposure to oxygen, light, or temperature fluctuations, which can lead to oxidative damage [46]. Specifically, the amino acid composition of casein encompasses amino acids endowed with antioxidant properties, such as cysteine and methionine, integrated into its molecular structure [47]. Consequently, our investigation unequivocally elucidates that the SMC5 condition emerges as the preeminent choice for the development of microencapsulated *B. coagulans* products endowed with heightened antioxidative capabilities, equipping them with the capacity to counteract the deleterious effects of free radicals.

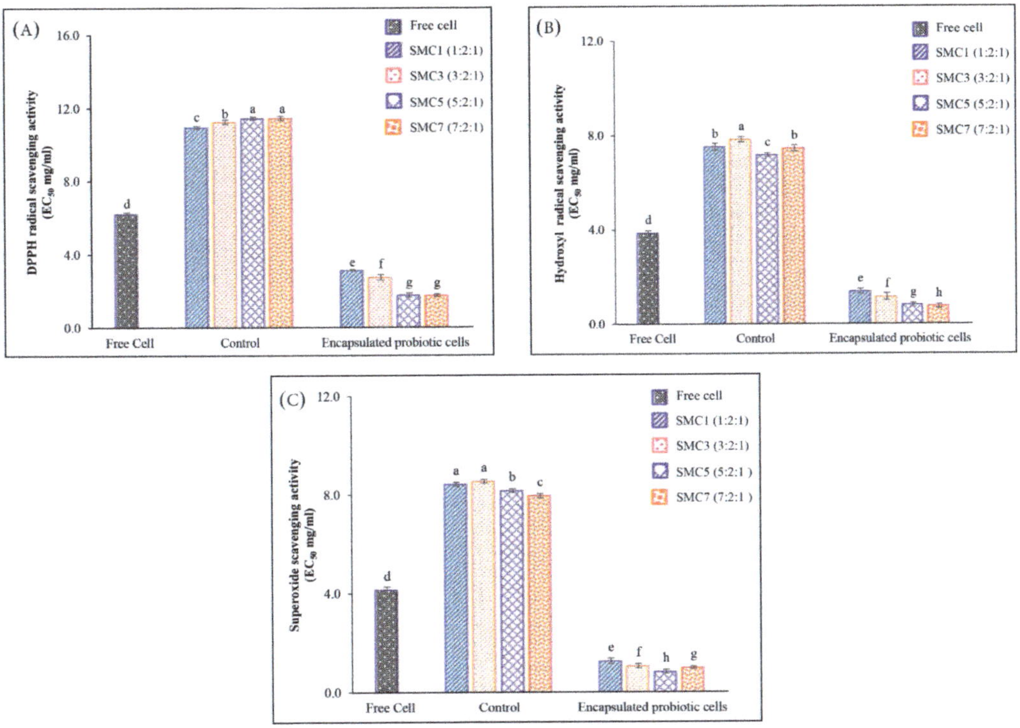

Figure 3. The graph illustrates the in vitro antioxidant activity of both free and encapsulated probiotic cells within different wall matrices, assessed using distinct methods: DPPH (**A**), hydroxyl radicals (**B**), and superoxide anion radicals (**C**). Values with notable differences ($p \leq 0.05$) are indicated by individual superscript letters (a–h).

3.2.2. α-Amylase and α-Glucosidase Inhibitory Activities

Diabetes is a chronic metabolic disease marked by elevated blood glucose levels, resulting in long-term damage to the heart, blood vessels, eyes, kidneys, and nerves. One primary approach to diabetes management involves inhibiting the digestion and absorption of carbohydrates, such as α-amylase, which gradually increases blood sugar levels after meals. It can effectively control post-meal blood glucose spikes [48]. Its primary function occurs within the small intestine, where it enzymatically breaks down disaccharides and complex carbohydrates, converting them into simpler sugars, such as glucose. The inhibition of α-glucosidase leads to a delay in the absorption of glucose from the gastrointestinal tract into the bloodstream, contributing to the reduction of post-meal blood sugar levels [49].

Within the ambit of our study, it was ascertained that all microencapsulated products exhibited a heightened degree of inhibitory activity against α-amylase in comparison to their free-cell counterparts. Among the various microcapsules investigated, SMC1 displayed the least inhibition of α-amylase activity, even when juxtaposed against all other microcapsules, as graphically represented in Figure 4A. Remarkably, SMC5 showcased the most robust inhibitory activity against α-amylase, as substantiated by an IC_{50} value of 1.32 mg/mL. Intriguingly, these observed activities appeared to remain unaffected by alterations in the ratio of skim milk within the microcapsules. Notably, SMC5's inhibitory potency against α-amylase surpassed that of acarbose, a pharmaceutical agent commonly employed for the management of diabetes via the inhibition of carbohydrate digestion. The IC_{50} value for acarbose stood at 1.42 mg/mL. Our findings harmonize with those delineated by Li et al. [50], who also documented a favorable correlation between the presence of *B. coagulans* and the inhibition of α-glucosidase activity.

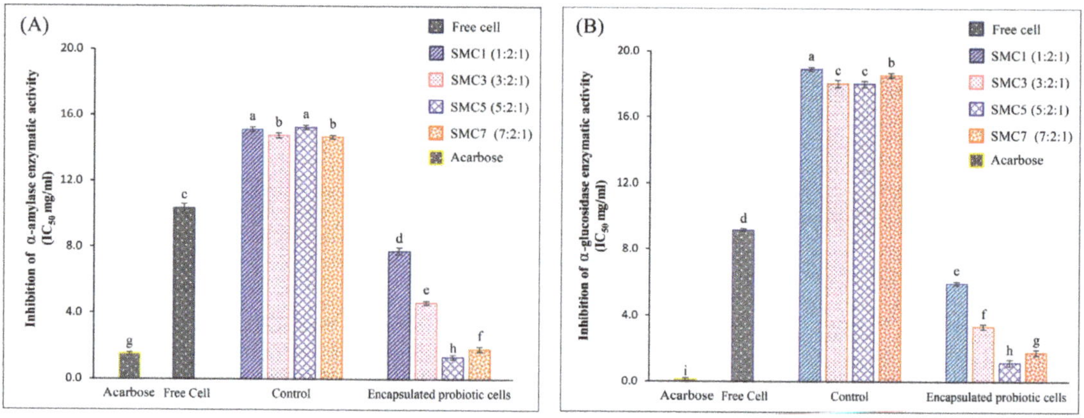

Figure 4. In vitro antidiabetic activity of free and encapsulated probiotic cells in different wall matrices via α-amylase (**A**) and α-glucosidase (**B**). Values with notable differences ($p \leq 0.05$) are indicated by individual superscript letters (a–i).

Among the diverse array of microencapsulated products examined, SMC5 distinctly emerges as the frontrunner, showcasing the most pronounced α-glucosidase-inhibitory activity, characterized by an exceptionally low IC_{50} value of 1.15 mg/mL, as graphically delineated in Figure 4B. These observations suggest that microcapsules possess the capacity to retard the absorption of glucose from the intestinal tract into the circulatory system. This phenomenon holds promise for the management of postprandial blood glucose levels. It is of paramount importance to underscore, however, that SMC5, despite its commendable efficacy, still exhibits a lower potential in comparison to acarbose, a synthetic α-glucosidase inhibitor boasting an IC_{50} value of 0.033 mg/mL. Studies have reported similar results regarding the inhibition of α-glucosidase activity in *B. coagulans* CC spores. Additionally, *B. subtilis* B2 from fermented food demonstrated the ability to produce an α-glucosidase inhibitor, as documented by Kim et al. [51] and Zhu et al. [52]. Probiotics, including *B. coagulans*, may produce inhibitory compounds such as organic acids, bacteriocins, short-chain fatty acids (SCFAs), or peptides that can interfere with the activities of α-amylase and α-glucosidase. These compounds may directly bind to the enzymes, rendering them less active [53]. Consequently, our research underscores the paramount suitability of SMC5 for the development of microencapsulated *B. coagulans* products with antidiabetic attributes, endowing them with the capacity to inhibit both α-amylase and α-glucosidase activities.

3.2.3. DPP-IV-Inhibitory Activity

DPP-IV is a crucial enzyme involved in the breakdown of incretin hormones like glucagon-like peptide-1 (GLP-1), which plays a key role in regulating blood sugar levels.

The inhibition of DPP-IV results in elevated levels of active GLP-1, which, in turn, promotes increased insulin secretion and a reduction in blood sugar levels [54].

In our empirical findings, it was discerned that among the gamut of microcapsules under scrutiny, SMC7 notably exhibited the highest degree of DPP-IV-inhibitory activity, as underscored by its IC_{50} value of 0.25 mg/mL (Figure 5). Subsequently, SMC5 emerged as the second-most efficacious, yielding an IC_{50} value of 0.37 mg/mL. When assessing the overarching proficiency in enzyme inhibition across diverse wall matrices, the hierarchy of efficacy was delineated as follows: SMC7 > SMC5 > SMC3 > SMC1 > free cells > control samples (comprising encapsulated materials bereft of *B. coagulans* spores). This study found a direct improvement in the DPP-4 inhibition of the encapsulated products with an increasing ratio of skim milk. This suggests that skim milk might safeguard the bioactive material responsible for DPP-IV-inhibitory activity. Interestingly, this trend was not observed in the control samples, indicating a unique effect specific to the encapsulated products. However, the control samples exhibited some degree of inhibitory activity, which was attributed to the presence of casein components in skim milk powder. These casein components contain small peptides that have been extensively studied for their various bioactive properties, including potential health benefits such as reducing hypertension (high blood pressure) and modulating the immune system [55]. Our study's findings align with previous research by Wu et al. [56], who reported similar results for *B. amyloliquefaciens* demonstrating DPP-4-inhibitory activity. Moreover, probiotic bacteria have the ability to produce a range of bioactive compounds, such as peptides and metabolites, which are capable of inhibiting DPP-4 activity. These compounds can directly interact with the enzyme, diminishing its capacity to break down incretin hormones like GLP-1, as demonstrated by Yan et al. [10]. This interaction highlights the potential significance of probiotic compounds in regulating DPP-4 activity and preserving incretin hormones. Our findings align with the work conducted by Mudgil et al. [57], who reported the plausible DPP-IV-inhibitory activity of probiotic strains, suggesting potential improvements in blood glucose regulation. These insights hold implications for ameliorating fasting blood glucose levels in individuals afflicted with type 2 diabetes and enhancing glucose tolerance. Such discernments are poised to inform dietary choices for diabetic patients and contribute to the development of safer and more efficacious pharmaceutical agents in the foreseeable future.

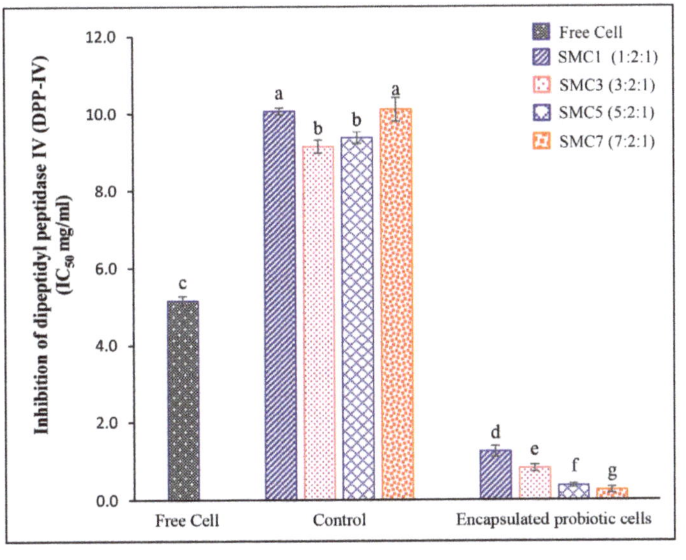

Figure 5. In vitro DPP-IV-inhibitory activity of free and encapsulated probiotic cells in different wall matrices. Values with notable differences ($p \leq 0.05$) are indicated by individual superscript letters (a–g).

The encapsulation processes of SMCs played a vital role in improving their biological activities. These bioactivities were found to correlate with survival rates, as depicted in Table 1. Notably, SMC5 and SMC7 exhibited the highest survival rates after the spray-drying process. This observation suggests that skim milk is a critical component for their effectiveness and demonstrates stronger inhibitory activity. As a result, these two conditions hold the potential to offer enhanced therapeutic benefits, especially in processes like the spray drying of *B. coagulans*.

3.3. *The Survival Rate during In Vitro Simulation of Gastrointestinal Digestion*

Figure 6 portrays the viability profiles of both free and microencapsulated *B. coagulans* cells, employing various wall materials denoted as SMCs, throughout a six-hour simulation mimicking gastrointestinal digestion. The free cells demonstrated a decline in viability amounting to 3.2 log CFU/g over the entire duration of the digestion process. Of particular note, all samples encapsulated within SMCs exhibited markedly superior survival rates subsequent to the simulated gastrointestinal digestion in comparison to their free-cell counterparts. This enhancement was substantiated by a diminution in viability of less than 1.0 log CFU/g. These findings underscore the efficacy of microencapsulation employing a protein-based coating augmented with prebiotic inulin in significantly bolstering the survivability of probiotic entities within the simulated gastric and intestinal environments, surpassing the resilience of unencapsulated spores.

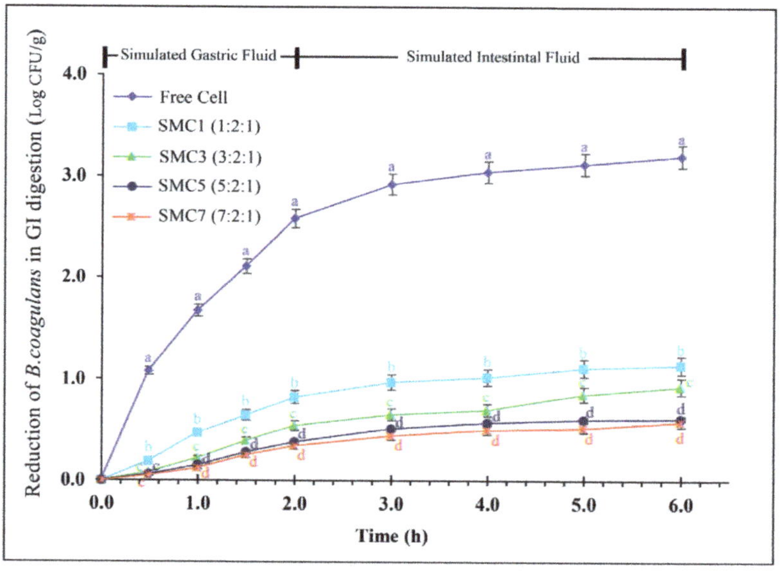

Figure 6. The reduction in *B. coagulans* during *in vitro*-simulated gastrointestinal digestion of free and encapsulated cells with various wall materials. Values with notable differences ($p \leq 0.05$) are indicated by individual superscript letters (a–d) at the same time point (h).

Notably, these rates still exceeded the recommended 6 log CFU/g set by the International Dairy Federation (Table 1). The SMC7 sample exhibited the highest cell survival, with only a decrease of 0.58 log CFU/g during gastrointestinal digestion, followed by the SMC5 sample, which had a nonsignificant difference from SMC7 with a loss of 0.93 log CFU/g, and the SMC1 sample, with a loss of 1.14 log CFU/g. Our study yielded more robust results compared to those presented by Yoha et al. [58]. They found that encapsulating *L. plantarum* using a combination of fructo-oligosaccharide, whey protein, and maltodextrin led to a significant loss of cell viability, with reductions of 2–3 log during exposure to gastric conditions and 4 log under small-intestinal conditions. In contrast, our research demonstrated that skim milk proteins and cellulose acetate create a protective barrier around probiotics. This barrier potentially reduces their exposure to proteolytic enzymes in the digestive system, enhancing the probiotics' survival and enabling them to maintain their activity within the gut [59].

Our experiments have conclusively confirmed that microencapsulation acts as a vital shield for the *B. coagulans* probiotic, safeguarding it from the harsh gastric environment characterized by low pH levels and the presence of digestive enzymes in the intestinal tract. To achieve this, we utilized cellulose acetate phthalate, a well-known polymer recognized for its ability to control pH-dependent releases. This material dissolves under mildly acidic or neutral intestinal conditions yet remarkably withstands highly acidic gastric fluids [26]. This exceptional quality has the potential to significantly slow the release of *B. coagulans* spores and metabolite materials during the gastric phase, ensuring a more controlled release in the intestinal phase. This protective mechanism not only guarantees the probiotic's efficacy but also holds the promise of delivering numerous health benefits upon oral consumption. Importantly, our findings shed light on the intricate effects of microencapsulation, revealing its dependence on the careful selection of coating materials and the specific physiological conditions encountered within the digestive tract.

Furthermore, our investigation sought to evaluate the bioactive activity of these encapsulated samples during GI digestion; however, it was observed that these samples contained a high concentration of gastrointestinal salts (sodium chloride and phosphate-buffered saline), gastrointestinal enzymes (pepsin and pancreatin), and bile salts (porcine bile salt). Consequently, our research focused exclusively on the assessment of survival rates, given the substantial interference posed by the factors mentioned above on the bioactive potential of the probiotics.

4. Conclusions

The optimal ratio for successful spray drying of *B. coagulans* was developed, resulting in the creation of a stable encapsulated synbiotic powder with improved survival rates and enhanced bioactive properties. The proportion of skim milk powder in the formulation was increased, proving beneficial as it bolstered the survival of *B. coagulans* spores during the spray-drying process, enabling them to withstand the rigors of gastrointestinal digestion. Particularly noteworthy was the robust survival of *B. coagulans* when exposed to simulated intestinal fluid, demonstrating its potential in the gut. Furthermore, heightened biological activities were observed in the encapsulated products. These included effective antioxidant activity against substances like DPPH, hydroxy radicals, and superoxide radicals, as well as noteworthy antidiabetic effects, including the inhibition of α-amylase and α-glucosidase enzymes. Among the various conditions tested, SMC5 exhibited the most impressive performance in terms of protecting *B. coagulans* cells and delivering multifunctional bioactive properties. The research-derived formula has the potential to benefit a wide range of probiotics, suggesting exciting prospects for the development of commercial products via further exploration and optimization.

Author Contributions: Conceptualization, S.S., P.S. and K.N.; methodology, S.S., P.S., N.N. and K.N.; formal analysis, S.S., P.S., N.N., S.K. and K.N.; investigation, S.S., P.S. and K.N.; resources, S.S. and P.S.; writing-original draft preparation, S.S. and P.S.; writing-review and editing, S.S., P.S., N.N., P.J., S.K. and K.N. All authors have read and agreed to the published version of the manuscript.

Funding: This work was supported by Maejo University (MJU), Suranaree University of Technology (SUT), Naresuan University (NU), Thailand Science Research and Innovation (TSRI), and the National Science, Research, and Innovation Fund (NSRF).

Informed Consent Statement: All study participants provided informed consent.

Data Availability Statement: The article contains the data and materials supporting the conclusions of this study.

Conflicts of Interest: The authors declare no conflict of interest.

References

1. Poshadri, A.; Deshpande, H.W.; Khodke, U.M.; Katke, S.D. *Bacillus coagulans* and its Spore as Potential Probiotics in the Production of Novel Shelf-Stable Foods. *Curr. Res. Nutr. Food Sci.* **2022**, *10*, 858–870. [CrossRef]
2. Khedkar, S.; Carraresi, L.; Bröring, S. Food or pharmaceuticals? Consumers' perception of health-related borderline products. *PharmaNutrition* **2017**, *5*, 133–140. [CrossRef]
3. Lebeer, S.; Bron, P.A.; Marco, M.L.; Van Pijkeren, J.-P.; Motherway, M.O.C.; Hill, C.; Pot, B.; Roos, S.; Klaenhammer, T. Identification of probiotic effector molecules: Present state and future perspectives. *Curr. Opin. Biotechnol.* **2018**, *49*, 217–223. [CrossRef] [PubMed]
4. Kieps̈, J.; Dembczyński, R. Current trends in the production of probiotic formulations. *Foods* **2022**, *11*, 2330. [CrossRef]
5. Majeed, M.; Natarajan, S.; Sivakumar, A.; Ali, F.; Pande, A.; Majeed, S.; Karri, S. Evaluation of anti-diarrhoeal activity of Bacillus coagulans MTCC 5856 and its effect on gastrointestinal motility in Wistar rats. *Int. J. Pharma Bio Sci.* **2016**, *7*, 311–316.
6. Altun, G.K.; Erginkaya, Z. Identification and characterization of *Bacillus coagulans* strains for probiotic activity and safety. *LWT* **2021**, *151*, 112233. [CrossRef]
7. Jiménez-Villeda, B.E.; Falfán-Cortés, R.N.; Rangel-Vargas, E.; Santos-López, E.M.; Gómez-Aldapa, C.A.; Torres-Vitela, M.R.; Villarruel-López, A.; Castro-Rosas, J. Synbiotic Encapsulation: A Trend towards Increasing Viability and Probiotic Effect. *J. Food Process. Preserv.* **2023**, *2023*, 7057462.
8. Inatsu, Y.; Nakamura, N.; Yuriko, Y.; Fushimi, T.; Watanasiritum, L.; Kawamoto, S. Characterization of *Bacillus subtilis* strains in Thua nao, a traditional fermented soybean food in northern Thailand. *Lett. Appl. Microbiol.* **2006**, *43*, 237–242. [CrossRef]
9. Al-Hindi, R.R.; Abd El Ghani, S. Production of functional fermented milk beverages supplemented with pomegranate peel extract and probiotic lactic acid bacteria. *J. Food Qual.* **2020**, *2020*, 4710273. [CrossRef]
10. Yan, F.; Li, N.; Yue, Y.; Wang, C.; Zhao, L.; Evivie, S.E.; Li, B.; Huo, G. Screening for potential novel probiotics with dipeptidyl peptidase IV-inhibiting activity for type 2 diabetes attenuation in vitro and *in vivo*. *Front. Microbiol.* **2020**, *10*, 2855. [CrossRef]
11. Jauhiainen, T.; Collin, M.; Narva, M.; Cheng, Z.J.; Poussa, T.; Vapaatalo, H.; Korpela, R. Effect of long-term intake of milk peptides and minerals on blood pressure and arterial function in spontaneously hypertensive rats. *Milchwiss.—Milk Sci. Int.* **2005**, *60*, 358–362.
12. Razmpoosh, E.; Javadi, A.; Ejtahed, H.S.; Mirmiran, P.; Javadi, M.; Yousefinejad, A. The effect of probiotic supplementation on glycemic control and lipid profile in patients with type 2 diabetes: A randomized placebo controlled trial. *Diabetes Metab. Syndr. Clin. Res. Rev.* **2019**, *13*, 175–182. [CrossRef] [PubMed]
13. Huang, Y.; Wang, J.; Cheng, Y.; Zheng, Y. The hypocholesterolaemic effects of *Lactobacillus acidophilus* American Type Culture Collection 4356 in rats are mediated by the down-regulation of Niemann-Pick C1-Like 1. *Br. J. Nutr.* **2010**, *104*, 807–812. [CrossRef] [PubMed]
14. Thirabunyanon, M.; Boonprasom, P.; Niamsup, P. Probiotic potential of lactic acid bacteria isolated from fermented dairy milks on antiproliferation of colon cancer cells. *Biotechnol. Lett.* **2009**, *31*, 571–576. [CrossRef]
15. Bravo, J.A.; Forsythe, P.; Chew, M.V.; Escaravage, E.; Savignac, H.M.; Dinan, T.G.; Bienenstock, J.; Cryan, J.F. Ingestion of Lactobacillus strain regulates emotional behavior and central GABA receptor expression in a mouse via the vagus nerve. *Proc. Natl. Acad. Sci. USA* **2011**, *108*, 16050–16055. [CrossRef]
16. O'mahony, L. A randomized, placebo-controlled, double-blind comparison of the probiotic bacteria lactobacillus and bifidobacterium in irritable bowel syndrome (IBS): Symptom responses and relationship to cytokine profiles. *Gastroenterology* **2005**, *128*, 541–551. [CrossRef]
17. Kirjavainen, P.V.; Salminen, S.J.; Isolauri, E. Probiotic bacteria in the management of atopic disease: Underscoring the importance of viability. *J. Pediatr. Gastroenterol. Nutr.* **2003**, *36*, 223–227. [CrossRef]
18. Cian, R.E.; Campos-Soldini, A.; Chel-Guerrero, L.; Drago, S.R.; Betancur-Ancona, D. Bioactive Phaseolus lunatus peptides release from maltodextrin/gum arabic microcapsules obtained by spray drying after simulated gastrointestinal digestion. *Int. J. Food Sci. Technol.* **2019**, *54*, 2002–2009. [CrossRef]
19. Bevilacqua, A.; Campaniello, D.; Speranza, B.; Racioppo, A.; Altieri, C.; Sinigaglia, M.; Corbo, M.R. Microencapsulation of saccharomyces cerevisiae into alginate beads: A focus on functional properties of released cells. *Foods* **2020**, *9*, 1051. [CrossRef]
20. Singh, S.; Gupta, R.; Chawla, S.; Gauba, P.; Singh, M.; Tiwari, R.K.; Upadhyay, S.; Sharma, S.; Chanda, S.; Gaur, S. Natural sources and encapsulating materials for probiotics delivery systems: Recent applications and challenges in functional food development. *Front. Nutr.* **2022**, *9*, 971784. [CrossRef]
21. Misra, S.; Pandey, P.; Mishra, H.N. Novel approaches for co-encapsulation of probiotic bacteria with bioactive compounds, their health benefits and functional food product development: A review. *Trends Food Sci. Technol.* **2021**, *109*, 340–351. [CrossRef]
22. Yao, H.; Liu, B.; He, L.; Hu, J.; Liu, H. The incorporation of peach gum polysaccharide into soy protein based microparticles improves probiotic bacterial survival during simulated gastrointestinal digestion and storage. *Food Chem.* **2023**, *413*, 135596. [CrossRef] [PubMed]
23. Chandramouli, V.; Kailasapathy, K.; Peiris, P.; Jones, M. An improved method of microencapsulation and its evaluation to protect Lactobacillus spp. in simulated gastric conditions. *J. Microbiol. Methods* **2004**, *56*, 27–35. [CrossRef] [PubMed]
24. Razavi, S.; Janfaza, S.; Tasnim, N.; Gibson, D.L.; Hoorfar, M. Microencapsulating polymers for probiotics delivery systems: Preparation, characterization, and applications. *Food Hydrocoll.* **2021**, *120*, 106882.

25. Kothakota, A.; Kumar, A.; Kumar, M.; Juvvi, P.; Rao, S.; Kautkar, S. Characteristics of spray dried dahi powder with maltodextrin as an adjunct. *Int. J. Agric. Environ. Biotechnol.* **2014**, *7*, 849–865. [CrossRef]
26. Rowe, R.; Sheskey, P.; Quinn, M. *Handbook of Pharmaceutical Excipients*, 6th ed.; American Pharmacists Association: Washington, DC, USA, 2009; 917p.
27. Behboudi-Jobbehdar, S.; Soukoulis, C.; Yonekura, L.; Fisk, I. Optimization of spray-drying process conditions for the production of maximally viable microencapsulated *L. acidophilus* NCIMB 701748. *Dry. Technol.* **2013**, *31*, 1274–1283. [CrossRef]
28. Cao, G.; Tao, F.; Hu, Y.; Li, Z.; Zhang, Y.; Deng, B. Positive effects of a *Clostridium butyricum*-based compound probiotic on growth performance, immune responses, intestinal morphology, hypothalamic neurotransmitters, and colonic microbiota in weaned piglets. *Food Funct.* **2019**, *10*, 2926–2934. [CrossRef]
29. Russell, B.; Jelley, S.; Yousten, A. Selective medium for mosquito pathogenic strains of *Bacillus sphaericus* 2362. *J. Appl. Environ. Microbiol.* **1989**, *55*, 294–297. [CrossRef]
30. Arslan-Tontul, S.; Erbas, M. Single and double layered microencapsulation of probiotics by spray drying and spray chilling. *Lwt—Food Sci. Technol.* **2017**, *81*, 160–169.
31. Gomez-Mascaraque, L.G.; Morfin, R.C.; Pérez-Masiá, R.; Sanchez, G.; Lopez-Rubio, A. Optimization of electrospraying conditions for the microencapsulation of probiotics and evaluation of their resistance during storage and in-vitro digestion. *LWT—Food Sci. Technol.* **2016**, *69*, 438–446. [CrossRef]
32. Suwanangul, S.; Sangsawad, P.; Alashi, M.A.; Aluko, R.E.; Tochampa, W.; Chittrakorn, S.; Ruttarattanamongkol, K. Antioxidant activities of sacha inchi (*Plukenetia volubilis* L.) protein isolate and its hydrolysates produced with different proteases. *Maejo Int. J. Sci. Technol.* **2021**, *15*, 48–60.
33. Zhang, B.; Deng, Z.; Ramdath, D.D.; Tang, Y.; Chen, P.X.; Liu, R.; Liu, Q.; Tsao, R. Phenolic profiles of 20 Canadian lentil cultivars and their contribution to antioxidant activity and inhibitory effects on α-glucosidase and pancreatic lipase. *Food Chem.* **2015**, *172*, 862–872. [CrossRef] [PubMed]
34. Mudgil, P.; Jobe, B.; Kamal, H.; Alameri, M.; Al Ahbabi, N.; Maqsood, S. Dipeptidyl peptidase-IV, α-amylase, and angiotensin I converting enzyme inhibitory properties of novel camel skin gelatin hydrolysates. *LWT* **2019**, *101*, 251–258.
35. Sangsawad, P.; Katemala, S.; Pao, D.; Suwanangul, S.; Jeencham, R.; Sutheerawattananonda, M. Integrated Evaluation of Dual-Functional DPP-IV and ACE Inhibitory Effects of Peptides Derived from Sericin Hydrolysis and Their Stabilities during *In Vitro*-Simulated Gastrointestinal and Plasmin Digestions. *Foods* **2022**, *11*, 3931. [CrossRef]
36. Xie, S.; Qu, P.; Luo, S.; Wang, C. Potential uses of milk proteins as encapsulation walls for bioactive compounds: A review. *J. Dairy Sci.* **2022**, *105*, 7959–7971. [CrossRef]
37. Paulo, F.; Santos, L. Microencapsulation of caffeic acid and its release using aw/o/w double emulsion method: Assessment of formulation parameters. *Dry. Technol.* **2019**, *37*, 950–961. [CrossRef]
38. Soukoulis, C.; Behboudi-Jobbehdar, S.; Yonekura, L.; Parmenter, C.; Fisk, I. Impact of milk protein type on the viability and storage stability of microencapsulated *Lactobacillus acidophilus* NCIMB 701748 using spray drying. *Food Bioprocess Technol.* **2014**, *7*, 1255–1268. [CrossRef]
39. Sharifi-Rad, M.; Anil Kumar, N.V.; Zucca, P.; Varoni, E.M.; Dini, L.; Panzarini, E.; Rajkovic, J.; Tsouh Fokou, P.V.; Azzini, E.; Peluso, I. Lifestyle, oxidative stress, and antioxidants: Back and forth in the pathophysiology of chronic diseases. *Front. Physiol.* **2020**, *11*, 694.
40. Kodali, V.P.; Sen, R. Antioxidant and free radical scavenging activities of an exopolysaccharide from a probiotic bacterium. *Biotechnol. J. Healthc. Nutr. Technol.* **2008**, *3*, 245–251. [CrossRef]
41. Zhang, B.; Zhang, H.; Yu, Y.; Zhang, R.; Wu, Y.; Yue, M.; Yang, C. Effects of *Bacillus coagulans* on growth performance, antioxidant capacity, immunity function, and gut health in broilers. *Poult. Sci.* **2021**, *100*, 101168. [CrossRef]
42. Jeong, H.J.; Park, J.H.; Lam, Y.; de Lumen, B.O. Characterization of lunasin isolated from soybean. *J. Agric. Food Chem.* **2003**, *51*, 7901–7906. [CrossRef] [PubMed]
43. Sui, L.; Zhu, X.; Wu, D.; Ma, T.; Tuo, Y.; Jiang, S.; Qian, F.; Mu, G. In vitro assessment of probiotic and functional properties of Bacillus coagulans T242. *Food Biosci.* **2020**, *36*, 100675. [CrossRef]
44. Zhou, R.; Xu, Y.; Dong, D.; Hu, J.; Zhang, L.; Liu, H. The effects of microcapsules with different protein matrixes on the viability of probiotics during spray drying, gastrointestinal digestion, thermal treatment, and storage. *eFood* **2023**, *4*, e98. [CrossRef]
45. Şengül, N.; Aslím, B.; Uçar, G.; Yücel, N.; Işık, S.; Bozkurt, H.; Sakaoğullarí, Z.; Atalay, F. Effects of exopolysaccharide-producing probiotic strains on experimental colitis in rats. *Dis. Colon Rectum* **2006**, *49*, 250–258. [CrossRef]
46. Vivek, K.; Mishra, S.; Pradhan, R.C.; Nagarajan, M.; Kumar, P.K.; Singh, S.S.; Manvi, D.; Gowda, N.N. A comprehensive review on microencapsulation of probiotics: Technology, carriers and current trends. *Appl. Food Res.* **2022**, *3*, 100248.
47. Suetsuna, K.; Ukeda, H.; Ochi, H. Isolation and characterization of free radical scavenging activities peptides derived from casein. *J. Nutr. Biochem.* **2000**, *11*, 128–131. [CrossRef] [PubMed]
48. Awosika, T.O.; Aluko, R.E. Inhibition of the in vitro activities of α-amylase, α-glucosidase and pancreatic lipase by yellow field pea (*Pisum sativum* L.) protein hydrolysates. *Int. J. Food Sci. Technol.* **2019**, *54*, 2021–2034. [CrossRef]
49. Kumari, V.; Huligere, S.S.; Ramu, R.; Naik Bajpe, S.; Sreenivasa, M.; Silina, E.; Stupin, V.; Achar, R.R. Evaluation of probiotic and antidiabetic attributes of lactobacillus strains isolated from fermented beetroot. *Front. Microbiol.* **2022**, *13*, 911243. [CrossRef]
50. Li, S.; Ke, W.; Zhang, Q.; Undersander, D.; Zhang, G. Effects of Bacillus coagulans and Lactobacillus plantarum on the Fermentation Quality, Aerobic Stability and Microbial Community of Triticale Silage. *Chem. Biol. Technol. Agric.* **2023**, *10*, 79. [CrossRef]

51. Kim, H.-W.; Choi, S.-Y.; Lee, D.-C.; Rhee, H.-I. Intestinal Production of Alpha-Glucosidase Inhibitor by Bacillus coagulans Spores. *Microorganisms* **2023**, *11*, 1462. [CrossRef]
52. Zhu, Y.; Li, X.; Teng, C.; Sun, B. Enhanced production of α-glucosidase inhibitor by a newly isolated strain of Bacillus subtilis B2 using response surface methodology. *Food Bioprod. Process.* **2013**, *91*, 264–270. [CrossRef]
53. Vieco-Saiz, N.; Belguesmia, Y.; Raspoet, R.; Auclair, E.; Gancel, F.; Kempf, I.; Drider, D. Benefits and inputs from lactic acid bacteria and their bacteriocins as alternatives to antibiotic growth promoters during food-animal production. *Front. Microbiol.* **2019**, *10*, 57.
54. Nongonierma, A.B.; FitzGerald, R.J. Features of dipeptidyl peptidase IV (DPP-IV) inhibitory peptides from dietary proteins. *J. Food Biochem.* **2019**, *43*, e12451.
55. Tidona, F.; Criscione, A.; Guastella, A.M.; Zuccaro, A.; Bordonaro, S.; Marletta, D. Bioactive peptides in dairy products. *Ital. J. Anim. Sci.* **2009**, *8*, 315–340.
56. Wu, S.; Lu, S.; Liu, J.; Yang, S.; Yan, Q.; Jiang, Z. Physicochemical properties and bioactivities of rice beans fermented by *Bacillus amyloliquefaciens*. *Engineering* **2021**, *7*, 219–225. [CrossRef]
57. Mudgil, P.; Aldhaheri, F.; Hamdi, M.; Punia, S.; Maqsood, S. Fortification of Chami (traditional soft cheese) with probiotic-loaded protein and starch microparticles: Characterization, bioactive properties, and storage stability. *LWT* **2022**, *158*, 113036. [CrossRef]
58. Yoha, K.; Moses, J.; Anandharamakrishnan, C. Effect of different drying methods on the functional properties of probiotics encapsulated using prebiotic substances. *Appl. Microbiol. Biotechnol.* **2023**, *107*, 1575–1588.
59. Kil, B.J.; Yoon, S.J.; Yun, C.-H.; Huh, C.-S. The effect of milk protein on the biological and rheological properties of probiotic capsules. *J. Microbiol. Biotechnol.* **2020**, *30*, 1870. [CrossRef]

foods

MDPI

Article

Kombucha and Water Kefir Grains Microbiomes' Symbiotic Contribution to Postbiotics Enhancement

Marina Pihurov [1], Bogdan Păcularu-Burada [1], Mihaela Cotârleț [1], Leontina Grigore-Gurgu [1], Daniela Borda [1], Nicoleta Stănciuc [1], Maciej Kluz [2,*] and Gabriela Elena Bahrim [1,*]

1 Faculty of Food Science and Engineering, Dunărea de Jos University of Galați, Domnească Street No. 111, 800201 Galați, Romania; marina.pihurov@ugal.ro (M.P.); bogdan.pacularu@ugal.ro (B.P.-B.); mihaela.cotarlet@ugal.ro (M.C.); leontina.gurgu@ugal.ro (L.G.-G.); daniela.borda@ugal.ro (D.B.); nicoleta.stanciuc@ugal.ro (N.S.)

2 Department of Bioenergetics and Food Analysis and Microbiology, University of Rzeszow, 35601 Rzeszow, Poland

* Correspondence: mkluz@ur.edu.pl (M.K.); gabriela.bahrim@ugal.ro (G.E.B.); Tel.: +40-03-3613-0182 (G.E.B.)

Abstract: Wild artisanal cultures, such as a symbiotic culture of bacteria and yeasts (SCOBY) and water kefir grains (WKG), represent a complex microorganism consortia that is composed of yeasts and lactic and acetic acid bacteria, with large strains of diversity and abundance. The fermented products (FPs) obtained by the microbiome's contribution can be included in functional products due to their meta-biotics (pre-, pro-, post-, and paraprobiotics) as a result of complex and synergistic associations as well as due to the metabolic functionality. In this study, consortia of both SCOBY and WKG were involved in the co-fermentation of a newly formulated substrate that was further analysed, aiming at increasing the postbiotic composition of the FPs. Plackett–Burman (PBD) and Response Surface Methodology (RSM) techniques were employed for the experimental designs to select and optimise several parameters that have an influence on the lyophilised starter cultures of SCOBY and WKG activity as a multiple inoculum. Tea concentration (1–3%), sugar concentration (5–10%), raisins concentration (3–6%), SCOBY lyophilised culture concentration (0.2–0.5%), WKG lyophilised culture concentration (0.2–0.5%), and fermentation time (5–7 days) were considered the independent variables for mathematical analysis and fermentation conditions' optimisation. Antimicrobial activity against *Bacillus subtilis* MIUG B1, *Staphylococcus aureus* ATCC 25923, *Escherichia coli* ATCC 25922, and *Aspergillus niger* MIUG M5, antioxidant capacity (DPPH), pH and the total acidity (TA) were evaluated as responses. The rich postbiotic bioactive composition of the FP obtained in optimised biotechnological conditions highlighted the usefulness of the artisanal co-cultures, through their symbiotic metabolic interactions for the improvement of bioactive potential.

Keywords: water kefir grains; SCOBY membranes; co-fermentation; design of experiments; postbiotics

Citation: Pihurov, M.; Păcularu-Burada, B.; Cotârleț, M.; Grigore-Gurgu, L.; Borda, D.; Stănciuc, N.; Kluz, M.; Bahrim, G.E. Kombucha and Water Kefir Grains Microbiomes' Symbiotic Contribution to Postbiotics Enhancement. *Foods* **2023**, *12*, 2581. https://doi.org/10.3390/foods12132581

Academic Editor: Xiqing Yue

Received: 12 June 2023
Revised: 28 June 2023
Accepted: 30 June 2023
Published: 2 July 2023

1. Introduction

Artisanal cultures are known as wild consortia of microorganisms which can grow on unconventional fermentation substrates, resulting in valuable products rich in bioactive compounds (biotics).

Briefly, SCOBY-based membranes are made of a natural consortium of microorganisms which work in mutualistic symbiosis due to a wide variety of species comprising mostly acetic acid bacteria (*Gluconacetobacter* ssp., *Acetobacter* ssp., and *Gluconobacter* ssp.) and yeasts (*Zygosaccharomyces* ssp., *Brettanomyces* ssp., and *Saccharomyces* ssp.), but also some lactic acid bacteria strains (*Lactobacillus* ssp.). The bacterial strains generate a polysaccharide stroma in which the yeasts are attached, causing the development of a membrane with a thickness of several centimetres at the liquid–air border [1,2]. This SCOBY-based membrane usually ferments black or green tea supplemented with sucrose. The fermentation took place over 7 to 10 days, at 20–25 °C, and as such a beverage with a rich postbiotic composition was achieved. It contained organic acids (acetic, lactic, malic, tartaric, citric, gluconic, and glucuronic), and other compounds (substances with antibiotic properties, ethanol, water-soluble vitamins, hydrolytic enzymes, and amino acids) [3,4].

WKG culture is typically composed of lactic acid bacteria (*Lactobacillus* ssp., *Streptococcus* ssp., and *Leuconostoc* ssp.), yeasts (*Saccharomyces* ssp. and *Dekkera* ssp.), and acetic acid bacteria, encapsulated in the polysaccharide dextran (and a limited concentration of levan) matrix, looking like hard granules, small, translucent, and irregular [5,6]. These grains are usually cultured in water and sucrose supplemented with dried fruits (often raisins or figs), afterwards producing a useful beverage [7]. After 2–3 days of fermentation at 20–25 °C, the FP contains a wide range of synthesised metabolites, including lactic acid, esters, glycerol, carbon dioxide, and acetic acid isoamyl acetate, as well as ethanol, ethyl octanoate, ethyl decanoate, and ethyl hexanoate [1,8].

The microbial strains' symbiotic interaction in the artisanal consortium offers metabolite production in the FPs (pre-, pro-, post-, and para-probiotics). The diversity of the microbial community ensures the stability and safety of FPs against spoilage and pathogenic microorganisms [5]. The products fermented with artisanal cultures also exhibit valuable biological activities, including immunomodulatory, anti-inflammatory [9], antihypertensive, hepatoprotective, cholesterol-lowering, and antioxidant potential [10]. This study's goal was the co-cultivation of SCOBY and WKG (lyophilised cultures) in a new formulated fermentation medium and the optimisation of the biotechnological parameters to produce FP with improved bioactive content.

2. Materials and Methods

2.1. Multiplication and Freeze-Drying of Water Kefir Grains as Starter Cultures

WKG (Medicer Bios, Bucharest, Romania) were grown on a specific medium based on sterile tap water supplemented with 10% (*w*/*v*) sugar and 1% (*w*/*v*) raisins, at 25 ± 1 °C, for a period of 48 h, under aerobic conditions. Further, the grains were firstly washed using Milli Q water, and then immersed in a fresh medium. The mixtures were incubated under the same conditions as mentioned before, with successive cultivation steps to allow the granules to multiplicate [11]. Furthermore, in order to assure the cryoprotection of the viable cells, the granules were washed with ultrapure water and supplemented with 10% (*w*/*v*) inulin and lyophilised at −80 °C (Christ Alpha 1-4 LD plus, Osterode am Harz, Germany). The cryodesiccated water kefir grains were ground and further stored at a temperature of 4 °C.

2.2. Multiplication of Kombucha's Biofilm as a Starter Culture

The SCOBY-based membranes, purchased from a private household from the Republic of Moldova, were multiplied via successive cultivation in a sugar-based medium (7.5%, *w*/*v*) with the addition of 3% (*w*/*v*) infusion of black tea (Aaro Forstman Oy, Vantaa, Finland). The mixture was incubated for 10 days at room temperature (22 ± 2 °C). Then, 20% (*w*/*w*) sterile inulin powder was added to 100 g of biofilm that was previously divided in small pieces, the mixture being homogenised and freeze-dried [12]. The resulting biofilm was finally mortared and stored at 4 °C.

2.3. Experimental Design and Optimisation of WKG and SCOBY Co-Cultivation Process

2.3.1. The Plackett–Burman Design to Select the Factors Influencing the Synergism between WKG and SCOBY

For the screening of the most significant parameters that influence the co-cultivation of WKG and SCOBY, FPs were prepared according to the Plackett–Burman experimental Design (PBD), by varying the following parameters: 1–3% (*w*/*v*) black tea leaves (the tea mixture being infused for 5 min in boiled water and cooled down until 90 ± 1 °C), 3–6% (*w*/*v*) raisins, 5–10% (*w*/*v*) sugar, pH = 6.30 (Table 1). After the medium's sterilisation at 105 °C, for a period of 10 min, it was inoculated, after cooling, with the lyophilised starter cultures (0.2–0.5% *w*/*v*), and incubated under aerobic and stationary conditions for 5 to 7 days at 30 °C. After fermentation, the samples were analysed immediately.

Table 1. Independent variables and the PBD variation ranges.

Independent Variables	Minimum Value	Maximum Value
	−1	+1
A, black tea concentration, % (*w*/*v*)	1.0	3.0
B, sugar concentration % (*w*/*v*)	5.0	10.0
C, raisins concentration % (*w*/*v*)	3.0	6.0
D, time of fermentation, days	5.0	7.0
E, freeze-dried SCOBY concentration, % (*w*/*v*)	0.2	0.5
F, freeze-dried concentration WKG, % (*w*/*v*)	0.2	0.5

The Minitab software took into consideration 3 central points and 6 factorial and thus generated 15 experimental variants. The responses that were regarded were the pH and titratable acidity, the antioxidant capacity, and the antimicrobial activities (against above-mentioned microorganisms).

2.3.2. Optimisation of WKG and SCOBY Co-Cultivation via Response Surface Methodology (RSM)

After analysing the responses obtained from PBD experimental runs, the factors influencing the fermentation with WKG and SCOBY were identified, as follows: concentration of black tea and raisins, as well as the fermentation time. Subsequently, five variation levels were analysed for the independent variables (Table 2) included in the RSM analysis. For the statistical validation of the experimental models, a p value of <0.05 was regarded as being significant.

Table 2. Variation levels of the independent variables in RSM.

Independent Variables	Variation Levels				
	−1	0	+1	$-\alpha$	$+\alpha$
A, black tea concentration, % (*w*/*v*)	1.00	2.00	3.00	1.39	2.61
C, raisins concentration % (*w*/*v*)	3.00	4.50	6.00	3.58	5.42
D, fermentation time, days	5.00	6.00	7.00	5.39	6.61

The factors that were not optimised (B, E, F) remained constant; respectively, a 5% sugar concentration (*w*/*v*), a 0.2% lyophilised SCOBY concentration (*w*/*v*), and a concentration of WKG of 0.2% (*w*/*v*).

2.4. *The Evaluation of the Responses*

2.4.1. Acidifying Potential

The pH analysis was assessed with a digital pH meter (Mettler Toledo, FiveEasy F20, Greifensee, Switzerland).

The titratable acidity was expressed in Thörner degrees (°Th), using the AOAC method [13]. In brief, about 4 g of sample was weighed and distilled water was added to reach 50 mL, in volumetric flask. Aliquots of 10 mL were used for the NaOH 0.1N titration, using phenolphthalein, until the appearance of a weak pink colour.

2.4.2. Evaluation of the Antifungal Properties of the FPs

The antifungal activity assessment was carried out on the indicator strain of *Aspergillus niger* MIUG M5, belonging to the MIUG Collection, from the Faculty of Food Science and Engineering, "Dunărea de Jos" University, Galati, Romania. The incubation of the mould strain was achieved at 25 °C, for 96 h, using the Yeast Glucose Chloramphenicol

(YGC) Agar. The inoculum was obtained by the suspension of the spores in sterile saline solution (0.9% NaCl) at a concentration of 1×10^5 spores/mL, by counting with the Thoma chamber. From the fresh FP, a volume of 1 mL was taken and dispersed in a Petri dish and thoroughly mixed with a volume of 20 mL of Potato Dextrose Agar (PDA) medium (Oxoid, England). After solidification, from the spore's suspension a volume of 10 µL was inoculated in the centre of the plate and left for incubation at 25 °C for 4 days (96 h). The control was assessed under the above-mentioned conditions but without the FP addition. Subsequent to the incubation time, the diameters of the mould's growth were determined and the inhibition ratio (*RI*) was determined using Equation (1) [14]:

$$RI = \frac{A_c - A_t}{A_c} \times 100 \tag{1}$$

where *RI* represents the growth inhibition ratio, A_c represents the mould growth's diameter of the control sample, and A_t the mould strain's diameter on the FP-supplemented medium.

2.4.3. Assessment of the FPs' Antibacterial Activity

Antibacterial activity was tested against the indicator strains *Bacillus subtilis* MIUG B1, *Escherichia coli* ATCC 25922, and *Staphylococcus aureus* ATCC 25923, strains that were cultivated on Plate Count Agar (PCA) (Scharlau, Barcelona, Spain), respectively, and Mueller II Hinton agar (Biolab, Hungary) for 24 h, at 37 °C. Then, the colony was placed into the Nutrient Broth (for *B. subtilis*) or, respectively, Muller Hinton broth (for *E. coli* and *S. aureus*), with an overnight incubation at 37 °C. Afterwards, the bacterial inoculum was dimensioned spectrophotometrically (OD_{600nm}) at 0.3, corresponding to a concentration around 2.4×10^8 CFU/mL. Subsequently, a volume of 500 µL of bacterial suspension was added in the Petri dishes with the specific media for each bacterial strain in the wells made (8 mm diameter), and 100 µL of FP was added. Afterwards, the incubation of the plates took place at 37 °C, for 48 h, the inhibition zone being determined and expressed in mm [15,16].

2.4.4. Evaluation of the Antioxidant Properties of the FPs

In order to extract the bioactive from the fermented medium, an ultrasound-assisted method (MRC. Ltd., Holon, Israel) was applied, considering an extraction time of 30 min at 40 °C, followed by centrifugation at 7000 rpm and 4 °C for 15 min. To obtain the 2,2-diphenyl-1-picrilhydrasyl (DPPH) radical scavenging potential (DPPH) solution, 4 mg of DPPH were transferred in 100 mL of HPLC-grade methanol (≥99.9%) and allowed to dissolve. The DPPH solution was prepared daily and stored in dark conditions [17,18]. From the supernatant, an aliquot of 0.1 mL sample was homogenised with 3.9 mL DPPH solution and kept in the dark for 90 min, and the absorbance was read at 515 nm [19,20].

Antioxidant activity was performed by adding the 2,2-diphenyl-1-picrilhydrasyl (DPPH) radical scavenging potential, allowing the antioxidant activity in µM TE/mL to be calculated on a calibration curve based on 6-hydroxy-2,5,7,8-tetramethylcroman-2-carboxylic acid (Trolox). The antioxidant activity was calculated according to the Formulas (2) and (3):

$$\text{RSA, \%} = [(A_m - A_p)/A_m] \times 100 \tag{2}$$

$$\mu\text{M TE/mL} = (\text{RSA, \%} - 3.3672)/0.3483 \tag{3}$$

where A_m was the absorbance of the control and A_p the absorbance of the sample analysed.

In brief, 4 mg of DPPH was transferred in 100 mL of HPLC-grade methanol (≥99.9%) and allowed to dissolve. The DPPH solution was prepared daily and stored in dark conditions [17,18].

2.5. Analysis of Organic Acids and Polyphenolic Content in the FP Obtained in Optimised Fermentation Conditions

2.5.1. The Assessment of the Organic Acids

The determination was achieved using an HPLC system, Agilent 1200 (Agilent Technologies, Santa Clara, CA, USA), with a multi-wavelength detector (MWD) and a quaternary pump, autosampler, degasser, and a thermostat. The column was a Hamilton RPR X300 (250 × 4.1 mm, particle size 7 µm, Hamilton, Bonaduz, Switzerland), with a gradient elution composed of mobile phase A (KH_2PO_4, 20 mM, pH 2.4) and phase B—acetonitrile 90% (*v*/*v*) (ACN) [21]. The mobile phases' mixtures had the following steps: min 0—80% A, min 10—40% A, min 12.5—40% A, min 12.6—80% A. The organic acids' separation profile was achieved at 210 nm, injection volume of 20 µL, at 30 °C, with a 1.5 mL/min flow-rate [22]. The data acquisition was assessed with the ChemStation software.

The organic acids were identified and quantified based on external calibration curves using HPLC-grade organic acids' standard solutions (Sigma Aldrich, Schnelldorf, Germany).

2.5.2. Evaluation of Polyphenols

The Agilent 1200 high-performance liquid chromatography system (Agilent Technologies, Santa Clara, CA, USA) was applied to determine the polyphenolic composition of the analysed samples. Consequently, the compounds of interest were separated with a Synergi Max-RP-80Å column with a guard column (250 × 4.6 mm, particle size of 4 µm, Phenomenex, Torrance, CA, USA) using mobile phase A (ultrapure water: acetonitrile: formic acid = 87:3:10) and mobile phase B (ultrapure water: acetonitrile: formic acid = 40:50:10), with the following elution program: min 0—94% A, min 20—80% A, min 35—60% A, min 40—40% A, min 45—10% A. For the polyphenolic compounds' separation, at 30 °C, a 20 µL volume was deployed into the column, at a 0.5 mL/min flow rate. The time of the method was 80 min and then the data were processed using ChemStation program version B.04.03 [14,23,24].

The polyphenols were identified and quantified simultaneously at the wavelengths of 280 nm and 320 nm, based on external calibration curves for the available polyphenolic HPLC-grade standards (Sigma Aldrich, Schnelldorf, Germany).

2.6. Statistical Analysis

The design of the experiments was assessed with the Minitab 17 software (v. 1.0, LLC, Pennsylvania State University, Centre County, PA, USA). One-way ANOVA and Tukey tests considering a 95% confidence interval ($p < 0.05$) were used to analyse the experimental results, which were considered as averages of triplicate measurements followed by standard deviation.

3. Results and Discussions

3.1. The Selection of the Most Important Parameters That Influenced the Fermentation Process with Artisanal Consortia via PBD Analysis

This strategy was applied for the fermentation process, aiming at designing the appropriate culture medium by adjusting the carbon (C) and nitrogen (N) sources, the C/N ratio, minerals, trace elements, growth factors, and the fermentation parameters. In the customised formulas for fermentation, the main source of C was considered, whereas the fresh or dried fruits provided the nitrogen [25]. The analysed parameters and the interactions between them could be evaluated objectively using statistical methods [26].

The statistical modelling with PBD generated 15 experimental combinations using the ranges of variation for the chosen factors, as follows: black tea concentration 1–3% (*w*/*v*), sugar concentration 5–10% (*w*/*v*), raisins concentration 3–6% (*w*/*v*), 5–7 days of fermentation, lyophilised SCOBY ranging from 0.2 to 0.5% (*w*/*v*), and WKG ranging from 0.2 to 0.5 % (*w*/*v*), respectively (Table 3).

Table 3. The PBD of experiments and the corresponding responses obtained based on the independent variables' variation.

Run	Independent Variables						Responses						
	A *	B	C	D	E	F	pH	Total Acidity, °Th	Antioxidant Activity, µM TE/mL	Antibacterial Activity against *E. coli*, mm	Antibacterial Activity against *S. aureus*, mm	Antibacterial Activity against *B. subtilis*, mm	Antifungal Activity against *A. niger*, RI %
1	1.0	10.0	6.0	7.0	0.2	0.5	3.85	37.50	2.411	0.00	0.00	5.00	100.00
2	3.0	5.0	6.0	7.0	0.2	0.5	3.94	40.00	2.404	0.00	0.00	7.00	82.96
3	2.0	7.5	4.5	6.0	0.35	0.35	3.85	51.25	2.412	0.00	0.00	5.83	100.00
4	1.0	5.0	6.0	7.0	0.5	0.2	3.87	50.00	2.409	0.00	0.00	3.83	100.00
5	2.0	7.5	4.5	6.0	0.35	0.35	3.83	43.75	2.412	0.00	0.00	5.33	100.00
6	1.0	10.0	6.0	5.0	0.5	0.2	3.85	41.25	2.398	0.00	14.00	4.17	88.34
7	1.0	5.0	3.0	7.0	0.5	0.5	3.82	45.00	2.401	0.00	0.00	1.50	100.00
8	3.0	10.0	6.0	5.0	0.5	0.5	3.96	38.75	2.393	0.00	0.00	7.33	100.00
9	2.0	7.5	4.5	6.0	0.35	0.35	3.88	36.25	2.401	0.00	0.00	5.50	100.00
10	1.0	10.0	3.0	5.0	0.2	0.5	3.87	20.00	2.394	0.00	11.67	2.50	100.00
11	1.0	5.0	3.0	5.0	0.2	0.2	3.46	225.00	2.398	12.67	12.33	14.33	82.06
12	3.0	5.0	3.0	5.0	0.5	0.5	3.96	56.25	2.388	0.00	0.00	6.00	100.00
13	3.0	10.0	3.0	7.0	0.2	0.2	3.91	62.50	2.388	0.00	0.00	7.17	100.00
14	3.0	10.0	3.0	7.0	0.5	0.2	3.88	56.25	2.390	0.00	0.00	6.33	83.86
15	3.0	5.0	6.0	5.0	0.2	0.2	3.92	58.75	2.388	0.00	0.00	6.00	100.00

* A—tea concentration, % (*w*/*v*); B—sugar concentration, % (*w*/*v*); C—raisins concentration % (*w*/*v*); D—time of fermentation, days; E—SCOBY culture's concentration % (*w*/*v*); F—WKG culture's concentration, % (*w*/*v*).

Thus, the following results were obtained; 3.46–3.96 for pH and 20–225 °Th for titratable acidity, while for the antioxidant activity a value of 2.388–2.412 µM TE/mL was found, and 0.00–12.67 mm for the antibacterial activity inhibition zone against *E. coli*, 0.00–14.00 mm against *S. aureus*, 1.50–14.33 mm against *B. subtilis*, and an 82.06–100% inhibition zone for the antifungal activity.

The statistical models, based on some analysed responses, respectively, the antioxidant activity and the antibacterial activity against *S. aureus*, were validated in accordance with the regression coefficients higher than 80% and at a $p < 0.05$ value.

The main parameters that influenced the FPs' antioxidant activity were the tea's concentration (A), raisins' concentration (C), and the time of fermentation (D), and their impact on the studied response variables is shown in the Pareto diagram (Figure 1a).

Analysing the results comparatively, it can be stated that the FPs that were involved in the co-culture were characterised by a higher bioactive capacity compared to the unfermented medium, due to the polyphenols present in the tea and the metabolites of yeasts and bacteria, including vitamins, organic acids, and extracellular enzymes that contribute to the structural and compositional changes during the fermentation of kombucha [27].

The product fermented with WKG had a high antioxidant potential on account of the presence of lactic and acetic acid bacteria, as well as yeasts, their metabolites, and cell lysis' products that occurred during fermentation [28].

For the antibacterial activity against *S. aureus*, the significant factors were the concentration of tea (A) and the fermentation time (D), as the Pareto diagram in Figure 1a shows.

It is known that due to the production of post-biotics, FPs (including black tea substrate) have shown antibacterial activity of an 12–30.2 mm inhibition zone against several pathogens [29].

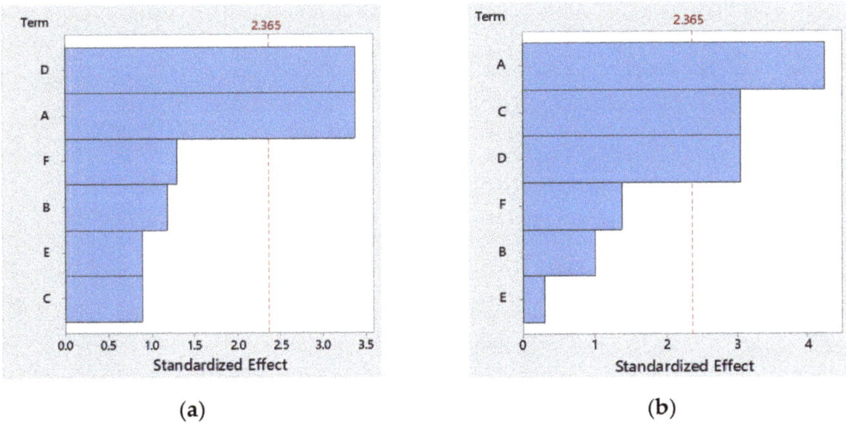

Figure 1. Pareto diagram of the independent variables' effect studied on (**a**) antibacterial capacity against *S. aureus* strain and (**b**) antioxidant activity of FPs.

For the rest of the analysed responses, no validation of the model was achieved.

The ANOVA results from Table 4 showed the significant contributions for the concentration of tea (p = 0.004), the concentration of raisins, and for the fermentation time (p = 0.019). Furthermore, this statistical model can be validated based on in the non-significant lack of fit (p = 0.923) [30].

Table 4. Antioxidant activity based on the ANOVA test.

Source	DF	Adj SS	Adj MS	F Value	p Value
Pattern	7	0.001001	0.000143	8.38	0.006
Linear	6	0.000677	0.000113	6.61	0.013
Concentration of tea, g%	1	0.000308	0.000308	18.03	0.004
Sugar concentration, g%	1	0.000018	0.000018	1.03	0.343
Concentration of raisins, g%	1	0.000159	0.000159	9.30	0.019
Time of fermentation, days	1	0.000159	0.000159	9.30	0.019
Concentration of kombucha, g%	1	0.000002	0.000002	0.09	0.774
Concentration of water kefir granules, g%	1	0.000032	0.000032	1.89	0.211
Curvature	1	0.000324	0.000324	19.01	0.003
Error	7	0.000119	0.000017		
Lack-of-Fit	5	0.000043	0.000009	0.22	0.923
Pure error	2	0.000077	0.000038		
Total	14	0.001120			

3.2. *Optimisation of the Fermentation Process with Artisanal Co-Culture to Increase the FPs' Functional Potential*

The statistical results from the PBD allowed the selection of three important parameters: tea concentration, raisin concentration, and fermentation time. The other factors, namely, the concentration of SCOBY lyophilised culture (0.2%) and the concentration of WKG lyophilised culture (0.2%) at 30 °C for 5 days of fermentation, remained constant.

The amount of inoculum, the amount of sugar and fruit added, the medium composition, the amount of oxygen, and the time and temperature of fermentation were also mentioned as factors that determined the best fermentation of WKG and had an impact on the composition and properties of the FP [28].

Table 5 presents the experimental matrix obtained by the Central Composite Design (CCD) model that generated 20 experimental variants, with the corresponding analysed responses: pH, titratable acidity, antioxidant activity, and antibacterial and antifungal activities against the targeted strains.

Table 5. CCD with the analysed responses correlating with the independent variables' variation in the RSM analysis.

Independent Variables					Responses					
Run	A *	C	F	pH	Total Acidity, °Th	Antioxidant Activity, µM TE/mL	Antibacterial Activity against *E. coli*, mm	Antibacterial Activity against *S. aureus*, mm	Antibacterial Activity against *B. subtilis*, mm	Antifungal Activity against *A. niger*, RI %
1	2.00	4.50	6.00	3.92	62.50	2.384	0.00	0.00	7.33	100.00
2	1.39	5.42	5.39	3.42	637.50	2.385	6.67	5.50	18.50	70.12
3	2.00	4.50	6.00	3.90	75.00	2.395	0.00	0.00	7.17	100.00
4	2.61	5.42	5.39	3.94	158.75	2.381	0.00	0.00	7.00	100.00
5	2.61	3.58	5.39	3.91	100.00	2.391	0.00	0.00	8.50	89.24
6	1.39	3.58	5.39	3.67	137.50	2.381	5.67	7.67	10.33	100.00
7	1.39	3.58	6.61	3.83	87.50	2.388	0.00	0.00	7.83	100.00
8	2.00	4.50	6.00	3.92	63.75	2.379	0.00	0.00	6.50	100.00
9	1.39	5.42	6.61	3.82	112.50	2.381	0.00	0.00	5.17	100.00
10	2.61	5.42	6.61	3.91	143.75	2.389	0.00	0.00	6.67	100.00
11	2.00	4.50	6.00	3.89	113.75	2.386	0.00	0.00	7.33	85.66
12	2.61	3.58	6.61	3.91	140.00	2.372	0.00	0.00	7.83	75.30
13	2.00	4.50	7.00	3.90	57.50	2.372	0.00	0.00	7.17	89.24
14	2.00	3.00	6.00	3.95	51.25	2.354	0.00	0.00	6.17	100.00
15	2.00	4.50	5.00	3.95	56.25	2.373	0.00	0.00	7.00	100.00
16	2.00	4.50	6.00	3.97	53.75	2.357	0.00	0.00	7.17	89.64
17	1.00	4.50	6.00	3.80	138.75	2.372	0.00	0.00	5.33	88.84
18	2.00	6.00	6.00	3.92	140.00	2.006	0.00	0.00	8.00	100.00
19	2.00	4.50	6.00	3.92	112.50	2.375	0.00	0.00	8.50	100.00
20	3.00	4.50	6.00	3.92	87.50	2.372	0.00	0.00	7.83	100.00

* A—black tea's concentration, % (*w*/*v*); C—raisins' concentration, % (*w*/*v*); F—fermentation time, days.

Following the statistical analysis of the obtained results, two mathematical models, for pH and total acidity, were validated, with a probability value of 0.003, thus highlighting the factors with a significant interaction for each validated response (Table 6).

Table 6. Validation of the interaction of significant independent variables.

Response	Variables with Significant Interaction	*p* Value
pH	1. Black tea's concentration, g%, and Fermentation time, days	0.003
Total acidity, °Th	1. Black tea's concentration, g%, and Raisins' concentration, g%	0.013
	2. Black tea's concentration, g%, and Time of fermentation, days	0.003
	3. Raisins' concentration, g%, and Time of fermentation, days	0.006

The interactions between the variables, as well as their impact on the response, can be visualised in the contour and surface graphs (Figure 2a–c), which highlight the correlation between the tea concentration, the fermentation time, the concentration of raisins, and the responses obtained for the validated models.

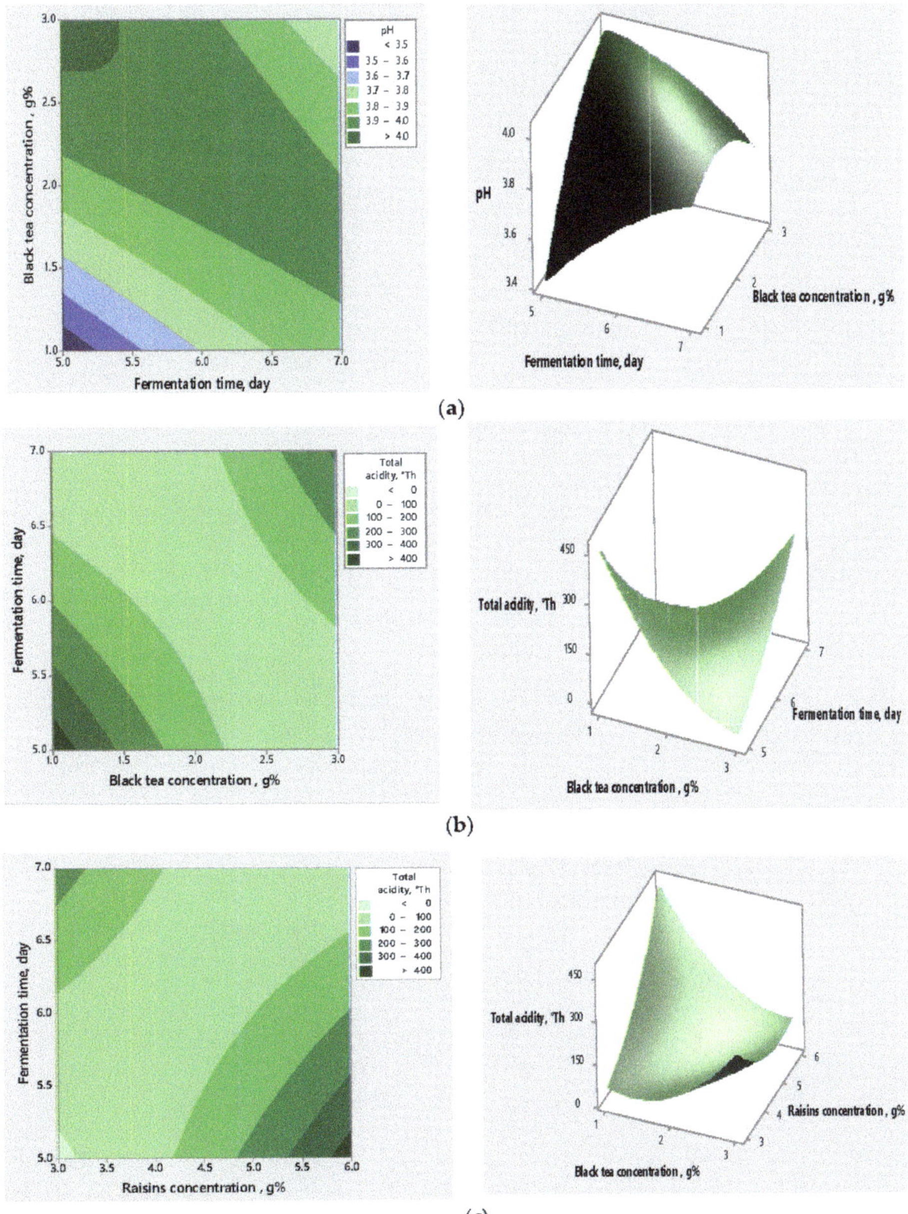

Figure 2. Contour graphs (**left**) and surface graphs (**right**) for correlative effect of the significant variables on pH (**a**) and total acidity (**b**,**c**).

Analysing the above graphs, the acidification potential increased when increasing the concentration of raisins and decreasing the concentration of tea and the time of fermentation.

According to the CCD experimental data, the following values for the analysed responses were obtained: 3.42–3.97 for pH, 51.25–637.5 °Th, 2.006–2.395 μM TE/mL for the antioxidant potential, 0–6.67 mm antibacterial activity inhibition zone against *E. coli*, 0–7.67 mm against *S. aureus*, 5.17–18.5 mm against *B. subtilis*, and 70.12–100% inhibition for antifungal activity against *A. niger*.

Following the analysis of the validated models and the significant factors, an optimised FP with an increased bioactive activity was designed by the formulated medium based on 3.52% (*w*/*v*) raisins, 1.0% (*w*/*v*) black tea, and 5% (*w*/*v*) sugar, inoculated with 0.2% (*w*/*v*) WKG lyophilised culture and 0.2% (*w*/*v*) SCOBY lyophilised culture co-fermentation at 30 °C, for 5 days, under stationary aerobic conditions.

The validation models for pH and total acidity (titratable acidity) were then analysed. The experimental values ranged between the predicted values for a 95% confidence level. Also, the desirability of the model was 0.901, close to 1, which indicates that by following the chosen parameters favourable results for the analysed responses can be achieved (Table 7).

Table 7. Validation of the models.

Response	95% Confidence Level Range	Experimental Value
pH	3.30–3.70	3.30 ± 0.01
Titratable acidity	55.5–444.6	375.83 ± 0.02
Desirability of the model	0.901	

The FP obtained in optimised conditions was characterised by a high acidity of 375.83 °Th and a low pH of 3.30 compared to the control (unfermented sample), which had an acidity potential of 9.27 °Th and a pH value of 4.69. Also, the control showed no antibacterial activity for the antifungal inhibition calculated with a ratio of 2.68%. Therefore, the antioxidant activity was higher (2.507 μM TE/mL) due to the polyphenols from the tea. The analysis of the validated models showed that the optimisation of bio-processes improved the FPs' acidification capacity.

3.3. *Organic Acids and Polyphenols Content in the Optimised FP*

3.3.1. Organic Acids Content

Using the high-performance liquid chromatography technique, the compounds present in the unfermented medium (control) and the fermented one were quantified (Table 8).

Table 8. Content of organic acids in FP vs. unfermented medium.

Organic Acid	Concentration	
	Unfermented Medium, mg/mL	Fermented Product, mg/mL
Acetic acid	4.34 ± 0.01 ^{B}	8.72 ± 0.12 ^{A}
Butyric acid	37.90 ± 0.42 ^{B}	45.81 ± 0.17 ^{A}
Isovaleric acid	ND *	0.88 ± 0.12 ^{A}

^{A,B}—significant differences in the obtained values compared to the control ($p < 0.05$). * ND—not determined.

Due to the symbiosis between yeasts and lactic acid bacteria, the development of the homopolysaccharide matrix from WKG and organic acid production were achieved. In this regard, yeasts helped bacteria by providing nitrogen as simple assimilable compounds (dipeptides, tripeptides and amino acids) through their proteolytic activity. Also, the carbon source had a key signification in the fermentative capacity of the WKG [25]. The association and competition between the bacteria and yeasts in kombucha were unique, leading to chained reactions from various metabolites, including up to 6.4 g/L in acetic and lactic acids, and up to 0.5 g/L in citric, gluconic, malic, and succinic acids [31]. The consortium members' cooperative association is well established. Less than 30% of the consortium is made of lactic acid bacteria strains, which are recognised for producing both lactic and also gluconic acids, which contribute to the antibacterial and antioxidant characteristics of the FP [32].

Among the identified short-chain fatty acids, acetic acid is characteristic for SCOBY fermentation, also being produced in small amounts as a postbiotic of the WKG consortium. As such, the acetic acid concentration increased from 4.34 mg/mL to 8.72 mg/mL. Butyric acid is not frequently found in kombucha-based drinks, but still its presence may occur, as Uţoiu et al., 2018 reported; after 5 days of fermentation, 0.14 g/L butyric acid was determined [33], compared to the present study, where the amount of butyric acid increased from 37.90 mg/mL to 45.81 mg/mL. Isovaleric acid is a volatile compound that contributes to the flavour of the FP, being the result of the interaction between acetic acid bacteria and yeasts (e.g., *Acetobacter indonesiensis* with *Brettanomyces bruxellensis*). The literature highlighted a concentration of up to 0.007 mg/mL; instead, the present study reported an amount of 0.88 mg/mL in the FP.

Previously, in an FP obtained by co-fermentation with milk kefir grains and SCOBY, some organic acids such as lactic acid, acetic acid, citric acid, isovaleric acid, and butyric acid, which presented the following concentrations, respectively, of 24.39 mg/mL, 25.21 mg/mL, 5.77 mg/mL, 4.36 mg/mL, and 67.33 mg/mL, were synthesised by the artisanal cultures [12]. Therefore, the lactic and citric acids were not identified in the optimised fermented product's composition obtained by fermentation of the formulated medium with a multiple starter culture, based on WKG and SCOBY microbiota; the result can be attributed to the synergistic functionality of the consortia in tested conditions, in correlation with the chemical composition of the fermentation substrate.

3.3.2. Content of Polyphenols and Flavonoids

The major bioactive compounds identified in the product obtained in optimised conditions were caffeic acid, 255.64 μg/mL; rutin trihydrate, 568.93 μg/mL; and epicatechin, 1135.69 μg/mL, whereas the caffeic acid was found in a lower concentration in the control, respectively, 16.80 μg/mL, this bioactive compound being specific to black tea (Table 9). Gallic acid and isorhamnetin, ferulic, and chlorogenic acids were present in smaller concentrations. Some compounds have been identified by Vázquez-Cabral et al., 2017, in a kombucha beverage, e.g., myricetin, 0.184 mg/L; gallic acid, 54.396 mg/L; caffeic acid, 16.213 mg/L; chlorogenic acid, 0.539 mg/L; epicatechin, 142.62 mg/L; and rutin, 4.245 mg/L. Our experimental data were confirmed by other results from similar works, that reported, in a fermented product with WKG, compounds such as chlorogenic acid, caffeic acid, tannins, vitamins C and D, glucosides, and various enzymes including lipase, amylase, and protease [34].

Table 9. Content of bioactive compounds in the FP vs. unfermented medium.

Bioactive Compound	Unfermented Medium, μg/mL	Fermented Product, μg/mL
Gallic acid	ND	39.68 ± 2.56^{a}
Caffeic acid	16.80 ± 0.02^{b}	255.64 ± 54.77^{a}
Chlorogenic acid	2 ± 0.00^{a}	0.25 ± 0.02^{b}
Ferulic acid	ND	0.36 ± 0.00^{a}
Rutin trihydrate	ND	568.92 ± 50.06^{a}
Epicatechin	ND	1135.59 ± 44.76^{a}
Isorhamnetin	ND	11.04 ± 0.96^{a}

The results are expressed as mean ± standard deviation. Values with different superscript letters in the same row indicate significant differences between samples ($p < 0.05$). $^{a, b}$—significant differences between the control and fermented sample.

Previously, in our research regarding co-fermentation with SCOBY and milk kefir grains, in the sample obtained under optimised conditions, several compounds were quantified: gallic acid ≅ 71 μg/mL, epicatechin ≅ 1063 μg/mL, caffeic acid ≅ 315 μg/mL, quercetin ≅ 18 μg/mL, apigenin ≅ 0.22 μg/mL, and isorhamnetin ≅ 3 μg/mL [12].

Following the statistical and mathematical modelling analysis, an FP with improved bioactive properties was obtained. In the tested biotechnological conditions, the main independent variables with an influence on the quality of the FP turned out to be the concentration of tea, the fermentation time, and the concentration of raisins. Thus, the optimised fermentation conditions were: (i) composition of the medium: 3.52% (w/v)

raisins, 1.0% (*w/v*) black tea, 5% (*w/v*) sugar in sterilised tap water; (ii) inoculum: 0.2% (*w/v*) lyophilised culture of WKG and 0.2% (*w/v*) lyophilised culture of SCOBY; and (iii) fermentation process: aerobic conditions, in a stationary system, at a temperature of 30 °C, for 5 days.

According to these biotechnological conditions, the obtained FP presented a high acidity potential of 375.83 °Th, and a 3.25 pH value. The organic acids were also highlighted in different concentrations; acetic—8.72 mg/mL, butyric—45.81 mg/mL, isovaleric—0.88 mg/mL, respectively; polyphenolic compounds such as phenolic acids: caffeic—255.64 µg/mL, gallic—39.68 µg/mL, ferulic—0.36 µg/mL, and chlorogenic—0.25 µg/mL; and flavonoids derived from quercetin: rutin trihydrate—568.93 µg/mL, isorhamnetin—11.94 µg/mL, and epicatechin—1135.69 µg/mL. The presence of these compounds demonstrates the functional potential of the FP.

4. Conclusions

The obtained results confirmed the possibility of using multiple SCOBY and WKG starter cultures to ferment a newly formulated black tea, raisin, and sugar-based medium to improve the postbiotic composition of the FP. The variant of FP with the increased bioactive potential was obtained following statistical techniques for the selection of the parameters (independent variables) and the optimisation of the process. The analysed responses demonstrated the ability of the bacteria and yeasts from the SCOBY and WKG microbiome to work in symbiosis. The preservation and usage of the artisanal cultures as freeze-dried cultures ensured the stability of the strain's functionality. This study demonstrated the versatile metabolism and synergism of the wild microorganisms (bacteria and yeasts) from the multiple consortia. The idea of using these artisanal cultures for the co-fermentation of the unconventional substrates demonstrated their employment for multiple applications. Thus, by variation of the fermentation parameters and exploitation of PBD and CCD tools it is possible to obtain FPs with different compositions and bioactive properties to be used as ingredients for food and feed formulation.

Author Contributions: Conceptualisation, G.E.B., M.P. and M.K.; methodology, M.C., M.P., B.P.-B. and N.S.; software, D.B., M.P. and B.P.-B.; validation, D.B., M.C. and B.P.-B.; formal analysis, M.P., L.G.-G. and B.P.-B.; investigation, M.P., M.C., B.P.-B. and N.S.; resources, G.E.B.; data curation, M.P., M.C., L.G.-G. and D.B.; writing—original draft preparation, M.P.; writing—review and editing, M.P., M.C., M.K., and G.E.B.; visualisation, M.C., D.B., N.S. and G.E.B.; supervision, M.C. and G.E.B.; project administration, G.E.B.; funding acquisition, G.E.B. All authors have read and agreed to the published version of the manuscript.

Funding: This work was supported by a grant from the Ministry of Research, Innovation and Digitisation, CNCS/CCCDI—UEFISCDI, project number PCE 159/2021, within PNCDI III.

Institutional Review Board Statement: Not applicable.

Data Availability Statement: Data are contained within the article.

Conflicts of Interest: The authors declare no conflict of interest. The funders had no role in the design of the study; in the collection, analyses, or interpretation of data; in the writing of the manuscript; or in the decision to publish the results.

References

1. Pihurov, M.; Păcularu-Burada, B.; Cotârleţ, M.; Vasile, M.A.; Bahrim, G.E. Novel Insights for Metabiotics Production by Using Artisanal Probiotic Cultures. *Microorganisms* **2021**, *9*, 2184. [CrossRef]
2. Nikolaev, Y.A.; Plakunov, V.K. Biofilm-"city of Microbes" or an Analogue of Multicellular Organisms? *Microbiology* **2007**, *76*, 125–138. [CrossRef]
3. Jayabalan, R.; Waisundara, V.Y. Kombucha as a Functional Beverage. In *Functional and Medicinal Beverages: Volume 11: The Science of Beverages*; Elsevier Inc.: Amsterdam, The Netherlands, 2019; pp. 413–446. ISBN 9780128163979.
4. Oliveira, Í.A.C.L.D.; Rolim, V.A.D.O.; Gaspar, R.P.L.; Rossini, D.Q.; de Souza, R.; Bogsan, C.S.B. The Technological Perspectives of Kombucha and Its Implications for Production. *Fermentation* **2022**, *8*, 185. [CrossRef]

5. Moretti, A.F.; Moure, M.C.; Quiñoy, F.; Esposito, F.; Simonelli, N.; Medrano, M.; León-Peláez, Á. Water Kefir, a Fermented Beverage Containing Probiotic Microorganisms: From Ancient and Artisanal Manufacture to Industrialized and Regulated Commercialization. *Futur. Foods* **2022**, *5*, 100123. [CrossRef]
6. Wang, X.; Wang, P. Red Beetroot Juice Fermented by Water Kefir Grains: Physicochemical, Antioxidant Profile and Anticancer Activity. *Eur. Food Res. Technol.* **2022**, *249*, 939–950. [CrossRef]
7. Havva¸safak, H.H.; Gün, I.; Kalit, M.T.; Kalit, S. Physico-Chemical, Microbiological and Sensory Properties of Water Kefir Drinks Produced from Demineralized Whey and Dimrit and Shiraz Grape Varieties. *Foods* **2023**, *12*, 1851. [CrossRef]
8. Spizzirri, U.G.; Loizzo, M.R.; Aiello, F.; Prencipe, S.A.; Restuccia, D. Non-Dairy Kefir Beverages: Formulation, Composition, and Main Features. *J. Food Compos. Anal.* **2023**, *117*, 105130. [CrossRef]
9. Moreira, M.E.C.; Santos, M.H.D.; Zolini, G.P.P.; Wouters, A.T.B.; Carvalho, J.C.T.; Schneedorf, J.M. Anti-Inflammatory and Cicatrizing Activities of a Carbohydrate Fraction Isolated from Sugary Kefir. *J. Med. Food* **2008**, *11*, 356–361. [CrossRef]
10. Alsayadi, M.; Al Jawfi, Y.; Belarbi, M.; Sabri, F.Z. Antioxidant Potency of Water Kefir. *J. Microbiol. Biotechnol. Food Sci.* **2013**, *2*, 2444–2447.
11. Laureys, D.; Aerts, M.; Vandamme, P.; De Vuyst, L. The Buffer Capacity and Calcium Concentration of Water Influence the Microbial Species Diversity, Grain Growth, and Metabolite Production During Water Kefir Fermentation. *Front. Microbiol.* **2019**, *10*, 2876. [CrossRef]
12. Pihurov, M.; Păcularu-Burada, B.; Cotârleț, M.; Bahrim, G.E. Tailoring the Optimized Fermentation Conditions of SCOBY-Based Membranes and Milk Kefir Grains to Promote Various Functional Properties. *Foods* **2022**, *11*, 3107. [CrossRef] [PubMed]
13. Mirani, A.; Goli, M. Optimization of Cupcake Formulation by Replacement of Wheat Flour with Different Levels of Eggplant Fiber Using Response Surface Methodology. *Food Sci. Technol.* **2022**, *42*, 1–9. [CrossRef]
14. Păcularu-Burada, B.; Turturică, M.; Rocha, J.M.; Bahrim, G.E. Statistical Approach to Potentially Enhance the Postbiotication of Gluten-Free Sourdough. *Appl. Sci.* **2021**, *11*, 5306. [CrossRef]
15. Balouiri, M.; Sadiki, M.; Ibnsouda, S.K. Methods for in Vitro Evaluating Antimicrobial Activity: A Review. *J. Pharm. Anal.* **2016**, *6*, 71. [CrossRef] [PubMed]
16. Cotârleț, M.; Stănciuc, N.; Bahrim, G.E. Yarrowia Lipolytica and Lactobacillus Paracasei Solid State Fermentation as a Valuable Biotechnological Tool for the Pork Lard and Okara's Biotransformation. *Microorganisms* **2020**, *8*, 1098. [CrossRef]
17. Moraes Filho, M.L.; Busanello, M.; Garcia, S. Optimization of the Fermentation Parameters for the Growth of Lactobacillus in Soymilk with Okara Flour. *LWT-Food Sci. Technol.* **2016**, *74*, 456–464. [CrossRef]
18. Vohra, B.M.; Fazry, S.; Sairi, F.; Babul-Airianah, O. Effects of Medium Variation and Fermentation Time on the Antioxidant and Antimicrobial Properties of Kombucha. *Malays. J. Fundam. Appl. Sci.* **2019**, *15*, 298–302. [CrossRef]
19. Diguță, C.F.; Nițoi, G.D.; Matei, F.; Luță, G.; Cornea, C.P. The Biotechnological Potential of *Pediococcus* Spp. Isolated from Kombucha Microbial Consortium. *Foods* **2020**, *9*, 1780. [CrossRef]
20. Matei, B. Studies On The Valorisation Potential Of Kombucha Type Microbial Consortia. Doctoral Thesis, University of Agronomic Sciences and Veterinary Medicine of Bucharest, Bucharest, Romania, 2021. Available online: https://usamv.ro/images/Programe_de_studii/Doctorat/Teze_de_doctorat/Arhiva_2021/Matei_Bogdan_rezumat_EN.pdf (accessed on 21 February 2022).
21. Cawthray, G.R. An Improved Reversed-Phase Liquid Chromatographic Method for the Analysis of Low-Molecular Mass Organic Acids in Plant Root Exudates. *J. Chromatogr. A* **2003**, *1011*, 233–240. [CrossRef]
22. Neffe-Skocińska, K.; Sionek, B.; Ścibisz, I.; Kołożyn-Krajewska, D. Acid Contents and the Effect of Fermentation Condition of Kombucha Tea Beverages on Physicochemical, Microbiological and Sensory Properties. *CYTA-J. Food* **2017**, *15*, 601–607. [CrossRef]
23. Khaleil, M.M. A Bioprocess Development Study of Polyphenol Profile, Antioxidant and Antimicrobial Activities of Kombucha Enriched with *Psidium guajava* L. *J. Microbiol. Biotechnol. Food Sci.* **2020**, *9*, 1204–1210. [CrossRef]
24. Mizzi, L.; Chatzitzika, C.; Gatt, R.; Valdramidis, V. HPLC Analysis of Phenolic Compounds and Flavonoids with Overlapping Peaks. *Food Technol. Biotechnol.* **2020**, *58*, 12–19. [CrossRef]
25. Zannini, E.; Lynch, K.M.; Nyhan, L.; Sahin, A.W.; O' Riordan, P.; Luk, D.; Arendt, E.K. Influence of Substrate on the Fermentation Characteristics and Culture-Dependent Microbial Composition of Water Kefir. *Fermentation* **2022**, *9*, 28. [CrossRef]
26. Sitanggang, A.B.; Wu, H.S.; Wang, S.S.; Lan, J.C.W. Fermentation Strategies: Nutritional Requirements. In *Industrial Fermentation: Food Processes, Nutrient Sources and Production Strategies*; Nova Science Pub Inc.: Hauppauge, NY, USA, 2010; pp. 217–247. ISBN 1608765504.
27. Ivanišová, E.; Meňhartová, K.; Terentjeva, M.; Godočíková, L.; Árvay, J.; Kačániová, M. Kombucha Tea Beverage: Microbiological Characteristic, Antioxidant Activity, and Phytochemical Composition. *Acta Aliment.* **2019**, *48*, 324–331. [CrossRef]
28. Dolores, M.; Ana, P.; Bengoa, A.; Iraporda, C.; Medrano, M.; Garrote, G.L.; Abraham, A.G. Water Kefir: Factors Affecting Grain Growth and Promoting Properties of the Fermented Beverage. *J. Appl. Microbiol.* **2022**, *133*, 162–180. [CrossRef]
29. Subardjo, M.V.K. *Black Tea Water Kefir Beverage*; Massey University: Albany, New Zealand, 2017.
30. Atalar, I.; Dervisoglu, M. Optimization of Spray Drying Process Parameters for Kefir Powder Using Response Surface Methodology. *LWT-Food Sci. Technol.* **2015**, *60*, 751–757. [CrossRef]
31. Andreson, M.; Kazantseva, J.; Kuldjärv, R.; Malv, E.; Vaikma, H.; Kaleda, A.; Kütt, M.L.; Vilu, R. Characterisation of Chemical, Microbial and Sensory Profiles of Commercial Kombuchas. *Int. J. Food Microbiol.* **2022**, *373*, 109715. [CrossRef]
32. Ferremi Leali, N.; Binati, R.L.; Martelli, F.; Gatto, V.; Luzzini, G.; Salini, A.; Slaghenaufi, D.; Fusco, S.; Ugliano, M.; Torriani, S.; et al. Reconstruction of Simplified Microbial Consortia to Modulate Sensory Quality of Kombucha Tea. *Foods* **2022**, *11*, 3045. [CrossRef] [PubMed]

33. Uțoiu, E.; Matei, F.; Toma, A.; Diguță, C.F.; Ștefan, L.M.; Mănoiu, S.; Vrăjmașu, V.V.; Moraru, I.; Oancea, A.; Israel-Roming, F.; et al. Bee Collected Pollen with Enhanced Health Benefits, Produced by Fermentation with a Kombucha Consortium. *Nutrients* **2018**, *10*, 1365. [CrossRef] [PubMed]
34. Vázquez-Cabral, B.D.; Larrosa-Pérez, M.; Gallegos-Infante, J.A.; Moreno-Jiménez, M.R.; González-Laredo, R.F.; Rutiaga-Quiñones, J.G.; Gamboa-Gómez, C.I.; Rocha-Guzmán, N.E. Oak Kombucha Protects against Oxidative Stress and Inflammatory Processes. *Chem. Biol. Interact.* **2017**, *272*, 1–9. [CrossRef]

foods

MDPI

Article

Anti-Obesity Effects of SPY Fermented with *Lactobacillus rhamnosus* BST-L.601 via Suppression of Adipogenesis and Lipogenesis in High-Fat Diet-Induced Obese Mice

Taewook Kang [1,2], Jin Ree [1,2], Joo-Woong Park [2], Hyewon Choe [2,3] and Yong Il Park [1,*

1 Department of Biotechnology, Graduate School, The Catholic University of Korea, Bucheon 14662, Republic of Korea; twkang@biostream.co.kr (T.K.); jree@biostream.co.kr (J.R.)
2 Biostream Co., Ltd., Suwon 10442, Republic of Korea; pjwrnds@biostream.co.kr (J.-W.P.); hwchoe@biostream.co.kr (H.C.)
3 Graduate School of Genetics and Biotechnology, Kyung Hee University, Yongin 17104, Republic of Korea
* Correspondence: yongil382@catholic.ac.kr; Tel.: +82-2-2164-4512; Fax: +82-2-2164-4846

Abstract: In this research, the potential anti-obesity efficacy of *Lactobacillus rhamnosus* BST-L.601 and its fermented product (named SPY) with mashed sweet potato paste were investigated using 3T3-L1 preadipocytes and high-fat diet (HD)-induced obese mice. SPY (0–0.5 mg/mL) dose-dependently and significantly reduced lipid accumulation and TG content and the expression of adipogenic markers (C/EBPα, PPAR-γ, and aP2) and fatty acid synthetic pathway proteins (ACC and FAS) in 3T3-L1 adipocytes, demonstrating that SPY suppresses adipocyte differentiation and lipogenesis. Oral administration of SPY (4×10^7 CFU/kg body weight) to HD-induced obese mice for 12 weeks significantly reduced the body and liver weight, the size of adipocytes, and the weight of epididymal, visceral, and subcutaneous fat tissues. SPY was more effective in decreasing body weight gain in HD mice than in treatment with BST-L.601 alone. Administration of SPY or BST-L.601 also reduced the serum level of total cholesterol and LDL cholesterol and leptin secretion at a similar level. These results revealed that both SPY and BST-L.601 effectively suppress HD-induced adipogenesis and lipogenesis, suggesting that these materials would be useful in the functional foods industry to ameliorate and/or prevent obesity.

Keywords: *Lactobacillus rhamnosus*; probiotics; prebiotics; fermented sweet potato; anti-obesity

Citation: Kang, T.; Ree, J.; Park, J.-W.; Choe, H.; Park, Y.I. Anti-Obesity Effects of SPY Fermented with *Lactobacillus rhamnosus* BST-L.601 via Suppression of Adipogenesis and Lipogenesis in High-Fat Diet-Induced Obese Mice. *Foods* **2023**, *12*, 2202. https://doi.org/10.3390/foods12112202

Academic Editors: Xiqing Yue and Diego A. Moreno

Received: 17 April 2023
Revised: 19 May 2023
Accepted: 27 May 2023
Published: 30 May 2023

1. Introduction

Obesity, one of the major metabolic diseases, is being magnified as a worldwide health problem related to various fatal diseases such as cardiac dysfunction, diabetes, hypertension, osteoarthritis, and cancer [1]. Recently, the incidence rate of obesity has rapidly increased, as reported by the OECD in 2017. Moreover, a serious problem with this situation is childhood obesity, with an estimated 15.5% of OECD infants being obese [2]. Obesity is defined as an abnormal accumulation of body fat, resulting in excessive expansion and growth of adipose tissue due to an imbalance between energy intake and expenditure [3]. The development of obesity is defined by an increased adipose tissue mass that can be driven by either an unusually larger number or expanded fat cells (adipocytes) [4]. The expanded size of adipocytes (hypertrophy) is predominantly attributed to the accumulation of lipids (lipogenesis), and the increased cell number of adipocytes (hyperplasia) leads to the proliferation and differentiation of adipocyte precursor cells to mature adipocytes, which is a cellular process called adipogenesis [4]. Therefore, the mass of adipose tissue can be controlled by inhibiting adipogenesis, reducing the accumulation of lipids, improving lipolysis, and/or guiding the apoptotic death of adipose cells [5]. During the differentiation of adipocytes, several adipogenic transcription factors such as sterol regulatory element-binding protein-1c (SREBP-1c), peroxisome proliferator-activator receptor-γ

(PPAR-γ), and CCAAT/enhancer binding protein-α (C/EBPα), are essential regulators of adipogenesis. The expression of C/EBPα, PPAR-γ, and several lipogenic enzymes, including fatty acid synthase (FAS) and acetyl-CoA carboxylase (ACC), is stimulated by SREBP-1c products [4,6]. Activation of lipogenic enzymes converts acetyl-CoA to fatty acids and triglycerides and induces tissue uptake into plasma [7].

To solve the obesity issue, studies on how to change the intestinal microbiome composition with probiotics such as Lactobacillus have been conducted in recent decades [8]. These results suggested that when the composition of the intestinal microbiome is changed, intestinal microorganisms (mainly lactic acid bacteria, LAB) influence energy consumption and lipid accumulation; thus, microorganisms of human intestines are useful for obesity control [9]. It was also reported that the composition of the intestinal microbial community was adjusted through oral administration of probiotics to induce obesity suppression, and the ingestion of LAB caused a change in the human intestinal microbial community, especially Lactobacillus species, for a long period of time, which could be different from obese people [10]. Probiotic strains, especially Lactobacillus genera, have proven anti-obesity effects by reducing fat mass and body weight [11]. The anti-obesity efficacy of probiotics can be demonstrated by increasing satiety and lowering insulin resistance [12]. As it has been recently suggested that intestinal microbes could cause obesity, the anti-obesity effect of controlling intestinal microbes through probiotics is being studied [13]. These mechanisms include modifying the composition of the gut microflora, improving barrier integrity in the intestine, producing beneficial metabolites, and regulating the host immune system [14].

Meanwhile, studies on obesity suppression using prebiotics along with probiotics have been conducted [15]. Prebiotics are indigestible, fermentable raw materials that can promote the proliferation of beneficial gut bacteria, or they are "A substrate that is selectively utilized by host microorganisms conferring a health benefit", as suggested by the International Scientific Association for Probiotics and Prebiotics [16]. Nondigestible polysaccharides, such as inulin, galacto-oligosaccharides (GOS), fructo-oligosaccharides (FOS), lactulose, and resistant starch (RS), are recognized as prebiotics [17]. Prebiotics suppresses the proliferation of harmful bacteria and encourages the growth of healthful bacteria, such as Lactobacilli, to promote the production of short-chain fatty acids (SCFAs) such as acetate, butyrate, and propionate [18]. Among others, resistant starch is one of the fractions thoroughly or partly fermented in the large intestine but cannot be digested in the small intestine of robust people. Suppressing the glycemic response, working as healthy probiotics, lowering cholesterol levels, and boosting the generation of SCFAs in the large intestine is reported as a function of resistant starch [19]. Sweet potato is a good and appropriate resource for the supplement of resistant starch [20]. It was demonstrated that the anti-obesity effects of the gut microbiome are related to lactic acid bacteria [21], and the dietary fiber of sweet potato helps the gut microflora profile in a healthy way [22]. Indeed, the possible correlation between probiotics and obesity has been studied by several researchers, suggesting that a particular phylum or probiotic species can regulate energy metabolism [23]. Lactobacillus is known to induce the degradation (fermentation) of indigestible complex polysaccharides and to promote the efficiency of metabolism in our body. In previous studies, models based on these features proved that these two LAB have remarkable anti-obesity effects [24]. Among LAB, *Lactobacillus rhamnosus* is a strain of Lactobacillus species that are regarded as GRAS strains (generally regarded as safe). In this context, it would be quite probable to obtain a synergistic outcome with enhanced biological activities, such as anti-obesity activity, of Lactobacillus probiotics by using sweet potato as a source of prebiotics simultaneously [25].

To address this hypothesis and to develop a new or better probiotic and/or probiotic-prebiotic combined composition for healthy functional food ingredient or a remedial agent to treat or prevent obesity, The purpose of this study is to compare and make an evaluation of the potential anti-obesity effects of a newly isolated probiotic strain, *Lactobacillus rhamnosus* BST-L.601 (deposited in KCTC under accession number KCTC13517BP), and the fermented product (named SPY) of mashed sweet potato paste (MSPP) with this strain, using 3T3-L1 preadipocyte cells and a high-fat diet (HD)-induced obese C57BL/6 mouse model.

2. Materials and Methods

2.1. Materials

Dulbecco's modified Eagle's medium (DMEM), fetal bovine serum (FBS), and Dulbecco's phosphate-buffered saline (DPBS) were purchased from Welgene (Gyeongsan, Republic of Korea). Trypsin-ethylenediaminetetraacetic acid (EDTA) and penicillin and streptomycin were purchased from Gibco-BRL (Grand Island, NY, USA). Isobutylmethylxanthine (IBMX), dexamethasone, insulin, Oil red O, and dimethyl sulfoxide (DMSO) were purchased from Sigma-Aldrich (St. Louis, MO, USA). The murine 3T3-L1 preadipocyte cell line (ATCC® CL-173™) was provided by the American Type Culture Collection (Manassas, VA, USA). 3-(4,5-Dimethylthiazolyl)-diphenyl tetrazoliumbromide (MTT) was obtained from DUCHEPA Biochemie (Haarlem, The Netherlands). Antibodies specific to β-actin, CCAAT/enhancer binding protein-α (C/EBPα), peroxisome proliferator-activated receptor-γ (PPAR-γ), adipocyte protein 2 (aP2), fatty acid synthase (FAS), acetyl-CoA carboxylase (ACC) and anti-rabbit IgG-HRP were purchased from Cell Signaling Technology (Danvers, MA, USA). Primers specific to β-actin, PPAR-γ, aP2, FAS, and ACC were purchased from Cosmo Genetech (Seoul, Republic of Korea).

2.2. Preparation of *Lactobacillus rhamnosus* BST-L.601 and Mashed Sweet Potato Medium

Lactobacillus rhamnosus BST-L.601 was isolated and identified from the stool of 20 randomly selected Korean people. Human stools were dissolved in germ-free PBS (pH 7.4) with a decimal method, inoculated into De Man, Rogosa, and Sharpe (MRS) agar broth (Becton-Dickinson, Franklin Lakes, NJ, USA), and cultured at 37 °C under anaerobic conditions for 24 to 48 h. After cultivation, the largest colonies were selected, inoculated into MRS broth medium, and cultured at 37 °C under anaerobic and static conditions for 24 to 48 h. From the culture broth showing confluent growth with over 1.0 absorbance at 600 nm, strains were selected and inoculated into skim milk medium (Becton-Dickinson, Franklin Lakes, NJ, USA). The strains showing smooth curd formation were selected and smeared again onto the skim milk agar medium. Finally, a single strain was isolated from a single largest colony. This strain was identified as a strain of *L. rhamnosus* by analyzing the 16S rRNA sequence and deposited in KCTC under accession number KCTC 13517 BP (Korea Patent No. 10-2020-0012236).

A fermented sample (SPY) was prepared by fermentation of mashed sweet potato paste (MSPP) with *L. rhamnosus* BST-L.601. Sweet potato paste was prepared using domestic sweet potato cultivated and collected in Kangwon Province, Republic of Korea. After washing the sweet potatoes, the skin of the sweet potatoes was removed, cut randomly into smaller chips, and finely ground with equal amounts (by weight) of water using a grinder (HR3752/00, Philips, The Netherlands) to make the mashed sweet potato paste (MSPP). Yeast extract (20 g/L, Becton-Dickinson, Franklin Lakes, NJ, USA) was blended with MSPP, and the mixture was finely ground at 10,000 rpm for 10 min using a homogenizer (Daihan Scientific, Wonju, Republic of Korea). The pH of MSPP was adjusted from 6.5 to 8.0 using 1 M HCl before sterilization (15 min, 121 °C), and 40 g/L glucose solution was fortified to the paste after sterilization. *L. rhamnosus* BST-L.601 (1×10^6) CFU was inoculated into sweet potato paste with 40 g/L glucose solution to 10% (v/v) and fermented at 37 °C for 3 days. Additional fermentation was performed at 4 °C for 3 additional days after the first fermentation.

2.3. Cell Culture and Differentiation of Pre-Adipocytes

Murine 3T3-L1 preadipocyte cells (ATCC® CL173) were cultured in preadipocyte medium [PM; a mixture of Dulbecco's modified Eagle medium (DMEM) added with 10% fetal bovine serum (FBS)] and a 1% penicillin–streptomycin mixture at 37 °C in a humidified atmosphere of 5% CO_2 until reaching approximately 90% confluence. Unless stated otherwise, preadipocytes were seeded onto 6-well plates at a density of 24,000 cells/well and cultured until 80% confluence was reached. The differentiation of 3T3-L1 preadipocyte cells into mature adipocytes was achieved by culturing cells in differentiation media I [DMEM, 10% FBS, 1 mM dexamethasone, 0.5 mM 3-isobutyl-1-methylxanthine (IBMX), and 10 µg/mL insulin] and differentiation media II [DMEM, 10% fetal bovine serum (FBS), and 10 µg/mL insulin] for 48 h. After that, adipocytes were cultured in DMEM with 10% FBS for original growth and subcultured every 48 h until use.

2.4. Cell Viability Determination

Whether L. rhamnosus BST-L.601 and SPY are toxic to 3T3-L1 preadipocyte cells and differentiated mature adipocyte cells was assessed by measuring cell viability using an MTT assay. Cells were exposed to increasing concentrations of each sample (L.601 or SPY). 3T3-L1 preadipocytes (4000 cells/well) were cultured in 96-well plates with SPY at various concentrations (0.05, 0.1, 0.25, and 0.5 mg/mL) dissolved in PM culture medium at 37 °C in a humidified atmosphere of 5% CO_2 for 48 h. Cell viability was determined by the addition of 50 mL MTT solution (1 mg/mL in phosphate-buffered saline; PBS) to each well and incubation at 37 °C for 4 h. After the culture medium is removed, DMSO was added to each well and incubated at room temperature for 30 min. Absorbance was measured at 570 nm on a microplate reader (Molecular Devices, Seoul, Republic of Korea).

2.5. Oil Red O Staining

To examine the effect of SPY on differentiation and lipogenesis, cells were cultured in MDI differentiation medium in 6-well plates, treated with SPY for 8 days, and stained with Oil red O dye. After incubation, cells were washed gently with PBS, fixed with 4% paraformaldehyde for 30 min at room temperature, rinsed with PBS, and then stained with freshly prepared 0.5% (*w*/*v*) Oil red O solution at 37 °C for 1 h. The stained cells were photographed using an inverted microscope (X100, Olympus, Tokyo, Japan) to visualize lipid droplets. To determine the lipid content, the retained dye in adipocytes was extracted with 100% isopropanol and quantified at 517 nm using a microplate reader. Compared to the control, the relative lipid content of each sample was expressed.

The intracellular TG contents were verified by a Cayman Chemical Triglyceride Assay kit (Ann Arbor, MI, USA), as a method of the manufacturer's instructions. Differentiated adipocytes (Day 8) were treated with increasing amounts of SPY (0.05, 0.1, 0.25, and 0.5 mg/mL) in 6-well plates. The cells were washed and scraped with 200 mL of PBS, and homogenized by sonication for 2 min. Total TG in cell lysates was assayed after that.

2.6. Total RNA Preparation and Reverse Transcription-Polymerase Chain Reaction (RT–PCR)

To determine the mRNA expression levels of inducible FAS, ACC, PPAR-γ, and aP2, total RNA from SPY-treated cells was prepared using a total RNA extraction kit (Intron Biotechnology, Republic Korea). RT-PCR was performed using the One-step RT–PCR Pre-Mix kit (Intron Biotechnology, Seongnam-si, Republic of Korea) with appropriate sense and antisense primers for FAS (sense 5′-CGGCTGCAGGTGGTCGATAGG-3′ and antisense 5′-TGTAGGGGTTGCCGCAATGTC-3′), PPAR-γ (sense 5′-GTCTGTGGGGATAAAGCATC-3′ and antisense 5′- CTGATGGCATTGTGAGACAT-3′), ACC (sense 5′-GAAGAGAACAAAA GCGACATG-3′ and antisense 5′-AATGGCTGATAGGAAGATAGA-3′), and β-actin (sense 5′-AGG+TATCCTGACCCTGAAGTACC-3′ and antisense 5′-GTTGCCAATAGTGATGACC TGGC-3′). Primers were amplified under incubation conditions of 95 °C predenaturation for 5 min and 30 cycles of 95 °C denaturation for 30 s, 58 °C annealing for 30 s, 72 °C extension for 40 s, and a final elongation step for 10 min at 72 °C. The products obtained by RT-PCR were separated on a 1.5% agarose gel and stained with ethidium bromide. The gels were then viewed under UV transillumination, and the relative levels of mRNA against β-actin were quantified using ImageJ software from NIH (Bethesda, MD, USA).

2.7. Western Blot Analysis

To determine the expression level of adipogenic-related proteins, western blot analysis was performed using the lysates from 3T3-L1 adipocytes cultured in a differentiation medium within or without SPY for 8 days. On Day 8, PBS buffer was used to wash the cells. Cold lysis buffer (pH 7.4) containing 20 mM Tris-HCl, 150 mM NaCl, 1 mM Na2EDTA, 1 mM EGTA, 1% NP-40, 1% sodium deoxycholate, 2.5 mM sodium pyrophosphate, 1 mM β-glycerophosphate, 1 mM Na_3VO_4, and 1 mg/mL leupeptin was used to resuspend the cells. The cell lysates were centrifuged at 17,700× *g* and 4 °C for 10 min. The Bradford method (Bio-Rad, Hercules, GA, USA) was applied to determine the protein concentrations. An equal quantity of protein was divided on a 10% SDS-polyacrylamide gel and transferred to PVDF membranes (Bio-Rad, USA). Tris-buffered saline (TBS, pH 7.4) containing 5% nonfat dry milk was used to block the membrane, and primary anti-mouse FAS, aP2, PPAR-γ, and ACC antibodies were incubated with the blocked membrane overnight at 4 °C. The membranes were incubated with anti-mouse IgG-HRP (Cell Signaling Technology) for 2 h at room temperature after washing blocked membranes with TBS containing 0.1% Tween 20. Protein bands were visualized through the ECL system (ABclon, Seoul, Republic of Korea). The ImageJ Program (National Institute of Health, Bethesda, MD, USA) was utilized to quantify the band intensities and to normalize the levels of PPAR-γ, aP2, FAS, and ACC compared to β-actin.

2.8. Animal Care and Diets

The Catholic University of Korea approved the animal protocols followed in the present research (Approval Number: 2019-019). Male C57BL/6 mice purchased from Orient Bio Inc. (Gyeonggi-do, Republic of Korea) were kept under controlled temperatures (22–23 °C) on a 12/12-h light-dark cycle. After the acclimatization period, the male C57BL/6 mice were divided into seven groups (n = 8 in each group) and given a normal diet (ND) or high-fat diet (HD) with the indicated doses of samples for 12 weeks, normal diet control group (2018C, Envigo, Indianapolis, IN, USA, 18% calories from fat); HD, high-fat diet (TD.06414, Envigo, USA, 60.5% calories from fat); GAR, HD + 1% *Garcinia cambogia* extract (100 mg/kg); SPY-L, HD + SPY (30 mg, 4×10^6 CFU/kg); SPY-H, HD + SPY (300 mg, 4×10^7 CFU/kg); BST-L.601-L, HD + BST-L.601 (4×10^8 CFU/kg); BST-L.601-H, HD + BST-L.601 (4×10^9 CFU/kg). Each dose of the sample was administered. The *Garcinia cambogia* extract was used as a positive reference. Body weights were measured every week. After 12 weeks of administration of SPY or L.601, the mice were fasted for 24 h and anesthetized with a mixture of alfaxalone (0.15 mL/25 g/mice) and xylazine hydrochloride (0.01 mL/25 g/mice) before sacrifice. Blood serum, adipose tissues, and liver were individually taken for further analyses.

2.9. Serum Biochemical Analysis

To collect blood samples, mice were fasted. Blood serum was collected by cardiac puncture and stored at −80 °C before use. To determine levels of total cholesterol (Total-c), high-density lipoprotein cholesterol (HDL-c), and low-density lipoprotein cholesterol (LDL-c) in collected blood serum, LH-1500 Automatic Analyzer (LH-1500, Incheon, Republic of Korea) was utilized.

2.10. Chemical Composition Analysis

The general chemical composition, total carbohydrate, protein, polyphenols, and caffeic acid, in the mashed sweet potato paste (MSPP) powder and SPY (the fermented MSPP with BST-L.601), were determined and compared. Total carbohydrate was measured by the phenol-sulfuric acid method at 490 nm using glucose as a reference [26]. Bradford method was operated to quantify total protein and BSA was used as a standard for Bradford method at 595 nm [27]. Total polyphenols were determined by the Folin-Ciocalteu reagent method using gallic acid as a reference at 750 nm [28]. Each powder sample was blended with distilled water to make a 1.0% (*w*/*v*) solution (10 mg/mL) for 1 h using a nutator mixer

(FinePCR, Gunpo-si, Republic of Korea) while homogenizing with a sonicator (Branson, MO, USA) by 1 min sonication and 5 min cooling on ice-cold water. The sonication was repeated 10 times. After sonication, each suspension was diluted with distilled water to an appropriate concentration for analysis. The detection and quantification of the amount of caffeic acid in SPY and MSPP were performed by using an HPLC system equipped with a UV detector (1260 Infinity II, Agilent, Santa Clara, CA, USA). Samples were dissolved in methanol, fractionated on an XDB-C18 column (150 × 4.6 mm, 5 µm column, Agilent), and eluted at 1.0 mL/min in gradient mode with a mobile phase composed of water (pH 3.15 by formic acid) and acetonitrile.

2.11. Statistical Analysis

The results were shown in the means ± SDs for each treatment group in each experiment. Statistical analysis was carried out by using the Statistical Analysis System software package (SAS Institute, Cary, NC, USA). Significance was determined by a one-way analysis of variance, followed by Dunnett's range test for multiple comparisons, and data were analyzed using the SAS package program (SAS Institute Inc., Cary, NC, USA). Data were considered to be different at $p < 0.05$.

3. Results

3.1. Effects of SPY and BST-L.601 on the Viability of 3T3-L1 Pre-Adipocytes

To determine whether SPY and BST-L.601 are detrimental to 3T3-L1 preadipocytes, cell survival was evaluated by MTT assay with exposure to increasing doses of SPY (0.025, 0.05, 0.1, 0.25, and 0.5 mg/mL) containing BST-L.601 (2.5×10^7, 5×10^7, 1×10^8, 2×10^8, and 4×10^8 CFU/mL, respectively) (Figure 1A) or BST-L.601 (Figure 1B) for 24 h. As shown in Figure 1, both SPY (Figure 1A) and BST-L.601 (Figure 1B) were not significantly toxic to the growth of 3T3-L1 preadipocytes in the concentration range tested (up to 0.5 mg/mL for SPY and 4×10^8 CFU/mL for BST-L.601).

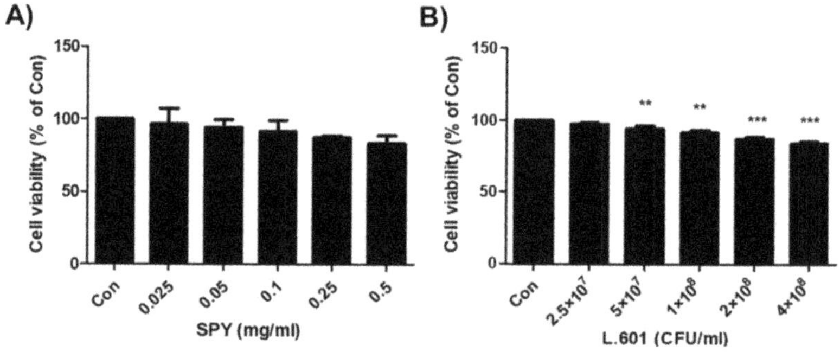

Figure 1. Effects of SPY and BST-L.601 on the viability of 3T3-L1 preadipocyte cells. Cytotoxicity of (**A**) SPY and (**B**) BST-L.601 was assessed by measuring cell viability using the MTT assay after 24 h of exposure to (**A**) SPY (0.025, 0.05, 0.1, 0.25, and 0.5 mg/mL) or (**B**) BST-L.601 (2.5×10^7, 5×10^7, 1×10^8, 2×10^8, and 4×10^8 CFU/mL). The data were expressed as a percentage normalized to sample-untreated control cells (Con). Values are the means ± SDs (n = 8). Values with different superscripts are significantly different among the groups by one-way ANOVA with Dunnett's multiple comparison test at ** $p < 0.01$; *** $p < 0.001$, compared to the Control group.

3.2. Effects of SPY on Adipocyte Differentiation

To assess the efficacy of SPY on inhibiting lipid accumulation, 3T3-L1 preadipocytes were exposed to varying levels of SPY for 8 days, and lipids in adipocytes were stained with Oil red O dye. Figure 2A, the representative images of Oil red O staining, shows that the undifferentiated control cells (UC) were not stained with dye, but huge numbers of stained spots (pink-colored) appeared in the differentiated control cells (Con). This indicates the preadipocytes were differentiated into mature adipocytes, actively synthesizing lipids (Figure 2A). However, these promoted lipids were substantially decreased by SPY (0.05, 0.1, 0.25, and 0.5 mg/mL) in a dose-dependent manner by 19.3%, 25.5%, 28.6%, and 47.4%, respectively, compared to the differentiated control cells (Con, 100%), showing that SPY effectively inhibited the differentiation of adipocyte cells (Figure 2A,B). Moreover, the markedly increased intracellular triglyceride (TG) contents in differentiating adipocytes (Con, 100%) were also effectively reduced to 23.2%, 28.8%, 33.3%, and 50.7% upon SPY (0.05, 0.1, 0.25, and 0.5 mg/mL) exposure (Figure 2C), indicating that SPY inhibited lipid accumulation, especially TG, during adipocyte differentiation.

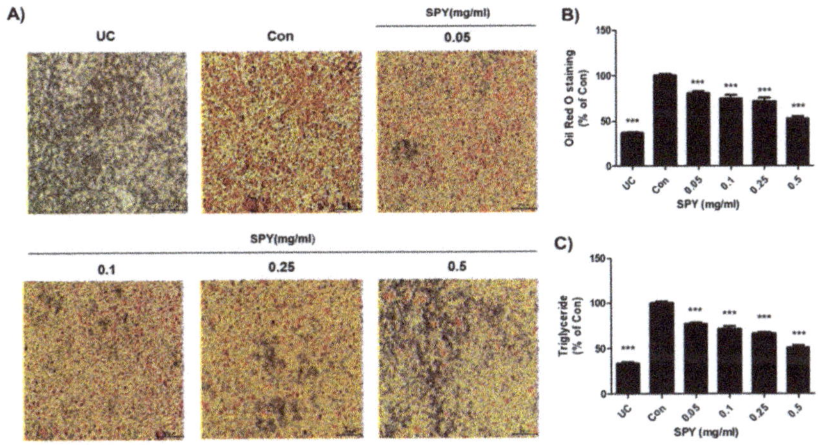

Figure 2. Effects of SPY on lipid accumulation and TG content in differentiating 3T3-L1 cells. Cells were cultured in an MDI differentiation medium and treated with varying concentrations of SPY (0.05, 0.1, 0.25, and 0.5 mg/mL) for 8 days. (**A**) Lipid droplets generated were stained with Oil red O dye and visualized under a microscope (×100). (**B**) Stained lipid droplets were solubilized with isopropanol and quantified at 517 nm on a microplate reader. (**C**) Intracellular triglyceride (TG) contents were measured using a triglyceride assay kit. Data are expressed as the means ± SDs (n = 3). Values with different superscripts are significantly different among the groups by one-way ANOVA with Dunnett's multiple comparison test at *** $p < 0.001$ compared to the Control group. UC, undifferentiated normal control cells; Con, sample-untreated differentiated control cells.

3.3. Inhibitory Effect of SPY on Differentiation and Lipogenesis-Related Protein Expression in 3T3-L1 Cells

Throughout adipogenesis, factors and proteins related to adipocytes such as C/EBPα, PPAR-γ, and aP2 are expressed [23]. C/EBPα and PPAR-γ stimulate the expression of lipogenic enzymes similar to FAS and ACC in various ways during adipogenesis [16,24]. To determine the effects of SPY on the expression of these adipogenic and lipogenic marker proteins, 3T3-L1 cells were cultured in an MDI differentiation medium for 8 days with the indicated concentration of SPY. Western blot analysis showed SPY treatment (0.05, 0.1, 0.25, and 0.5 mg/mL) resulted in predominant inhibition in the expression of these proteins (Figure 3). SPY (0.5 mg/mL) considerably inhibited the protein levels of PPAR-γ by 48.4% (Figure 3A,B), C/EBPα by 73.7% (Figure 3A,C), and aP2 by 75.7% (Figure 3A,D) when it is compared with the differentiated, Sample-untreated control cells (Con, 100%). SPY treatment (0.5 mg/mL) also significantly inhibited the expression of ACC by 50.0% (Figure 3E,F) and FAS by 80.9% (Figure 3E,G) compared to the differentiated but SPY-untreated group (Con, 100%). These results proved SPY effectively suppressed the differentiation of preadipocytes through the downregulation of diverse adipogenesis-specific transcription factors and lipogenesis marker proteins.

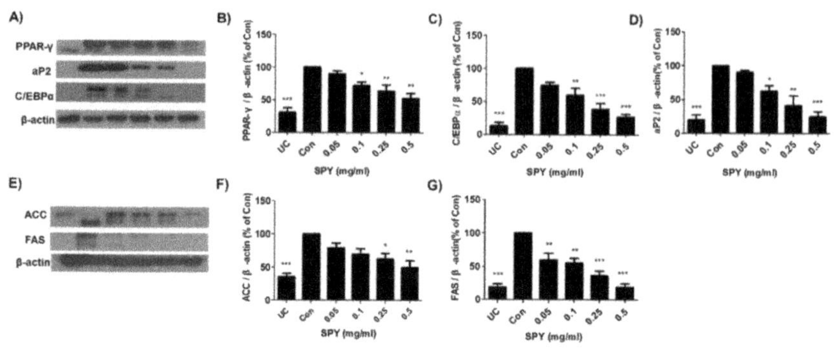

Figure 3. Inhibitory effect of SPY on differentiation and lipogenesis-related protein expression in 3T3-L1 adipocytes. (**A**) and (**E**) 3T3-L1 cells were cultured in MDI differentiation medium for 8 days with increasing concentrations of SPY (0.05, 0.1, 0.25, and 0.5 mg/mL), and cell lysates were used for Western blot analysis. The expression levels of (**B**) PPAR-γ, (**C**) C/EBPα, (**D**) aP2, (**F**) ACC, and (**G**) FAS were quantified by ImageJ software. β-Actin was used as a loading control. Each data point was expressed as the % of control cells (Con, 100%). Data are the means ± SDs (n = 3). Values with different superscripts are significantly different among the groups by one-way ANOVA with Dunnett's multiple comparison test at * $p < 0.05$; ** $p < 0.01$; *** $p < 0.001$, compared to the Control group. UC, undifferentiated control cells; Con, untreated sample, differentiated control cells.

3.4. Body and Organ Weight Changes of Mice Fed the Different Diets

C57BL/6 mice were fed a high-fat diet (HD) for 12 weeks and used as the HD-induced obese model. Mice were randomly divided into 7 groups (n = 8): ND, normal diet group; HD, high-fat diet (control); GAR, HD + 1% *Garcinia cambogia* extract-treated group (100 mg/kg); SPY-L, HD + SPY (4 × 10^6 CFU/kg); SPY-H, HD + SPY (4 × 10^7 CFU/kg); L.601-L, HD + BST-L.601 (4 × 10^8 CFU/kg); L.601-H, HD + BST-L.601 (4 × 10^9 CFU/kg). Oral administration of diet for mice with or without doses of samples proceeded for 12 weeks, and the body weight of each mouse was checked once a week. As shown in Figure 4B, HD-induced obese mice had an increased body weight by 48.0% after 12 weeks of feeding, and the SPY-H group and L.601-H group had decreased body weight by 19.6% and 14.8%, respectively, compared with the HD group. The relative weights of epididymal fat, visceral fat, abdominal fat, and liver tissue in the HD group were higher than those in the ND group (Table 1). The administration of SPY and BST-L.601 decreased the weight of total fat tissues (epididymal, visceral and subcutaneous, and abdominal fat tissues) by 29.8% (SPY-H) and 25.4% (L.601-H), respectively (Table 1).

Table 1. Changes in organ and fat tissue weights.

Group	ND	HD	GAR	SPY–	SPY–	L.601–	L.601–
Liver weight (g)	1.59 ± 0.17	1.93 ± 0.44	1.61 ± 0.16	1.71 ± 0.47	1.49 ± 0.19 *	1.52 ± 0.19 *	1.49 ± 0.17 *
Kidney weight (g)	0.45 ± 0.06	0.45 ± 0.04	0.48 ± 0.04	0.47 ± 0.08	0.49 ± 0.06	0.43 ± 0.06	0.48 ± 0.04
Spleen weight (g)	0.081 ± 0.016	0.083 ± 0.007	0.079 ± 0.012	0.078 ± 0.009	0.079 ± 0.007	0.072 ± 0.01	0.074 ± 0.012
Epididymis fat tissue (g)	1.029 ± 0.26 ***	3.246 ± 0.55	3.151 ± 0.21	2.920 ± 0.24	2.725 ± 0.41	3.446 ± 0.14	2.533 ± 0.48 **
Visceral fat tissue (g)	0.449 ± 0.11 ***	1.752 ± 0.17	1.394 ± 0.19 **	1.480 ± 0.17 *	1.321 ± 0.12 ***	1.578 ± 0.33	1.240 ± 0.11 ***
Abdominal fat tissue (g)	0.889 ± 0.19 ***	4.797 ± 0.37	3.512 ± 0.56 ***	3.855 ± 0.39 **	3.035 ± 0.56 ***	3.849 ± 0.34 **	3.373 ± 0.85 ***
Total fat (g)	2.48 ± 0.66 ***	10.14 ± 1.06	7.92 ± 0.67 **	8.35 ± 0.62 *	7.12 ± 1.11 ***	8.50 ± 0.46	7.57 ± 1.93 ***

ND, normal diet; HD, high-fat diet; GAR, HD supplemented with *Garcinia cambogia* extract (10 mg/kg); SPY–LHD supplemented with SPY (4 × 10^6 CFU/kg); SPY–H, HD supplemented with SPY (4 × 10^7 CFU/kg); L.601–L, HD supplemented with BST-L.601 (4 × 10^8 CFU/kg); L.601–H, HD supplemented with BST-L.601 (4 × 10^9 CFU/kg). L.601, *L. rhamnosus* BST.-L.601. Values are the means ± SDs (n = 8). Values with different superscripts are significantly different among the groups by one-way ANOVA with Dunnett's multiple comparison test at * $p < 0.05$; ** $p < 0.01$; *** $p < 0.001$, compared to the HD group.

3.5. SPY Prevents Hyperlipidemia in High-Fat Diet Mice

To assess the effects of SPY and BST-L.601 on the serum biochemical parameters in HD-induced obese mice, total cholesterol (Total-c), high-density lipoprotein cholesterol (HDL), low-density lipoprotein cholesterol (LDL), and leptin were analyzed by automated analyzer after various diets were orally administered to HD-induced obese mice. SPY (SPY-H) reduced the cholesterol and LDL levels by 27.9% and 32.6%, respectively, and BST-L.601 (L.601-H) reduced cholesterol and LDL levels by 35.7% and 36.4%, respectively (Figure 5A,C) compared with the HD group (100%). Oral administration of SPY and BST-L.601 (L.601-H) increased the ratio of HDL/Total-c level by 21.9% and 8.3%, respectively (Figure 5D). SPY also decreased the secretion of leptin, one of the hormones secreted from visceral and subcutaneous adipose tissue. Leptin secretion was decreased in the SPY-H group by 24.0% and in the L.601-H group by 27.8%, respectively (Figure 5E), when the secretion of leptin in the HD group is set at 100%. These results indicated that SPY and BST-L.601 effectively regulate cholesterol levels in blood serum.

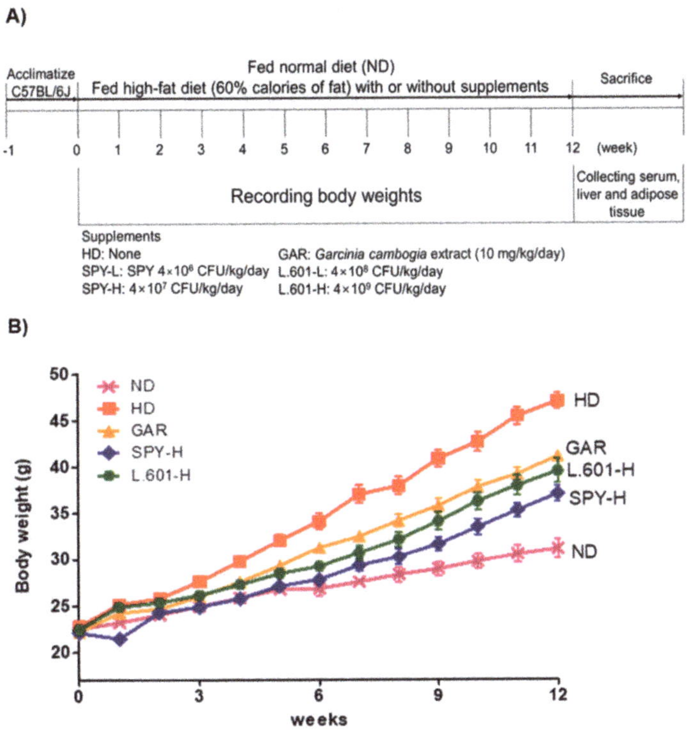

Figure 4. Protocol for animal treatment and body weight change of mice fed the different diets. (**A**) ND, normal diet; HD, high-fat diet; GAR, HD supplemented with *Garcinia cambogia* extract (100 mg/kg); SPY−L, a high-fat diet supplemented with SPY (4×10^6 CFU/kg); SPY− H, HD supplemented with SPY (4×10^7 CFU/kg); L.601− L, HD supplemented with BST-L.601 (4×10^8 CFU/kg); L.601− H, HD supplemented with BST-L.601 (4×10^9 CFU/kg). L.601, *L. rhamnosus* BST-L.601. (**B**) Body weight change of mice fed the different diets. Mice were given their diets and indicated doses of samples for 12 weeks, and the body weight was weighed every week. Values are the means ± SDs (n = 8).

3.6. Effects of SPY and BST-L.601 on the Size and Numbers of Adipocytes in Liver and Epididymal Fat Tissues

The increment of the mass and the number of adipocytes mainly depends on the amount of lipids accumulated in the fat cells. Therefore, the size and number of adipocytes accumulated in epididymal adipose tissue and liver tissue are habitually evaluated indicators to assess the potential anti-obesity efficacy of ingredients [24]. To determine the numbers and sizes of adipocytes in liver tissues and fat tissues, tissues were stained by hematoxylin and eosin staining (H&E). Sections of epididymal adipose tissue of the HD group revealed an increased number of expanded adipocytes (red-arrow) following H&E staining is completed (Figure 6A). However, administration of SPY (SPY-H) and BST-L.601 (L.601-H) significantly reduced these enlarged cell sizes of adipocytes by 25.0% and 11.0%, respectively, and the number of adipocytes per 1.48 μm^2 was increased by 28.5% and 7.6%, respectively (Figure 6B,C).

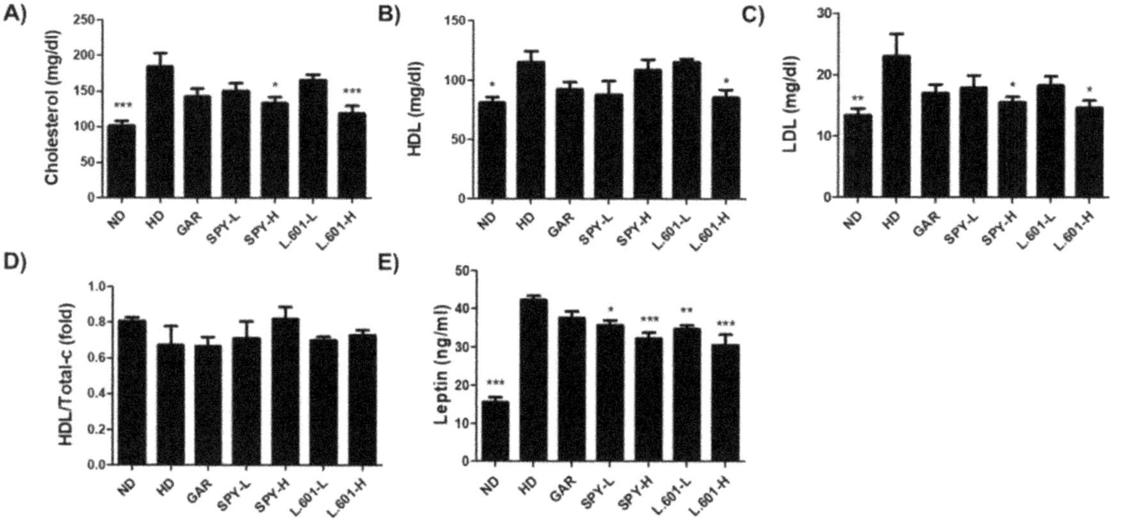

Figure 5. Effects of SPY and BST-L.601 on hyperlipidemia markers. The blood serum of HD-induced obese mice was analyzed by an automated analyzer after oral administration of various diets for 12 weeks. (**A**) Total cholesterol (Total-c), (**B**) high-density lipoprotein-cholesterol (HDL), (**C**) low-density lipoprotein-cholesterol (LDL), (**D**) HDL/Total-c, (**E**) Leptin. ND, normal diet; HD, high-fat diet; GAR, HD supplemented with *Garcinia cambogia* extract (10 mg/kg); SPY−L, HD supplemented with SPY (4×10^6 CFU/kg); SPY− H, HD supplemented with SPY (4×10^7 CFU/kg); L.601− L, HD supplemented with BST-L.601 (4×10^8 CFU/kg); L.601− H, HD supplemented with BST-L.601 (4×10^9 CFU/kg). L.601, *L. rhamnosus* BST-L.601. Values are the means ± SDs (n = 8). Values with different superscripts are significantly different among the groups by one-way ANOVA with Dunnett's multiple comparison test at * $p < 0.05$; ** $p < 0.01$; *** $p < 0.001$, compared to the HD group.

3.7. SPY and BST-L.601 Suppress Adipogenic and Lipogenic Marker Protein mRNA Expression in HD-Induced Obese Mice

RT-PCR analysis was performed to identify the effects of SPY and BST-L.601 on the expression levels of adipogenic markers in HD-induced obese mice. After 12 weeks of oral administration of SPY and BST-L.601, liver tissues were lysed, and the amount of mRNA was measured. Oral administration of SPY (SPY-H) and BST-L.601 (L.601-H) decreased the mRNA expression level of aP2 by 32.3% and 60.1% (Figure 7A), PPAR-γ by 31.0% and 25.5% (Figure 7B), FAS by 15.9% and 30.3% (Figure 7C), and ACC-1 by 13.3% and 41.0%, respectively (Figure 7D). These results suggested the exhibition of the anti-obesity ability of SPY and BST-L.601 through the downregulation of the expression of various adipogenesis-specific transcription factors and lipogenesis marker proteins.

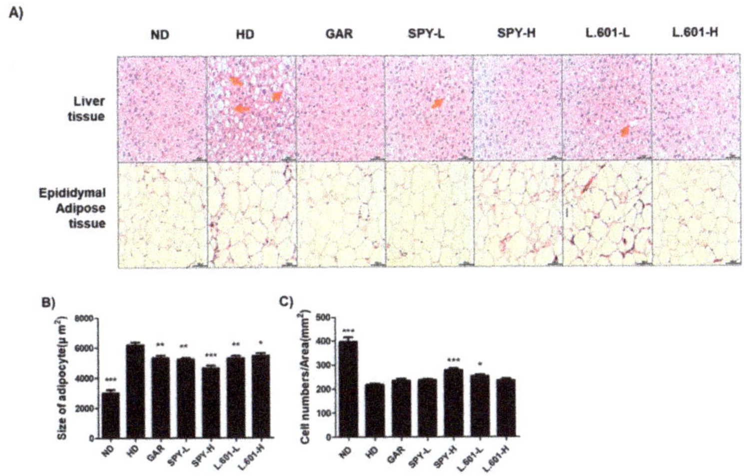

Figure 6. Effect of SPY and BST-L.601 on the size and numbers of adipocytes in liver and epididymal fat tissues in HD-induced obese mice. Liver tissues and adipose tissues were stained with hematoxylin and eosin (H&E). (**A**) Sections of stained epididymal adipose tissue and liver tissue were monitored under a light microscope (magnification, ×200 for epididymal adipose tissues, ×400 for liver tissues). (**B**) Size of epididymal adipose tissue and (**C**) cell numbers in the measured area (measured unit area = 1.48 μm^2). ND, normal diet; HD, high-fat diet; GAR, HD supplemented with *Garcinia cambogia* extract (100 mg/kg); SPY−L, HD supplemented with SPY (4×10^6 CFU/kg); SPY−H, HD supplemented with SPY (4×10^7 CFU/kg); L.601− L, HD supplemented with BST-L.601 (4×10^8 CFU/kg); L.601− H, HD supplemented with BST-L.601 (4×10^9 CFU/kg). L.601, *L. rhamnosus* BST-L.601. Values with different superscripts are significantly different among the groups by one-way ANOVA with Dunnett's multiple comparison test at * $p < 0.05$; ** $p < 0.01$; *** $p < 0.001$, compared to the HD group.

3.8. SPY and BST-L.601 Suppress Adipocyte Differentiation and Lipogenic Marker Protein Expression in Liver Tissue of HD-Induced Obese Mice

To determine the effect of SPY and BST-L.601 on adipogenic and lipogenic marker proteins in HD-induced obese mice, the levels of adipocyte differentiation transcription factors as PPAR-γ and aP2 and lipogenic enzymes similar to ACC and FAS were investigated by Western blotting. SPY (SPY-H group) and BST-L.601 (L.601-H group) were shown to reduce PPAR-γ by 60.8% and 41.9% (Figure 8A,B) and aP2 by 48.8% and 40.7% (Figure 8A,C) when the β-actin level in the HD group was set at 100%. Administration of SPY (SPY-H) and BST-L.601 (L.601-H) also decreased ACC by 48.5% and 53.5% (Figure 8A,D) and FAS by 43.8% and 41.7% (Figure 8A,E). These results showed that SPY and BST-L.601 effectively downregulated adipogenesis and lipogenesis.

3.9. Chemical Composition Profile of SPY and MSPP

The chemical composition of MSPP and SPY was determined as summarized in Table 2. The MSPP and SPY sample preparations contained total carbohydrates up to 922.4 mg/g (92.24%, *w/w*) and 810.5 mg/g (81.05%), protein up to 30.7 mg/g (0.0307%) and 16.3 mg/g (0.0163%), and total polyphenols up to 1.65 mg/g (0.00165%) and 1.99 mg/g (0.00199%), respectively, indicating that both samples consist mostly of carbohydrates (Table 2). One of the possible explanations for the decreased level of total carbohydrates in SPY, compared to that of MSPP, would be that some of the digestible carbohydrates were consumed for the growth and proliferation of BST-L.601 during fermentation.

Table 2. Chemical composition analysis.

	Total Carbohydrate [A] (mg GE/g)	Total Protein [B] (mg BE/g)	Total Polyphenols [C] (mg GAE/g)	Caffeic Acid (mg/mL)
MSPP [D]	922.4 (±7.69)	30.7 (±1.69)	1.65 (±0.01)	0.26
SPY	810.5 (±17.40)	16.3 (±1.00)	1.99 (±0.01)	0.18

[A] Total carbohydrate was measured by the phenol-sulfuric acid method and expressed as glucose equivalent (GE) in 1 g of dry sample. [B] Total protein was quantified by the Bradford method and expressed as BSA equivalent (BE) in 1 g of dry sample. [C] Total polyphenols were determined by the Folin-Ciocalteu reagent method and expressed as gallic acid equivalents (GAE) in 1 g of dry sample. [D] MSPP, mashed sweet potato paste. SPY, the fermented product of MSPP with *L. rhamnosus* BST-L.601.

The results from HPLC analysis for caffeic acid in MSPP (SP) and SPY confirmed the presence of caffeic acid in both samples up to 0.26 mg/g (MSPP) and 0.18 mg/g (SPY), respectively (Figure 9 and Table 2). One of the bioactive polyphenolic compounds of plant origin, caffeic acid (CFA) has been known to exert effects in anti-obesity by reducing body weight and regulating gut microbiota [29]. Therefore, in addition to BST-L.601 in SPY, CFA might contribute to the anti-obesity effects of SPY shown in this study, at least partly. Additionally, the presence of CFA in sweet potatoes was reported [30–32]. This suggests that CFA present in SPY can be utilized as an index compound for SPY and SPY-based products.

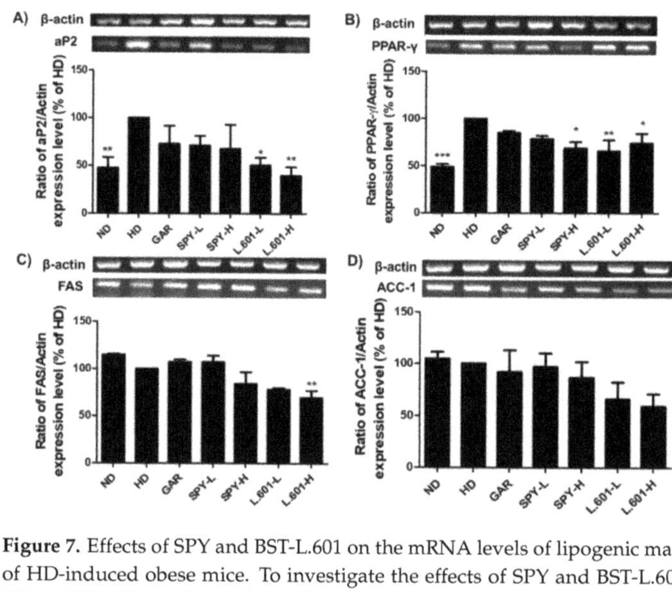

Figure 7. Effects of SPY and BST-L.601 on the mRNA levels of lipogenic markers in the liver tissue of HD-induced obese mice. To investigate the effects of SPY and BST-L.601 on the expression of (**A**) and (**B**) adipogenic markers (aP2 and PPAR-γ) and (**C**) and (**D**) lipogenic marker proteins (FAS and ACC-1), liver tissues of HD-induced obese mice treated with various diets for 12 weeks were homogenized and processed for RT-PCR and quantified by ImageJ software. The relative mRNA levels of lipogenic markers in the liver tissue were measured by RT-PCR. ND, normal diet; HD, high-fat diet; GAR, HD supplemented with *Garcinia cambogia* extract (100 mg/kg); SPY–L, HD diet supplemented with SPY (4×10^6 CFU/kg); SPY–H, HD supplemented with SPY (4×10^7 CFU/kg); L.601–L, HD supplemented with BST-L.601 (4×10^8 CFU/kg); L.601–H, HD supplemented with BST-L.601 (4×10^9 CFU/kg). L.601, *L. rhamnosus* BST-L.601. Values with different superscripts are significantly different among the groups by one-way ANOVA with Dunnett's multiple comparison test at * $p < 0.05$; ** $p < 0.01$, *** $p < 0.001$, compared to the HD group.

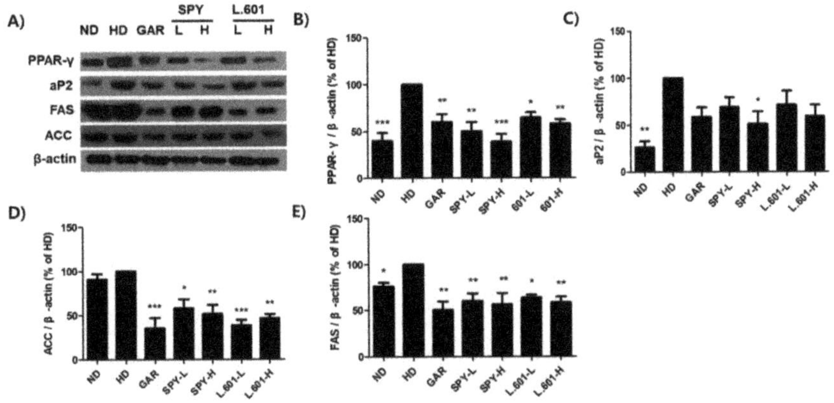

Figure 8. Effects of SPY and BST-L.601 on adipocyte differentiation and lipogenic marker proteins in the liver tissue of HD-induced obese mice. (**A**) Western blot analysis and ImageJ software were used to monitor and quantify the expression levels of adipocyte differentiation and lipogenic marker proteins, (**B**) PPAR-γ, (**C**) aP2, (**D**) ACC, and (**E**) FAS. β-Actin was used as a loading control. Each data point was expressed as the % of the HD group (100%). ND, normal diet; HD, high-fat diet; GAR, a high-fat diet supplemented with *Garcinia cambogia* extract (100 mg/kg); SPY−L, HD supplemented with SPY (4×10^6 CFU/kg); SPY− H, HD supplemented with SPY (4×10^7 CFU/kg); L.601− L, HD supplemented with BST-L.601 (4×10^8 CFU/kg); L.601− H, HD supplemented with BST.L-601 (4×10^9 CFU/kg). L.601, *L. rhamnosus* BST-L.601. Values with different superscripts are significantly different among the groups by one-way ANOVA with Dunnett's multiple comparison test at * $p < 0.05$; ** $p < 0.01$; *** $p < 0.001$, compared to the HD group.

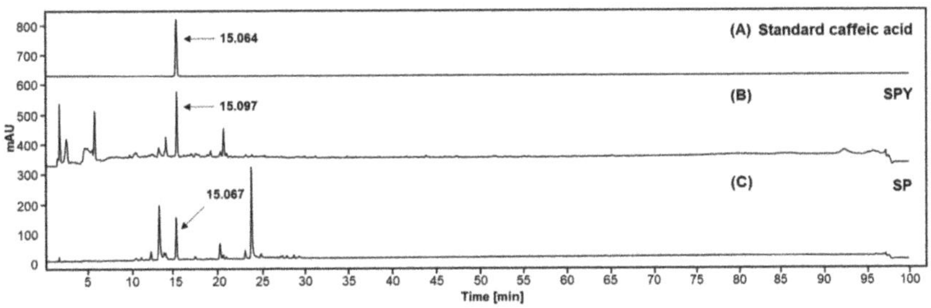

Figure 9. HPCL analysis for caffeic acid in SPY and MSPP. (**A**) Authentic caffeic acid (Sigma, San Francisco, CA, USA). (**B**) SPY and (**C**) SP (MSPP) samples dissolved in methanol were separately fractionated on an XDB-C18 column and detected by a UV detector at 330 nm and quantified by comparing the peak area on the chromatogram set to the peak area of standard caffeic acid at 100% as described in the manuscript.

4. Discussion

Several researchers have reported that oral medication of lactic acid bacteria (LAB) with prebiotics promotes the intestinal activity of LAB [22]. It was reported that the anti-obesity efficacy of the gut microbiome is related to LAB, and the dietary fiber from sweet potatoes helps a gut microbiome profile in a healthy way [21,22]. In this study, a newly isolated probiotic strain, *Lactobacillus rhamnosus* BST-L.601 (KCTC13517BP), and its fermented product (named SPY) with mashed sweet potato paste (MSPP) were evaluated to prove the efficacy of anti-obesity through using 3T3-L1 preadipocyte cells and an HD-induced obese mouse model.

Obesity is caused by the accumulation of triglycerides in adipocytes during the differentiation of 3T3-L1 preadipocytes [30]. Previous studies have shown that the first factors expressed after MDI treatment are C/EBPβ and C/EBPγ [33]. These two factors promote the expression of C/EBPα and PPAR-γ in the middle and end of preadipocyte differentiation [34]. Among them, C/EBPα is an essential factor inducing the inhibition of lipogenesis and adipogenesis. In addition, the decrease in the expression levels of C/EBPα and PPAR-γ induces a decrease in the expression of adipose protein 2 (aP2), which is known as a differentiation factor of preadipocytes, and PPAR-γ is a major factor involved in lipogenesis and adipogenesis [33]. The decrease in the expression level of these two factors leads to proteins that transport fatty acids to cells, causing lipid generation in cells [35]. This research showed that treatment of SPY decreased lipid droplets in 3T3-L1 preadipocytes and it means that SPY suppresses the adipogenesis of 3T3-L1 preadipocytes. Moreover, the expression of C/EBPα and PPAR-γ, which are the major adipogenesis factors, was significantly suppressed by SPY. Therefore, these results imply that SPY could negatively control the adipogenesis of 3T3-L1 preadipocytes.

Acetyl-CoA carboxylase (ACC) inhibits the activity of AMP-activated protein kinase (AMPK), reducing the production of malonyl-CoA [36,37]. Malonyl-CoA is synthesized as palmitate by fatty acid synthase (FAS), and palmitate is known to reduce fat accumulation and inhibit lipolysis [38,39]. In this research, oral administration of SPY reduced the expression of ACC and FAS, demonstrating its ability to downregulate lipogenesis. These results mean that SPY controls the expression of ACC and FAS, and also can be an effective agent to control adipogenic and lipogenic metabolism in 3T3-L1 preadipocytes. In previous research, it was proven that one of the *L. rhamnosus* strains controls the expression of transcription factors associated with adipogenic metabolism in an HD-induced obese mouse model [40].

The observed in vitro anti-obesity potential of SPY was further confirmed *in* an in vivo model system using HD-induced obese mice. The results showed that the significantly elevated body weight in HD mice was effectively reduced by oral administration of SPY or BST-L.601, with a more significant reduction by SPY than BST-L.601 treatment. The SPY used in this study was prepared by fermentation of mashed sweet potato paste (MSPP) with the BST-L.601 strain. Similar to our present observation, a combined preparation of prebiotics and probiotics was reported to reduce body fats [41]. LAB is demonstrated that it has efficacy in reducing hypercholesterolemia and preventing obesity [42]. Previously, it was reported that sweet potatoes contain significant amounts of resistant starch, caffeic acid, and many other compounds, including phenolic compounds, anthocyanins, and caffeoyl compounds [30–32]. One of the bioactive polyphenolic compounds of plant origin, caffeic acid (CFA) has been known to have various pharmacological activities such as anti-inflammatory, anti-cancer, anti-oxidant, and anti-obesity effects. Anti-obesity effectiveness of CFA is mediated by reducing body weight and regulating the gut microbiome in obese mice [29]. Resistant starch is known to express some beneficial effects related to metabolic syndrome such as inhibiting the increment of blood cholesterol and decreasing the glycemic response. Moreover, resistant starch acts as a functional prebiotic [19]. Therefore, although it is not proven right now, the observed results that SPY-H was more effective in decreasing the body weight gain in HD mice than treatment with BST-L.601 alone would be due to the presence of certain compounds contained in MSPP, such as resistant starch and CFA, which could synergistically contribute to the lowering of body weight gain in HD mice. Alternatively, it would also be possible that SPY may contain some other compounds produced during the fermentation of MSPP with BST-L.601 probiotics. On the other hand, hydroxycitric acid (HCA) is one of the key bioactive chemicals in *Garcinia cambogia*, and its anti-obesity effect has been explained unclearly for the last decades; thus, *Garcinia cambogia* has been widely added as a main raw material for anti-obesity functional foods [43]. In a previous study performed by another research group, it was confirmed that the administration of *Garcinia cambogia* powder (1%, *w*/*w*, 60% HCA) had an effect on inhibiting fat accumulation without toxicity [44,45]. For the above reasons, *Garcinia cambogia* was chosen as a positive control of anti-obesity efficacy for comparison with SPY. The results showed that SPY also exerted a significant positive effect on weight loss in fatty liver, epididymal adipose tissue, and visceral fat, which was comparable to that of the GAR (*Garcinia cambogia* extract, 100 mg/kg) group.

Meanwhile, Oil red O staining and TG content measurement in differentiating adipocyte cells showed that SPY significantly decreased lipid accumulation. Furthermore, histopathological analysis of adipose tissue located on the liver and epididymis in HD-induced obese mice by H&E staining demonstrated that administration of SPY-H and BST-L.601-H significantly reduced the enlarged cell size and increased the cell numbers of adipocytes in these tissues. One of the indicators for anti-obesity assessment is the size and number of adipocytes. Especially, adipocytes accumulated in epididymal and liver tissue are a useful index for verifying anti-obesity efficacy from chemicals or natural ingredients [46]. The important factor to define obesity is an increment of adipose tissue mass driven by either an explosive increment of the number or size of adipocytes [4]. Hypertrophy is mainly due to the accumulation of lipids and the hyperplasia caused by the proliferation and differentiation of adipocyte precursor cells into mature adipocytes [4]. This process is called adipogenesis. Thus, the regulation of the adipose tissue mass is conducted by suppressing adipogenesis, reducing lipogenesis, and enhancing lipolysis. An additional method to control adipose tissue is inducing apoptotic death of cells [5]. Therefore, the results of our study that SPY and BST-L.601 significantly reduced the enlarged cell size and increased the cell numbers of adipocytes in these tissues suggest that both SPY and BST-L.601 exert anti-obesity effects by effectively reducing adipose tissue mass by inhibiting lipogenesis of adipocytes.

On the other hand, as shown by H&E staining of liver tissues, oral administration of SPY and BST-L.601 prominently declined the immoderate generation and deposit of lipids in hepatocytes. This suggests that these materials may be potentially effective against hepatic steatosis, which is an important factor to define nonalcoholic fatty liver disease (NAFLD) [47]. Several animal studies using rodents have demonstrated that suppressing the accumulation of lipids in the liver tissue could be for hepatoprotection from HD-induced NAFLD [48,49]. The results of this study imply that SPY is useful for weight loss in fatty liver, epididymal adipose tissue, and visceral fat in HD-induced obese mice. Although it needs to be clarified through the following research, SPY, and BST-L.601 may exert protective effects to inhibit hepatic steatosis and thus NAFLD in HD-fed mice, which would be an additional health benefit of SPY and BST-L.601.

SPY was also shown to suppress the secretion of leptin, an adipocyte hormone that is secreted by adipocytes in response to their triglyceride levels, and is known to regulate energy expenditure and food intake [50,51]. Circulating leptin levels in the blood are associated with the extent of obesity; thus, leptin is a sensitive biomarker to indicate obesity [51,52]. In this study, oral administration of SPY and BST-L.601 to HD-induced obese mice for 12 weeks resulted in reduced levels of leptin in blood serum, suggesting that both materials, SPY, and BST-L.601, attenuated the secretion of leptin, thereby downregulating lipid metabolism in HD-fed mice.

5. Conclusions

This study revealed that, with no detectable level of cytotoxicity, SPY, a fermented product (named SPY) of *Lactobacillus rhamnosus* BST-L.601 in mashed sweet potato paste (MSPP) medium, significantly reduced the lipid accumulation and TG contents in 3T3-L1 adipocytes. Moreover, SPY also inhibited the differentiation of adipocytes and lipogenesis through suppression of the expression of adipogenesis-related markers such as C/EBPα, PPAR-γ, and aP2 and fatty acid synthetic pathway proteins such as ACC and FAS. In HD-induced obese mice, oral administration of SPY or BST-L.601 for 12 weeks significantly reduced the body and liver weight, the size of adipocytes, and the weight of epididymal, visceral, and subcutaneous fat tissues. Administration of SPY or BST-L.601 also reduced the serum levels of total cholesterol and LDL cholesterol and leptin secretion. These results indicated that both SPY and BST-L.601 effectively suppress HD-induced adipogenesis and lipogenesis through the downregulation of the expression of adipogenic marker proteins and lipogenesis-related marker proteins. SPY was more effective in decreasing body weight gain in HD mice than in treatment with BST-L.601 alone. The results of the present study suggest that SPY and BST-L.601 can be potential candidates as active ingredients to develop health-beneficial functional foods or new probiotic-prebiotic combined compositions to prevent or treat obesity.

Author Contributions: Conceptualization, J.-W.P., T.K. and Y.I.P.; methodology, T.K. and H.C.; validation, J.R. and J.-W.P.; formal analysis, T.K., H.C. and J.R.; investigation, T.K., J.-W.P. and J.R.; resources, T.K. and J.-W.P.; data curation, J.R. and H.C.; writing—original draft preparation, T.K.; writing—review and editing, Y.I.P. and J.R.; visualization, T.K. and J.R.; supervision, Y.I.P. and J.-W.P.; funding acquisition, Y.I.P. All authors have read and agreed to the published version of the manuscript.

Funding: This research was funded by Biostream Technologies Co., Ltd., grant number M-2019-D0321-00001, and supported by the Research Fund, 2020 of The Catholic University of Korea.

Data Availability Statement: The data presented in this study are available on request from the corresponding author.

Conflicts of Interest: This study was performed through the collaboration between University and Industrial company (The Catholic University of Korea- Biostream Technologies Co.) under the financial support from the company (grant number M-2019-D0321-00001). Authors, Joo-Woong Park, Hyewon Choe, and Taewook Kang, were (and till now) employed by the company Biostream Technologies Co. Taewook Kang and Jin Ree are affiliated to both company (Biostream Technologies Co.) and university (The Catholic University of Korea) as part-time graduate students. In addition to the financial support, the company research team (Dr. Joo-Woong Park, Hyewon Choe, and Taewook Kang) prepared the materials (probiotic strain, Lactobacillus rhamnosus BST-L.601, and the sweet potato yogurt, SPY, fermented with this strain) and performed composition analysis of these materials. Dr. Joo-Woong Park and Taewook Kang, as the company side, and Yong Il Park (the corresponding author) as the university side conceptualized, designed and supervised this study. The prepared materials were provided to university research team (Prof. Yong Il Park who is the corresponding author of this manuscript) for in vitro and in vivo bioactivity assay (Anti-obesity experiments). Taewook Kang, who is affiliated to both company and university as a part-time graduate student for Master degree performed most part of the bioactivity experiments and writing of the draft of this manuscript. Jin Ree and Hyewon Choe performed data curation and some instrumental analysis and data curation. The authors declare no conflict of interest.

References

1. Popkin, B.M.; Kim, S.; Rusev, E.R.; Du, S.; Zizza, C. Measuring the full economic costs of diet, physical activity and obesity-related chronic diseases. *Obes. Rev.* **2006**, *7*, 271–293. [CrossRef] [PubMed]
2. OECD. Obesity Update 2017. Available online: https://www.oecd.org/els/health-systems/Obesity-Update-2017.pdf (accessed on 22 December 2022).
3. Kim, J.H.; Kim, O.K.; Yoon, H.G.; Park, J.; You, Y.; Kim, K.; Lee, Y.H.; Choi, K.C.; Lee, J.; Jun, W. Anti-obesity effect of extract from fermented *Curcuma longa* L. through regulation of adipogenesis and lipolysis pathway in high-fat diet-induced obese rats. *Food Nutr Res.* **2016**, *60*, 30428. [CrossRef] [PubMed]
4. Gregoire, F.M.; Smas, C.M.; Sul, H.S. Understanding adipocyte differentiation. *Physiol. Rev.* **1998**, *78*, 783–809. [CrossRef] [PubMed]
5. Sorisky, A.; Pardasani, D.; Gagnon, A.; Smith, T.J. Evidence of adipocyte differentiation in human orbital fibroblasts in primary culture. *J. Clin. Endocrinol. Metab.* **1996**, *81*, 3428–3431.
6. Cristancho, A.G.; Lazar, M.A. Forming functional fat: A growing understanding of adipocyte differentiation. *Nat. Rev. Mol. Cell Biol.* **2011**, *12*, 722–734. [CrossRef]
7. Bazin, R.; Ferré, P. Assays of lipogenic enzymes. *Methods Mol. Biol.* **2001**, *155*, 121–127.
8. DiBaise, J.K.; Frank, D.N.; Mathur, R. Impact of the gut microbiota on the development of obesity: Current concepts. *Am. J. Gastroenterol. Suppl.* **2012**, *1*, 22. [CrossRef]
9. Conterno, L.; Fava, F.; Viola, R.; Tuohy, K.M. Obesity and the gut microbiota: Does up-regulating colonic fermentation protect against obesity and metabolic disease? *Genes Nutr.* **2011**, *6*, 241–260. [CrossRef]
10. Devaraj, S.; Hemarajata, P.; Versalovic, J. The human gut microbiome and body metabolism: Implications for obesity and diabetes. *Clin. Chem.* **2013**, *59*, 617–628. [CrossRef]
11. Bubnov, R.V.; Babenko, L.P.; Lazarenko, L.M.; Mokrozub, V.V.; Demchenko, O.A.; Nechypurenko, O.V.; Spivak, M.Y. Comparative study of probiotic effects of Lactobacillus and Bifidobacteria strains on cholesterol levels, liver morphology and the gut microbiota in obese mice. *Epma. J.* **2017**, *8*, 357–376. [CrossRef]

12. Pascale, A.; Marchesi, N.; Marelli, C.; Coppola, A.; Luzi, L.; Govoni, S.; Giustina, A.; Gazzaruso, C. Microbiota and metabolic diseases. *Endocrine* **2018**, *61*, 357–371. [CrossRef] [PubMed]
13. Tilg, H.; Kaser, A. Gut microbiome, obesity, and metabolic dysfunction. *J. Clin. Investig.* **2011**, *121*, 2126–2132. [CrossRef] [PubMed]
14. Mazloom, K.; Siddiqi, I.; Covasa, M. Probiotics: How effective are they in the fight against obesity? *Nutrients* **2019**, *11*, 258. [CrossRef] [PubMed]
15. Vallianou, N.; Stratigou, T.; Christodoulatos, G.S.; Tsigalou, C.; Dalamaga, M. Probiotics, prebiotics, synbiotics, postbiotics, and obesity: Current evidence, controversies, and perspectives. *Curr. Obes. Rep.* **2020**, *9*, 179–192. [CrossRef]
16. Choque Delgado, G.T.; Tamashiro, W. Role of prebiotics in regulation of microbiota and prevention of obesity. *Food Res. Int.* **2018**, *113*, 183–188. [CrossRef]
17. Gibson, G.R.; Hutkins, R.; Sanders, M.E.; Prescott, S.L.; Reimer, R.A.; Salminen, S.J.; Scott, K.; Stanton, C.; Swanson, K.S.; Cani, P.D.; et al. Expert consensus document: The International Scientific Association for Probiotics and Prebiotics (ISAPP) consensus statement on the definition and scope of prebiotics. *Nat. Rev. Gastroenterol. Hepatol.* **2017**, *14*, 491–502. [CrossRef]
18. Schwiertz, A.; Taras, D.; Schäfer, K.; Beijer, S.; Bos, N.A.; Donus, C.; Hardt, P.D. Microbiota and SCFA in lean and overweight healthy subjects. *Obesity* **2010**, *18*, 190–195. [CrossRef]
19. Topping, D.L.; Clifton, P.M. Short-chain fatty acids and human colonic function: Roles of resistant starch and nonstarch polysaccharides. *Physiol. Rev.* **2001**, *81*, 1031–1064. [CrossRef]
20. Zheng, Y.; Wang, Q.; Li, B.; Lin, L.; Tundis, R.; Loizzo, M.R.; Zheng, B.; Xiao, J. Characterization and prebiotic effect of the resistant starch from purple sweet potato. *Molecules* **2016**, *21*, 932. [CrossRef]
21. Tsai, Y.T.; Cheng, P.C.; Pan, T.M. Anti-obesity effects of gut microbiota are associated with lactic acid bacteria. *Appl. Microbiol. Biotechnol.* **2014**, *98*, 1–10. [CrossRef]
22. Liu, M.; Li, X.; Zhou, S.; Wang, T.T.Y.; Zhou, S.; Yang, K.; Li, Y.; Tian, J.; Wang, J. Dietary fiber isolated from sweet potato residues promotes a healthy gut microbiome profile. *Food Funct.* **2020**, *11*, 689–699. [CrossRef] [PubMed]
23. Bekaert, M.; Ouwens, D.M.; Hörbelt, T.; Van de Velde, F.; Fahlbusch, P.; Herzfeld de Wiza, D.; Van Nieuwenhove, Y.; Calders, P.; Praet, M.; Hoorens, A.; et al. Reduced expression of chemerin in visceral adipose tissue associates with hepatic steatosis in patients with obesity. *Obesity* **2016**, *24*, 2544–2552. [CrossRef] [PubMed]
24. LeBlanc, J.G.; Chain, F.; Martín, R.; Bermúdez-Humarán, L.G.; Courau, S.; Langella, P. Beneficial effects on host energy metabolism of short-chain fatty acids and vitamins produced by commensal and probiotic bacteria. *Microb. Cell Fact.* **2017**, *16*, 79. [CrossRef] [PubMed]
25. Frece, J.; Kos, B.; Svetec, I.K.; Zgaga, Z.; Beganović, J.; Lebos, A.; Susković, J. Synbiotic effect of *Lactobacillus helveticus* M92 and prebiotics on the intestinal microflora and immune system of mice. *J. Dairy Res.* **2009**, *76*, 98–104. [CrossRef] [PubMed]
26. Rasouli, M.; Ostovar-Ravari, A.; Shokri-Afra, H. Characterization and improvement of phenol-sulfuric acid microassay for glucose-based glycogen. *Eur. Rev. Med. Pharmacol. Sci.* **2014**, *18*, 2020–2024.
27. Bradford, M.M. A rapid and sensitive method for the quantitation of microgram quantities of protein utilizing the principle of protein-dye binding. *Anal. Biochem.* **1976**, *72*, 248–254. [CrossRef]
28. Hong, J.-I.; Kim, H.-J.; Kim, J.-Y. Factors affecting reactivity of various phenolic compounds with the Folin-Ciocalteu reagent. *J. Korean Soc. Food Sci. Nutr.* **2011**, *40*, 205–213. [CrossRef]
29. Xu, J.; Ge, J.; He, X.; Sheng, Y.; Zheng, S.; Zhang, C.; Xu, W.; Huang, K. Caffeic acid reduces body weight by regulating gut microbiota in diet-induced-obese mice. *J. Funct. Foods* **2020**, *74*, 104061. [CrossRef]
30. Meng, X.J.; Tan, C.; Feng, Y. Solvent extraction and in vitro simulated gastrointestinal digestion of phenolic compounds from purple sweet potato. *Int. J. Food Sci. Technol.* **2019**, *54*, 2887–2896. [CrossRef]
31. Philpott, M.; Ferguson, L.R.; Gould, K.S.; Harris, P.J. Anthocyanidin-containing compounds occur in the periderm cell walls of the storage roots of sweet potato (*Ipomoea batatas*). *J. Plant Physiol.* **2009**, *166*, 1112–1117. [CrossRef]
32. Zhao, J.G.; Yan, Q.Q.; Xue, R.Y.; Zhang, J.; Zhang, Y.Q. Isolation and identification of colourless caffeoyl compounds in purple sweet potato by HPLC-DAD-ESI/MS and their antioxidant activities. *Food Chem.* **2014**, *161*, 22–26. [CrossRef] [PubMed]
33. Steinberg, G.R.; Kemp, B.E.; Watt, M.J. Adipocyte triglyceride lipase expression in human obesity. *Am. J. Physiol. Endocrinol. Metab.* **2007**, *293*, E958–E964. [CrossRef] [PubMed]
34. Ntambi, J.M.; Young-Cheul, K. Adipocyte differentiation and gene expression. *J. Nutr.* **2000**, *130*, 3122s–3126s. [CrossRef]
35. Chawla, A.; Schwarz, E.J.; Dimaculangan, D.D.; Lazar, M.A. Peroxisome proliferator-activated receptor (PPAR) gamma: Adipose-predominant expression and induction early in adipocyte differentiation. *Endocrinology* **1994**, *135*, 798–800. [CrossRef] [PubMed]
36. Shin, J.; Li, B.; Davis, M.E.; Suh, Y.; Lee, K. Comparative analysis of fatty acid-binding protein 4 promoters: Conservation of peroxisome proliferator-activated receptor binding sites. *J. Anim. Sci.* **2009**, *87*, 3923–3934. [CrossRef] [PubMed]
37. Peterson, J.M.; Aja, S.; Wei, Z.; Wong, G.W. CTRP1 protein enhances fatty acid oxidation via AMP-activated protein kinase (AMPK) activation and acetyl-CoA carboxylase (ACC) inhibition. *J. Biol. Chem.* **2012**, *287*, 1576–1587. [CrossRef] [PubMed]
38. Lane, M.D.; Wolfgang, M.; Cha, S.H.; Dai, Y. Regulation of food intake and energy expenditure by hypothalamic malonyl-CoA. *Int. J. Obes.* **2008**, *32*, S49–S54. [CrossRef]
39. Jones, S.F.; Infante, J.R. Molecular pathways: Fatty acid synthase. *Clin. Cancer Res.* **2015**, *21*, 5434–5438. [CrossRef]
40. Saggerson, E.D. The regulation of glyceride synthesis in isolated white-fat cells. The effects of palmitate and lipolytic agents. *Biochem. J.* **1972**, *128*, 1057–1067. [CrossRef]

41. Khairani, A.F.; Islami, U.; Syamsunarno, M.; Lantika, U. Synbiotic purple sweet potato yogurt ameliorate lipid metabolism in high fat diet mice model. *Biomed. Pharmacol. J.* **2020**, *13*, 175–184. [CrossRef]
42. Choi, W.J.; Dong, H.J.; Jeong, H.U.; Jung, H.H.; Kim, Y.H.; Kim, T.H. Antiobesity effects of *Lactobacillus plantarum* LMT1-48 accompanied by inhibition of *Enterobacter cloacae* in the intestine of diet-induced obese mice. *J. Med. Food* **2019**, *22*, 560–566. [CrossRef]
43. Chuah, L.O.; Ho, W.Y.; Beh, B.K.; Yeap, S.K. Updates on antiobesity effect of Garcinia origin (-)-HCA. *Evid. Based Complem. Alternat. Med.* **2013**, *2013*, 751658. [CrossRef] [PubMed]
44. Hayamizu, K.; Ishii, Y.; Kaneko, I.; Shen, M.; Okuhara, Y.; Shigematsu, N.; Tomi, H.; Furuse, M.; Yoshino, G.; Shimasaki, H. Effects of *Garcinia cambogia* (hydroxycitric acid) on visceral fat accumulation: A double-blind, randomized, placebo-controlled trial. *Curr. Ther. Re.s Clin. Exp.* **2003**, *64*, 551–567. [CrossRef] [PubMed]
45. Semwal, R.B.; Semwal, D.K.; Vermaak, I.; Viljoen, A. A comprehensive scientific overview of *Garcinia cambogia*. *Fitoterapia* **2015**, *102*, 134–148. [CrossRef] [PubMed]
46. Lee, C.W.; Seo, J.Y.; Lee, J.; Choi, J.W.; Cho, S.; Bae, J.Y.; Sohng, J.K.; Kim, S.O.; Kim, J.; Park, Y.I. 3-O-Glucosylation of quercetin enhances inhibitory effects on the adipocyte differentiation and lipogenesis. *Biomed. Pharmacother.* **2017**, *95*, 589–598. [CrossRef]
47. Browning, J.D.; Horton, J.D. Molecular mediators of hepatic steatosis and liver injury. *J. Clin. Investig.* **2004**, *114*, 147–152. [CrossRef]
48. Yang, D.K.; Jo, D.-G. Mulberry fruit extract ameliorates nonalcoholic fatty liver disease (NAFLD) through inhibition of mitochondrial oxidative stress in rats. *Evid. Based Complem. Alternat. Med.* **2018**, *2018*, 8165716. [CrossRef]
49. Tran, N.K.S.; Kim, G.T.; Park, S.H.; Lee, D.; Shim, S.M.; Park, T.S. Fermented *Cordyceps militaris* extract prevents hepatosteatosis and adipocyte hypertrophy in high fat diet-fed mice. *Nutrients* **2019**, *11*, 1015. [CrossRef]
50. Van Harmelen, V.; Reynisdottir, S.; Eriksson, P.; Thörne, A.; Hoffstedt, J.; Lönnqvist, F.; Arner, P. Leptin secretion from subcutaneous and visceral adipose tissue in women. *Diabetes* **1998**, *47*, 913–917. [CrossRef]
51. Havel, P.J. Role of adipose tissue in body-weight regulation: Mechanisms regulating leptin production and energy balance. *Proc. Nutr. Soc.* **2000**, *59*, 359–371. [CrossRef]
52. Maffei, M.; Halaas, J.; Ravussin, E.; Pratley, R.E.; Lee, G.H.; Zhang, Y.; Fei, H.; Kim, S.; Lallone, R.; Ranganathan, S.; et al. Leptin levels in human and rodent: Measurement of plasma leptin and ob RNA in obese and weight-reduced subjects. *Nat. Med.* **1995**, *1*, 1155–1161. [CrossRef] [PubMed]

foods

MDPI

Article

Biochemical and Genomic Characterization of Two New Strains of *Lacticaseibacillus paracasei* Isolated from the Traditional Corn-Based Beverage of South Africa, Mahewu, and Their Comparison with Strains Isolated from Kefir Grains

Konstantin V. Moiseenko [1,*], Anna V. Begunova [2], Olga S. Savinova [1], Olga A. Glazunova [1], Irina V. Rozhkova [2] and Tatyana V. Fedorova [1]

1 A. N. Bach Institute of Biochemistry, Research Center of Biotechnology, Russian Academy of Sciences, Leninsky Ave. 33/2, 119071 Moscow, Russia
2 Federal State Budgetary Scientific Institution "All-Russian Research Institute of Dairy Industry", 115093 Moscow, Russia
* Correspondence: mr.moiseenko@gmail.com

Abstract: *Lacticaseibacillus paracasei* (formerly *Lactobacillus paracasei*) is a nomadic lactic acid bacterium (LAB) that inhabits a wide variety of ecological niches, from fermented foodstuffs to host-associated microenvironments. Many of the isolated *L. paracasei* strains have been used as single-strain probiotics or as part of a symbiotic consortium within formulations. The present study contributes to the exploration of different strains of *L. paracasei* derived from non-conventional isolation sources—the South African traditional fermented drink mahewu (strains MA2 and MA3) and kefir grains (strains KF1 and ABK). The performed microbiological, biochemical and genomic comparative analyses of the studied strains demonstrated correlation between properties of the strains and their isolation source, which suggests the presence of at least partial strain adaptation to the isolation environments. Additionally, for the studied strains, antagonistic activities against common pathogens and against each other were observed, and the ability to release bioactive peptides with antioxidant and angiotensin I-converting enzyme inhibitory (ACE-I) properties during milk fermentation was investigated. The obtained results may be useful for a deeper understanding of the nomadic lifestyle of *L. paracasei* and for the development of new starter cultures and probiotic preparations based on this LAB in the future.

Keywords: *Lacticaseibacillus paracasei*; mahewu; kefir grains; genome sequencing; antibacterial activity; milk fermentation; proteolytic activity; antioxidant activity; angiotensin-converting enzyme inhibitory activity (ACE-I)

Citation: Moiseenko, K.V.; Begunova, A.V.; Savinova, O.S.; Glazunova, O.A.; Rozhkova, I.V.; Fedorova, T.V. Biochemical and Genomic Characterization of Two New Strains of *Lacticaseibacillus paracasei* Isolated from the Traditional Corn-Based Beverage of South Africa, Mahewu, and Their Comparison with Strains Isolated from Kefir Grains. *Foods* **2023**, *12*, 223. https://doi.org/10.3390/foods12010223

Academic Editor: Xiqing Yue

Received: 17 November 2022
Revised: 27 December 2022
Accepted: 30 December 2022
Published: 3 January 2023

1. Introduction

Lactic acid bacteria (LAB) are a heterogeneous group of microorganisms that play a key role in various food fermentation processes [1,2]. In addition to advantageous storage and organoleptic properties, many foods fermented by LAB possess additional health-promoting benefits, such as imparting improvements to digestion and tolerance to lactose [3], hypocholesterolemic and antihypertensive effects [4,5] and antioxidant and anticarcinogenic activities [6,7]. Additionally, LAB itself can possess probiotic properties, such as the ability to impart improvements to the intestinal barrier and commensal microbial balance [8,9], production of beneficial enzymes (e.g., β-galactosidase and bile salt hydrolase) and neurochemicals [10,11], suppression of pathogenic microflora [12], and modulation of the immune system [13].

Among the LAB, *Lactobacillus* is the most well-known genus and currently comprises more than 200 species with extremely diverse phenotypes, genotypes and ecology [14].

Moreover, for each actively studied species of *Lactobacillus*, more than 100 strains are currently described and have genomes sequenced [15]. Although several researchers have proposed the idea that some beneficial properties of *Lactobacillus* spp. may be innate attributes of taxonomic ranks higher than the strain [16], the strain-specificity of such properties is still a cornerstone principle of probiotic science. Therefore, the isolation of individual strains of *Lactobacillus* spp. and the exploration of their beneficial properties are necessary, albeit tedious, undertakings to develop new associations of starter cultures for products with pronounced health benefits [17–19].

The present study contributes to the exploration of different strains of *Lacticaseibacillus paracasei* (formerly *Lactobacillus paracasei*). Many of the isolated *L. paracasei* strains have been used as single-strain probiotics or as part of a symbiotic consortium within formulations [20]. Examples actively marketed around the name "probiotic strains of *L. paracasei*" are: *L. paracasei* F19 from Chr. Hansen, which is mainly used as a part of the starter culture in the popular Scandinavian yogurt from the Arla Foods company; *L. paracasei* DG (*L. paracasei* CNCM I-1572), which is mainly used as a single-strain probiotic in the Enterolactis food supplement from the Italian company SOFAR; and *L. paracasei* Shirota, which is mainly used for the preparation of the Japanese sweetened probiotic milk beverage Yakult prom the Yakult Honsha company.

L. paracasei is closely related to such widely researched and used probiotic species as *Lacticaseibacillus casei* and *Lacticaseibacillus rhamnosus* and forms with these species a distinctive evolutionary/taxonomic group—the *L. casei* group [21,22]. The species of the *L. casei* group populate a wide variety of different niches, from fermented foodstuffs to host-associated microenvironments, and represent a paradigmatic "nomadic species" [23,24]. Although nomadic species are not permanent residents of intestinal ecosystems, they may persist in them for at least a limited time [23]. Additionally, when isolated from environments other than host-associated (e.g., commercial or artisanal dairy products and plants), the isolation sources of nomadic species are not well-correlated with the evolution of their genomes [25]. Consequently, no niche-specific adaptations common to the majority of strains isolated from the same source can be identified [26]. However, it is easily possible for the different strains of nomadic species to possess different independent adaptations to the same environment [24,27]. For example, Smocvina et al. [26] reported that dairy-derived strains of *L. paracasei* generally possessed reduced genome size with a smaller number of sugar cassettes. The inherent genetic flexibility of nomadic strains has made these strains a natural library of evolutionarily selected variations yet to be employed for biotechnological applications [28].

Currently, 310 strains of *L. paracasei* with sequenced genomes have been reported [15]; however, there are several pieces of evidence suggesting that not all the strain biodiversity of this species has been explored. Firstly, *L. paracasei* is considered to have an open pan-genome [29]; hence, the number of new gene families increases with the addition of the genomes from new strains [30]. Secondly, there is a limited number of strains isolated from non-conventional sources, examples of which would be non-dairy fermented foodstuff and symbiotic cultures of bacteria and yeast (SCOBY), such as kefir grains and kombucha [25]. The GeneBank database contains only eight genomes of *L. paracasei* isolated from kefir or kefir grains and only one from non-dairy fermented foodstuff (beer).

In this article, we report a comparative biochemical and genomic characterization of four strains of *L. paracasei* isolated from non-conventional sources: the strains *L. paracasei* MA2 and *L. paracasei* MA3 were isolated from the traditional, corn-based, nonalcoholic beverage of South Africa mahewu in the course of this work; and the strains *L. paracasei* KF1 and *L. paracasei* ABK were previously isolated from the SCOBY traditionally used in Commonwealth of Independent States (CIS) countries for preparation of kefir and kefir grains [31]. Additionally, for the studied strains, antagonistic activities against common pathogens and against each other were observed, and the ability to release bioactive peptides with antioxidant and angiotensin I-converting enzyme inhibitory (ACE-I) properties during milk fermentation was investigated.

2. Materials and Methods

2.1. Isolation, Identification and Profile of Enzymatic Activities

The strains *L. paracasei* MA2 and *L. paracasei* MA3 were isolated from mahewu, the samples of which were purchased in the distribution network of Durban (South Africa) and analytically characterized by Moiseenko et al. [32]. The isolation procedure was performed as described by Begunova et al. [31]. In brief, a series of tenfold dilutions were inoculated into MRS (De Man, Rogosa and Sharpe) broth with the addition of 10% ethanol and cultivated at 30 ± 1 °C for 3–5 days to enrich the medium with lactobacilli. An enriched culture of lactobacilli was inoculated on MRS agar (pH 5.4) and anaerobically incubated at a temperature of 30 ± 1 °C for 3–5 days. Morphologically identical colonies were selected for further cultivation in MRS broth.

The isolated lactobacilli were biochemically characterized using API 50CH (BioMerieux, Marcy l'Etoile, France) and API ZYM (BioMerieux) test systems according to the manufacturer's instructions. The results of the API 50CH test were analyzed with the APIWEB web server (https://apiweb.biomerieux.com, accessed on 24 September 2022). The genotyping of the obtained isolates was performed with colony polymerase chain reaction (PCR) from the MRS agar plates. The 16S rRNA gene was amplified according to [33] with a Taq DNA polymerase kit (Evrogen, Moscow, Russia) and the primers bak11w (5′-AGT TTG ATC MTG GCT CAG-3′) and bak4 (5′-AGG AGG TGA TCC ARC CGC A-3′). Successfully amplified PCR products were extracted from 2% agarose gel with a commercial QIAquick Gel Extraction Kit (Qiagen, Valencia, CA, USA). Sanger sequencing of the purified PCR products was carried out with the same primers as for PCR. The construction of the phylogenetic tree with obtained 16S rRNA sequences, as well as sequences from various *Lactobacillus* species type strains [14], was performed using the RAxML-HPC BlackBox (v 8.2.10) program [34] at the CIPRES Science Gateway [35].

The strains *L. paracasei* KF1 and *L. paracasei* ABK were obtained from the Microorganism Collection of the All-Russia Research Institute of the Dairy Industry (VNIMI, Moscow, Russia). Both strains were isolated from kefir grains and biochemically characterized using API 50CH and API ZYM test systems by Begunova et al. [31]. The sequences of the 16S ribosomal RNA genes of these strains can be found at the GenBank accessory numbers MW558119.1 and MN994625.1 for *L. paracasei* KF1 and *L. paracasei* ABK, respectively.

2.2. Inhibition of Pathogens and Antagonistic Interactions

The ability of the *L. paracasei* KF1, *L. paracasei* ABK, *L. paracasei* MA2 and *L. paracasei* MA3 to inhibit common pathogens was assessed according to Begunova et al. [36]. The pathogenic bacterium strains *Staphylococcus aureus* ATCC-6538 and *Escherichia coli* ATCC-25922 were purchased from the American Type Culture Collection (Manassas, VA, USA), and *Salmonella typhimurium* NCTC 00074 was purchased from the National Collection of Type Cultures (Salisbury, UK). In brief, the antagonistic activity was evaluated with the co-culture method. For the experimental samples, 20 mL of MRS broth was simultaneously inoculated with 1 mL (approximately 10^7 CFU·mL^{-1}) of the studied *L. paracasei* strain and 1 mL (approximately 10^7 CFU·mL^{-1}) of the pathogenic test-strain. The single-species cultivations of the pathogenic bacteria were used as a control. The incubation was carried out at 37 ± 2 °C, and samples were collected after 24 and 48 h. Pathogen cells were counted on commercial mediums based on pancreatic sprat hydrolysate, SPA agar medium (Mikrogen, Moscow, Russia), at 37 °C for 24–48 h.

The antagonistic interactions between the studied *L. paracasei* strains were assessed with the perpendicular streak test following Savinova et al. [37]. At the first stage, the pre-culture of the first LAB was streaked on the MRS agar and incubated under anaerobic conditions at 37 °C for 24 h. At the second stage, the pre-culture of the second LAB was streaked perpendicularly to the first LAB, and the plate was incubated under anaerobic conditions for another 24 h at 37 °C. The antagonistic interactions between LAB were assessed visually from the presence of a growth inhibition zone.

2.3. Fermentation of Milk

Growth characteristics and functional properties of the *L. paracasei* KF1, *L. paracasei* ABK, *L. paracasei* MA2 and *L. paracasei* MA3 were studied during fermentation of the skim milk. Sterile skim milk was inoculated with 1% of the corresponding strain and incubated at 30 °C for 72 h. Samples were collected under sterile conditions at 6, 16, 24, 48 and 72 h of fermentation, and the number of viable cells (colony-forming units (CFUs)) was counted on MRS agar and the pH was measured using a Seven Easy pH meter (Mettler Toledo, Greifensee, Switzerland).

For the measurements of proteolytic activity, antioxidant-capacity and ACE-I-activity protein-peptide fractions were isolated from fermented milk. For the samples with a pH above 4.6, the pH was adjusted to 4.6 by adding 0.75% trichloroacetic acid (TCA). The samples were centrifuged for 30 min at 4 °C and 10,000× *g* in a 5702R centrifuge (Eppendorf, Germany). The obtained supernatants were filtered through 0.45 μm PVDF syringe filters (Sartorius, Germany). The obtained protein-peptide fractions were frozen and stored at −80 °C until further analysis. Before the analysis, samples were thawed and additionally filtered with 0.45 μm polyvinylidene difluoride (PVDF) syringe filters.

The proteolytic activity was determined quantitatively as the amount of released amino groups using the 2,4,6-trinitrobenzenesulfonic acid (TNBS) method [38], as described by Torkova et al. [39]. The results were expressed as the amount of L-Leu molar equivalents, mM(L-Leu).

The antioxidant capacity of the samples was determined with the oxygen radical absorbance capacity fluorescence method (ORAC) with the generation of peroxyl radicals in the reaction medium, as described by Torkova et al. [39]. The results were expressed as the amount of Trolox molar equivalents, μM(TE).

The in vitro hypotensive effect of the fermented milks was assessed as the angiotensin-I-converting enzyme (ACE)-inhibiting activity of the samples (ACE-I activity), as described by Torkova et al. [39]. The measurements were performed with a BioTek Synergy 2 microplate photometer–fluorometer (BioTek). The results were expressed as the half maximal inhibitory concentration IC_{50} (reported as mg of protein per mL).

2.4. Genome Sequencing, Assembly and Annotation

The DNA isolation, genome sequencing and genome annotation of *L. paracasei* KF1, *L. paracasei* ABK, *L. paracasei* MA2 and *L. paracasei* MA3 were performed as described by Savinova et al. [37]. In brief, total DNA was extracted from liquid MRS cultures using a DNeasy mericon Food Kit (Qiagen) according to the manufacturer's protocol. The DNA library was prepared using the Ion AmpliSeq library kit 2.0 (Thermo Fisher Scientific, Waltham, MA, USA) and indexed with an Ion Xpress barcode adapters 1–16 kit (Thermo Fisher Scientific, MA, USA). Whole-genome sequencing was carried out using the Ion Torrent Personal Genome Machine (PGM) (Thermo Fisher Scientific). The obtained reads were pre-processed and assembled with CLC Genomics Workbench 11.0 (Qiagen). Upon submission, genome annotations were performed using NCBI Prokaryotic Genome Annotation Pipeline (PGAP) [40]. Additionally, annotation with ISfinder [41], PHASTER [42], PlasmidFinder [43] and BAGEL4 [44] was performed on the web. The comparative genome analysis was performed using the Anvi'o suite of programs [45,46].

2.5. Statistical Data Manipulations

All experiments were performed in three biological replicates. All statistical comparisons were firstly performed using a one-way ANOVA omnibus *F*-test. When a significant ($p < 0.05$) value for the *F*-statistics was found, differences between means were evaluated using Tukey's honestly significant difference (HSD) multiple comparison test ($p < 0.05$).

3. Results and Discussion

3.1. Isolation and Taxonomic Assignment of the Mahewu-Derived Strains

As a result of a series of successive subcultures on the media selective for lactobacilli, several rod-shaped (approximately 1 μkm in length) Gram-positive bacteria were isolated. The isolated bacteria formed small, round, creamy-yellow non-transparent colonies (approximately 1 mm in diameter) with smooth edges on MRS agar. To identify the isolated strains, their 16S rRNA genes were sequenced and compared with the sequences of the type strains registered in the GenBank database. As a result, two isolated strains demonstrated 99.9% similarity with the type strains of *L. paracasei*. These strains were named *L. paracasei* MA2 and *L. paracasei* MA3 and were deposited into the Microorganism Collection of the All-Russia Research Institute of the Dairy Industry (VNIMI, Moscow, Russia). The obtained sequences of 16S rRNA were deposited into GenBank under the accession numbers MW558121.1 and MW558122.1 for *L. paracasei* MA2 and *L. paracasei* MA3, respectively. The phylogenetic tree constructed using 16s rRNA sequences of various *Lactobacillus* species type strains [14] and 16S rRNA sequences of the mahewu-derived strains *L. paracasei* MA2 and *L. paracasei* MA3 is shown in Figure 1.

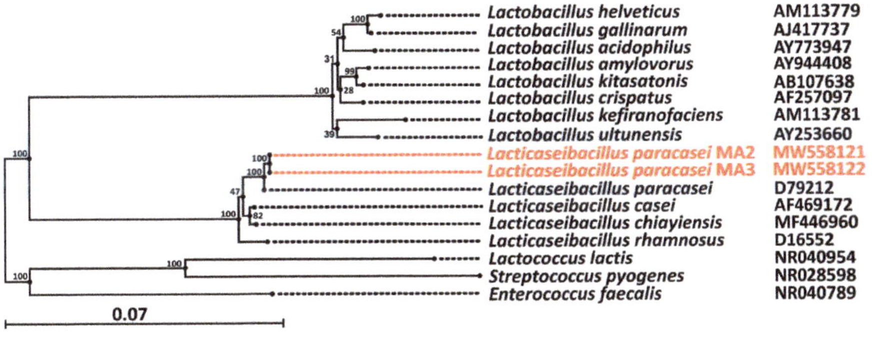

Figure 1. Phylogenetic tree constructed using 16S rRNA sequences of various *Lactobacillus* species type strains [14] and 16S rRNA sequences of strains isolated from mahewu in this work (marked in red). The evolutionary distance corresponding to one change every 100 nucleotides is shown by the scale. The number over the nodes corresponds to their bootstrap values.

3.2. Comparative Functional Characterization of the Mahewu- and Kefir-Derived Strains

3.2.1. Biochemical Characterization: Ability to Utilize Different Substrates and Profile of Enzymatic Activities

The ability of the mahewu-derived strains to utilize 49 different substrates was assessed using an API 50 test system. The obtained substrate-utilization patterns for both *L. paracasei* MA2 and *L. paracasei* MA3 were 91% identical to that typical for *L. paracasei* species, according to the APIWEB database. The comparison of the API 50 profile of the mahewu-derived strains with that of the kefir-derived strains, previously described by Begunova et al. [31], is presented on Figure 2.

As was expected for different strains of the same species, all four studied strains of *L. paracasei* were similar in their use of 39 out of 49 (i.e., 79.6%) tested substrates, 12 of which were used by all strains and 27 not by any strain. Notably, for four substrates (i.e., 8.2% of the total), the usage pattern clearly separated strains isolated from kefir and mahewu: only mahewu-derived strains were able to utilize glycerol, D-saccharose (sucrose) and potassium 2-ketogluconate, while D-melezitose was utilized only by kefir-derived strains. Three substrates (i.e., 6.1% of the total), arbutin, esculin ferric citrate and gentiobiose, were utilized by all strains except *L. paracasei* KF1. In addition, for three substrates (i.e., 6.1% of the total), strain-specific features of utilization were observed: only *L. paracasei* ABK demonstrated the ability to utilize D-melibiose and the inability to utilize D-tagatose, and only *L. paracasei* MA2 was not able to utilize D-fructose.

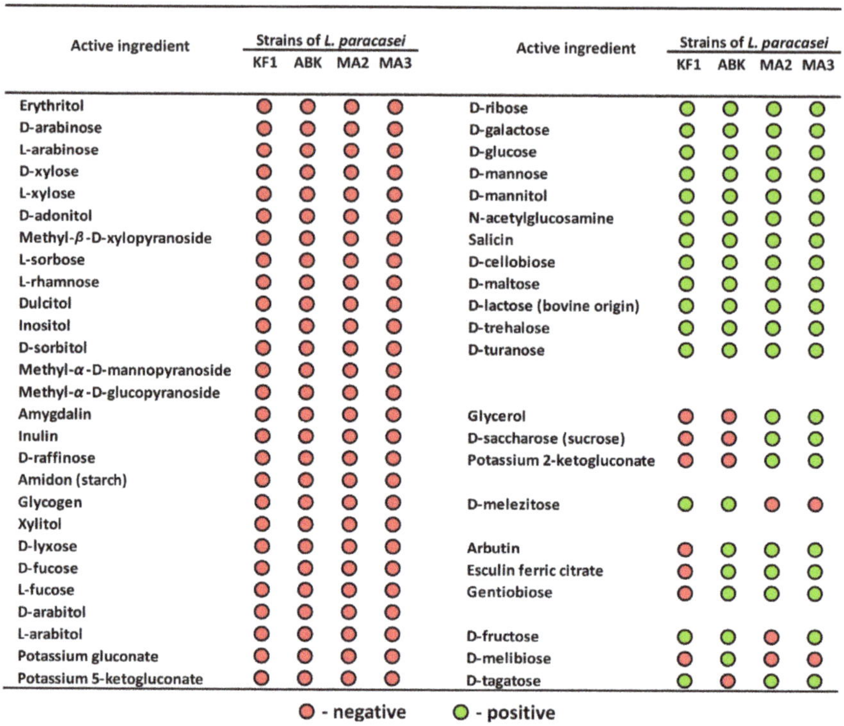

Figure 2. The API 50 substrate utilization profile of the kefir-derived (KF1 and ABK) and mahewu-derived (MA1 and MA2) strains of *L. paracasei*. The data for the kefir-derived strains were obtained from Begunova et al. [31].

The enzymatic activities of the mahewu-derived strains were assessed with a semi-quantitative API ZYM test system, which detects 19 enzymatic activities, including those of glycoside-hydrolases, proteases, phosphatases, and esterases. The comparison of the API ZYM profile of the mahewu-derived strains with that of the kefir-derived strains, previously described by Begunova et al. [31], is presented on Figure 3.

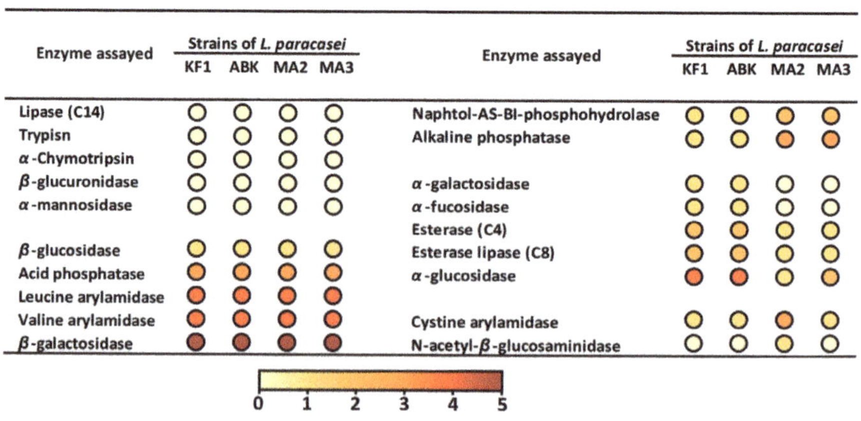

Figure 3. The API ZYM enzymatic activity profiles of kefir-derived (KF1 and ABK) and mahewu-derived (MA2 and MA3) strains of *L. paracasei*. The data for the kefir-derived strains were obtained from Begunova et al. [31].

All four studied *L. paracasei* strains demonstrated similar results for the majority of the tested enzymatic activities (10 out of 19 or 52.6%): five activities were absent for all strains, and five activities were detected at the same semi-quantitative level for all strains. For seven enzymatic activities (i.e., 36.8% of the total), the activity pattern clearly separated strains isolated from kefir and mahewu: while the mahewu-derived strains demonstrated higher activities of naphtol-AS-BI-phosphohydrolase and alkaline phosphatase, the activities of α-galactosidase, α-fucosidase, esterase (C4), esterase lipase (C8) and α-glucosidase were higher for kefir-derived strains. The *L. paracasei* MA2 demonstrated the highest cystine arylamidase activity among other strains and was the only strain with detectable N-acetyl-β-glucosaminidase activity.

It is worth comparing the results of the API 50 test with the results of the API ZYM test in terms of contrasting biochemical properties of strains derived from kefir and mahewu. In the API 50 tests, only 8.2% of all tested substrates demonstrated clear separation of these two groups of strains. Although this figure is small, it is arguably still larger than can be expected from random chance. Hence, based on the API 50 data, it was possible to hypothesize that specific differences between strains can be correlated with strains' isolation source. The data on the API ZYM test clearly substantiated this hypothesis, since 36.8% of all tested activities demonstrated the clear boundary between kefir- and mahewu-derived strains.

3.2.2. Inhibition of Pathogens and Antagonistic Interactions

To examine antagonistic activity of the studied *L. paracasei* strains against planktonic cells of the pathogenic bacteria, two-species co-cultivations were performed. The following strains of pathogenic bacteria were used: *S. typhimurium* NCTC 00074, *S. aureus* 2097 and *E. coli* B-125. The single-species cultivations of the pathogenic bacteria were used as a control. The dynamics of changes in the viable cell count of the pathogenic bacteria are shown in the Figure 4.

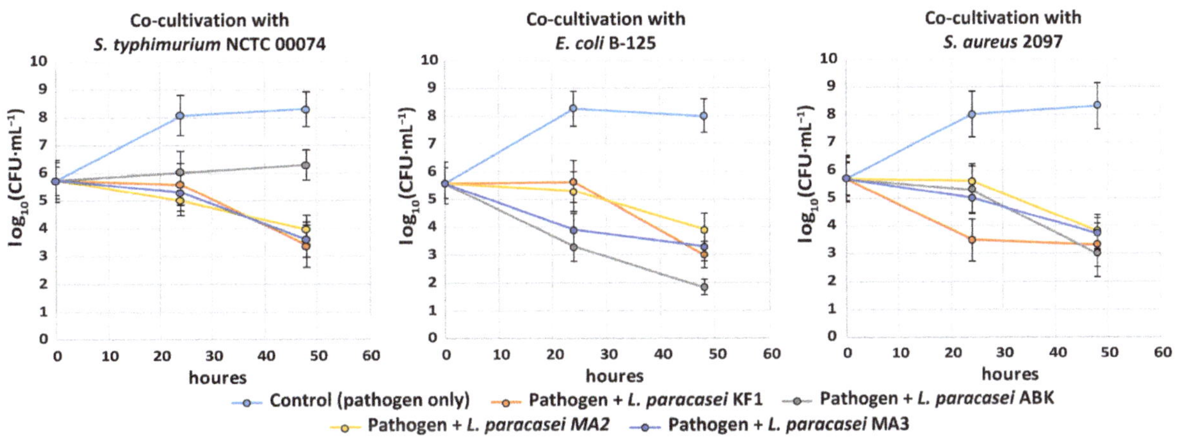

Figure 4. Changes in the viable cell count of *S. typhimurium* NCTC 00074, *E. coli* B-125 and *S. aureus* 2097 during single- and two-species cultivations with the kefir-derived (KF1 and ABK) and mahewu-derived (MA2 and MA3) strains of *L. paracasei*. The error bars represent the standard deviations from the mean.

While intensive growth of pathogenic bacterial strains was observed in their monoculture (control), in co-cultivations with the studied *L. paracasei* strains, their viable cell count constantly decreased. Generally, all the studied *L. paracasei* strains demonstrated similar abilities for inhibition of pathogens. The most prominent decrease, by three to four orders of magnitude in 48 h, was observed for *E. coli* B-125. The viable cell counts of other pathogens decreased by approximately two orders of magnitude. The most distinctive strain was *L. paracasei* ABK. Compared to other strains, this strain demonstrated the best inhibition for *E. coli* B-125, and the worst for *S. typhimurium* NCTC 00074.

The antagonistic interactions between studied strains of *L. paracasei* were tested during their solid-state co-cultivation with the perpendicular streak method (Table 1). Although strains did not demonstrate pronounced antagonism to each other, some weak antagonistic interactions were detected. Interestingly, while no antagonistic interactions were detected between the kefir-derived strains or between the mahewu-derived strains, the mahewu-derived strains were able to weakly suppress the growth of the kefir-derived strains.

Table 1. The antagonistic interactions between the studied strains of *L. paracasei*.

	Strains of *L. paracasei*		
	KF1	**ABK**	**MA2**
ABK	-		
MA2	+	+	
MA3	+	+	-

3.2.3. Growth Ability, Acidification Capability and Proteolytic Activity during Milk Fermentation

The dynamics of changes in the viable cell count are shown in Figure 5A. During the first 24 h of the fermentation process, the kefir-derived strains demonstrated lower growth rates compared to the mahewu-derived strains. After 24 h of fermentation, the kefir-derived strains continued to grow until 48 h, when they achieved the maximum viable cell count of $(2.35 \pm 1.07) \times 10^8$ CFU·mL^{-1}, and then their viable cell count decreased to $(1.10 \pm 1.06) \times 10^8$ CFU·mL^{-1} at 72 h of fermentation. In contrast, the mahewu-derived strains stopped their growth after 24 h of fermentation, when they achieved the maximum viable cell count of $(1.10 \pm 0.91) \times 10^8$ CFU·mL^{-1}, and then their viable cell count remained constant until 72 h of fermentation.

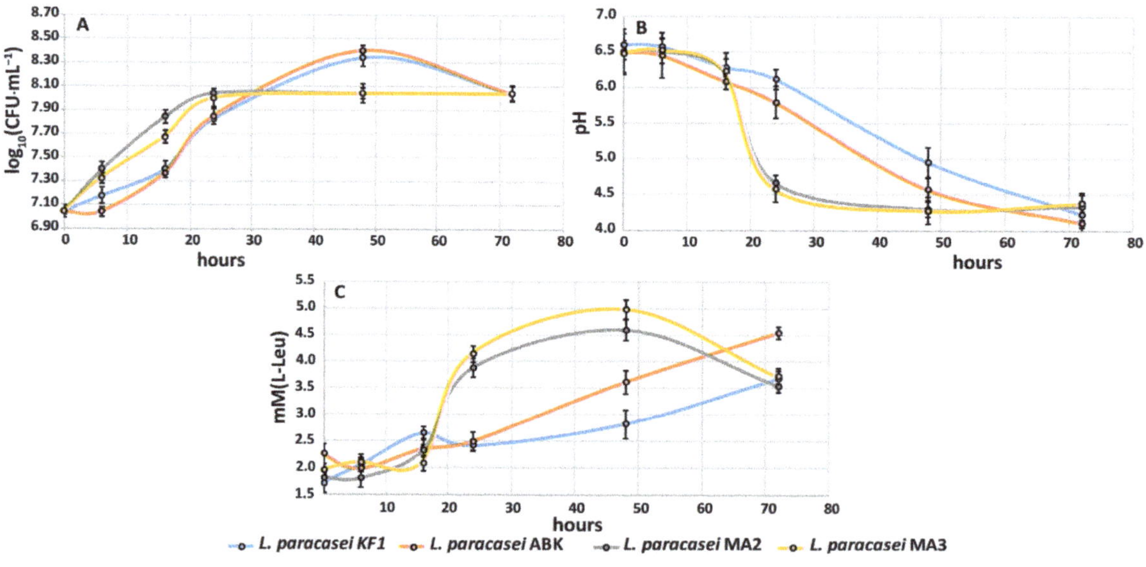

Figure 5. Fermentation characteristics of the kefir-derived (KF1 and ABK) and mahewu-derived (MA2 and MA3) strains of *L. paracasei* during milk fermentation: (**A**) the dynamics of change in the viable cell count; (**B**) the dynamics of change in the pH value; (**C**) the dynamics of change in the proteolytic activity. The error bars represent the standard deviations from the mean.

The dynamics of changes in the pH values are shown in Figure 5B. As expected, the acidification capabilities of all studied strains negatively correlated with their viable cell count. The faster the strain grew, the faster the acidification of its medium was observed to be. The mahewu-derived strains showed a pH decrease of approximately two units in a 24 h timespan and reached the final pH of 4.3 ± 0.2 at 72 h of fermentation. The kefir-derived strains gradually decreased their pH by approximately 0.04 pH units per hour, reaching the same final pH value as mahewu-derived strains at the end of the fermentation. While the mahewu-derived strains were already able to form clots after 24 h of fermentation, because of casein precipitation at pH lower than 4.6 [47], the clots in the milk fermented by kefir-derived strains were observed only after 72 h of fermentation.

The dynamics of changes of the proteolytic activity for the studied strains are shown in Figure 5C. For all strains, the proteolytic activity until 16 h of fermentation was almost the same and did not differ significantly from that at the beginning of fermentation. After 16 h of fermentation, the proteolytic activity of the mahewu-derived strains rose rapidly up to 3.88 ± 0.21 mM(L-Leu) at 24 h of fermentation and continued to increase up to 4.78 ± 0.22 mM(L-Leu) at 48 h. At 72 h of fermentation, the proteolytic activity of the mahewu-derived strains decreased down to 3.63 ± 0.15 mM(L-Leu). In contrast, the proteolytic activity of the kefir-derived strains gradually increased from 16 to 72 h of fermentation by approximately 0.04 and 0.02 mM(L-Leu) per hour, reaching the final values of 4.61 ± 0.15 and 3.68 ± 0.15 mM(L-Leu) at the end of the fermentations for *L. paracasei* KF1 and *L. paracasei* ABK, respectively.

Hence, all three studied parameters—growth ability, acidification capability and proteolytic activity—generally correlated with each other and with the origin of the strains during milk fermentation. The kefir-derived strains demonstrated slower growth, acidification capability and proteolytic activity compared to the mahewu-derived strains. Although slow growth and acidification of milk by the kefir-derived strain could present hindrances for the technological use of these strains, it could be an advantageous property from the probiotic perspective. Recently, Jung et al. [48] demonstrated that, in some cases of milk fermentation by *Lacticaseibacillus casei*, the slow-growing strains improved several of their probiotic characteristics (e.g., resistance to simulated gastrointestinal digestion and intestinal adhesion ability) after long-term fermentation.

3.2.4. Development of Antioxidant and Antihypertensive Properties during Milk Fermentation

The development of antioxidant activity during fermentation of milk by the studied strains is shown in Figure 6A. The antioxidant activity of milk fermented by the kefir-derived strains steadily increased over the entire fermentation time at approximately 2.7 and 6.9 μM(TE) per hour for *L. paracasei* KF1 *L. paracasei* ABK, respectively. For milk fermented by the mahewu-derived strains, the rapid growth of antioxidant activity up to 850 ± 23 μM(TE) was observed in the first 24 h of fermentation. From 24 to 48 h, the antioxidant activity stayed at almost the same level, after which it slightly decreased until the end of fermentation, reaching approximately 750 ± 33 μM(TE). Generally, the antioxidant activity of the fermented milk correlated with the strains' proteolytic activity. Previously, we have discussed a similar situation for *Lactobacillus helveticus*, *Lactobacillus rhamnosus* and *Lactobacillus reuteri* [49] and hypothesized that the main reason for this correlation is the not very stringent requirements that peptides must meet in order to possess reasonable antioxidant activity [50–52].

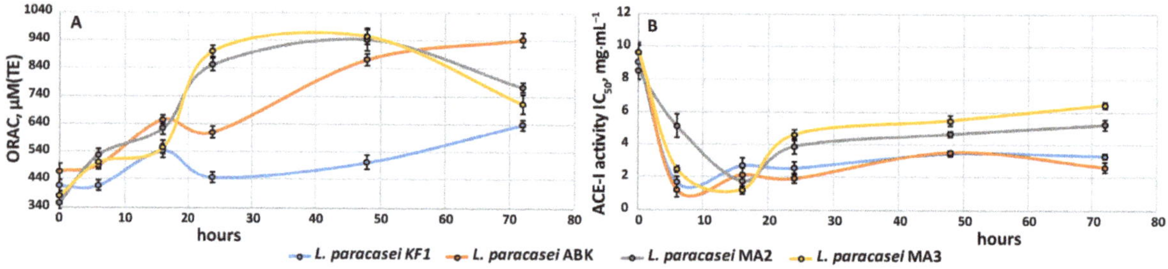

Figure 6. The development of antioxidant and ACE-I activities in the milk fermented by the kefir-derived (KF1 and ABK) and mahewu-derived (MA2 and MA3) strains of *L. paracasei*: (**A**) the development of antioxidant activity; (**B**) the development of ACE-I activity. The error bars represent the standard deviations from the mean.

The development of ACE-I activity is shown in Figure 6B. For milk fermented by all studied strains, the most active increase in ACE-I activity (i.e., decrease in the IC_{50}) was observed in the first 16 h of fermentation; at this time, the IC_{50} reached almost the same value for all strains at approximately 2.0 ± 0.4 mg·mL^{-1}. For milk fermented by the kefir-derived strains, the value of IC_{50} did not change until the end of the fermentation. For the mahewu-derived strains, the value of IC_{50} slightly increased at the end of fermentation, reaching approximately 5.5 ± 0.33 mg·mL^{-1}. The slower decrease in ACE-I activity in milk fermented by the kefir-derived strains may not only indicate the possibility of longer storage of this milk but also, once again, underline the possibility of using these strains in long-term fermentations without loss of ACE-I properties in the fermented products.

Currently, there is only one published article describing the ACE-I activity of the milk fermented by *L. paracasei* (strain L26) authored by Donkor et al. [53], in which an IC_{50} of 0.196 ± 0.008 mg·mL^{-1} was reported. Although the IC_{50} reported in our article (2.0 mg·mL^{-1}) is substantially higher, it is still in the range typical for milk fermented with *Lactobacillus* spp. [54].

3.3. *Comparative Genomic Characterization of the Mahewu- and Kefir-Derived Strains*

3.3.1. Genome Sequencing, Assembly and Annotation

Using Ion Torrent technology, the draft genomes of *L. paracasei* KF1, *L. paracasei* ABK, *L. paracasei* MA2 and *L. paracasei* MA3 were sequenced with overall coverage of 100× and ultimately assembled into 248, 246, 363 and 350 contigs, respectively (Table 2). For *L. paracasei* KF1, the N50 value was 36,612 bp, with the longest contig being 212,858 bp and the mean contig size 10,572 bp. For *L. paracasei* ABK, the N50 value was 36,610 bp, with the longest contig being 212,860 bp and the mean contig size 10,773 bp. For *L. paracasei* MA2, the N50 value was 37,017 bp, with the longest contig being 170,621 bp and the mean contig size 7622 bp. For *L. paracasei* MA3, the N50 value was 37,018 bp, with the longest contig being 170,654 bp and the mean contig size 8022 bp. The final size of the assemblies was 2.7 and 2.9 Mb for the kefir-derived (KF1 and ABK) and mahewu-derived (MA2 and MA3) strains, respectively. The Whole Genome Shotgun projects were deposited at DDBJ/ENA/GenBank under the accessions GCA_023470645, GCA_018967025, GCA_018966985 and GCA_023470655 for *L. paracasei* KF1, *L. paracasei* ABK, *L. paracasei* MA2 and *L. paracasei* MA3, respectively. The versions described in this paper are versions GCA_023470645.1, GCA_018967025.1, GCA_018966985.1 and GCA_023470655.1 for *L. paracasei* KF1, *L. paracasei* ABK, *L. paracasei* MA2 and *L. paracasei* MA3, respectively. All the sequenced genomes belong to two BioProgects—PRJNA824719 and PRJNA736961. In general, the obtained assemblies and annotations of the genomes of the two kefir-derived (KF1 and ABK) and two mahewu-derived (MA2 and MA3) strains of *L. paracasei* were of comparable quality to previously published genomes of other *L. paracasei* strains [15].

Table 2. Data on the genome sequencing of the kefir-derived (KF1 and ABK) and mahewu-derived (MA2 and MA3) strains of *L. paracasei*.

Kefir-Derived Strains							
***Lacticaseibacillus paracasei* KF1 (GB Accession: GCA_023470645.1)**				***Lacticaseibacillus paracasei* ABK (GB Accession: GCA_018967025.1)**			
Sequencing				**Sequencing**			
Sequencing technology	Ion Torrent	Number of reads	3,432,445	Sequencing technology	Ion Torrent	Number of reads	3,635,019
		Mean read size	208 bp			Mean read size	208 bp
Assembly		**Structural annotation**		**Assembly**		**Structural annotation**	
Assembly size, bp	2,697,398	Genes (total):	2791	Assembly size, Mb	2,698,106	Genes (total):	2796
Overall coverage	100×	- Protein coding	2517	Overall coverage	100×	- Protein coding	2524
Number of contigs	248	- RNA coding	78	Number of contigs	246	- RNA coding	78
Longest contig, bp	212,858	- Pseudogenes	196	Longest contig, bp	212,860	- Pseudogenes	194
N50 contig size, bp	36,612	CRISPR arrays	0	N50 contig size, bp	36,610	CRISPR arrays	0
Mean contig size, bp	10,572			Mean contig size, bp	10,773		
Mahewu-Derived Strains							
***Lacticaseibacillus paracasei* MA2 (GB Accession: GCA_018966985.1)**				***Lacticaseibacillus paracasei* MA3 (GB Accession: GCA_023470655.1)**			
Sequencing				**Sequencing**			
Sequencing technology	Ion Torrent	Number of reads	2,972,024	Sequencing technology	Ion Torrent	Number of reads	3,314,216
		Mean read size, bp	209 bp			Mean read size, bp	212 bp
Assembly		**Structural annotation**		**Assembly**		**Structural annotation**	
Assembly size, bp	2,878,977	Genes (total):	2977	Assembly size, Mb	2,870,266	Genes (total):	2965
Overall coverage	100×	- Protein coding	2651	Overall coverage	100×	- Protein coding	2650
Number of contigs	363	- RNA coding	79	Number of contigs	350	- RNA coding	79
Longest contig, bp	170,621	- Pseudogenes	247	Longest contig, bp	170,654	- Pseudogenes	236
N50 contig size, bp	37,017	CRISPR arrays	0	N50 contig size, bp	37,018	CRISPR arrays	0
Mean contig size, bp	7622			Mean contig size, bp	8022		

For all studied strains, the genome sizes and numbers of predicted CDSs were in the previously identified ranges—2.5–4 Mb and 2200–3200 CDSs—for free-living and nomadic *Lactobacillus* spp. [23]. It should be especially emphasized that the mahewu-derived strains possessed a 200 kb larger genome size (approximately 180 additional genes) compared to the kefir-derived strains. Additionally, the genomes of the mahewu-derived strains contained 50 more pseudogenes (i.e., the genes that have been silenced by one or more deleterious mutations).

Since pseudogenes can persist in bacterial genomes over a long evolutionary period, they can usually be thought of as "archaeological records" of pre-existing but now extinct proteins, enzymes or even entire biological pathways [55,56]. The accumulation of pseudogenes in the genomes of the mahewu-derived strains may be the result of relatively recent processes, such as niche change or weak selection towards corn-base substrates. In contrast, for kefir-derived strains, many of the protein-coding genes, and even the pseudogene "archaeological records" of them, may be lost due to the long stay of these strains in a stable SCOBY consortium of kefir grains.

3.3.2. Functional Annotation and Pan-Genomic Analysis

To reveal the specific genomic features of the studied *L. paracasei* strains that potentially could be associated with niche adaptations, functional annotations of the sequenced genomes and their comparative (i.e., pan-genomic) analysis were performed. For all sequenced genomes, approximately 83% of all predicted CDSs were functionally annotated by NCBI PGAP annotation pipeline [40], and approximately 76% were assigned to suitable clusters of orthologous groups of proteins (COG) by the Anvi'o [45,46] anvi-run-ncbi-cogs algorithm. Additionally, approximately 9% of all predicted CDSs were assigned to suitable KEGG pathway modules by the Anvi'o anvi-run-kegg-kofams algorithm, and the completeness of the modules (i.e., the completeness of the metabolic pathway encoded by the genes in these modules) was assessed by the Anvi'o anvi-estimate-metabolism algorithm. The results of the functional annotations are summarized in Supplementary Table S1. The comparison of the CDS content of the sequenced genomes and information about their main differences are shown in Figure 7.

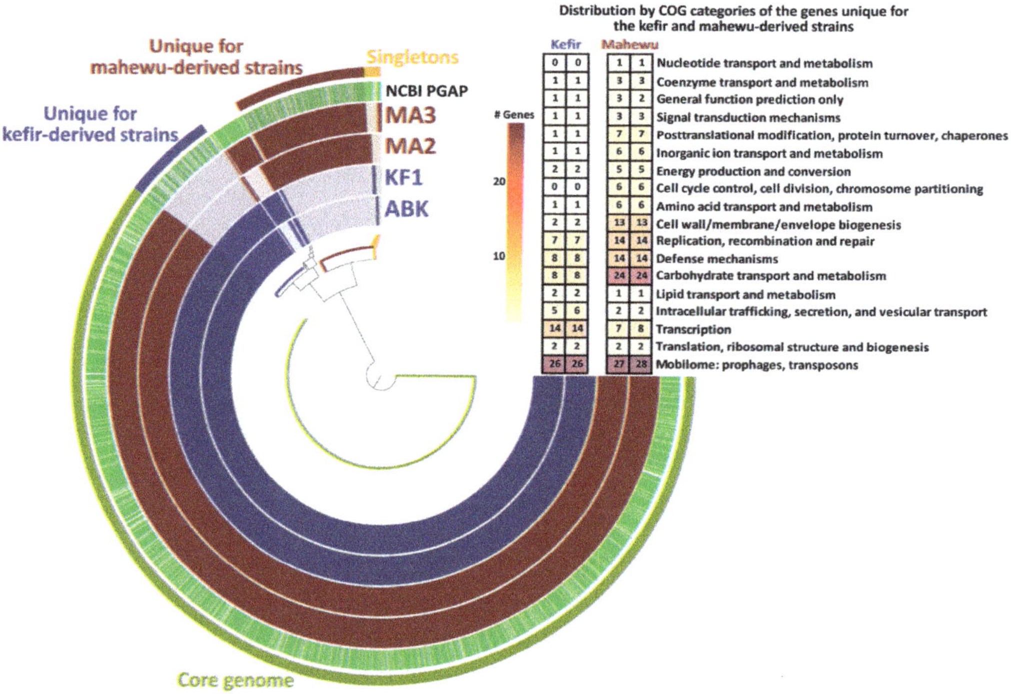

Figure 7. The Anvi'o diagram [45,46] representing the pan-genome analysis of kefir-derived (KF1 and ABK) and mahewu-derived (MA2 and MA3) strains of *L. paracasei*. Gene clusters (four inner rings) were ordered (inner dendrogram) according to their presence (solid color) or absence (grey color). The fifth ring depicts gene clusters that received functional annotation from the NCBI PGAP annotation pipeline [40], and the sixth ring depicts the binning of the pan-genome into core and accessory genomes. The later was further subdivided into gene clusters unique to either kefir- or mahewu-derived strains and strain-specific clusters (singletons). The heat map depicts the distribution by COG categories of the genes unique to either kefir- or mahewu-derived strains ("niche-specific genes"); niche-specific genes without an assigned COG category or assigned the "Unknown function" category were excluded from the heat map.

Recently, to discuss the genomes of strains belonging to the same species, the concepts of pan- and core genome have been introduced [57]. While the pan-genome is the union of sets of genes from all considered genomes, the core genome is the intersection of these sets.

All the genes from a pan-genome that do not belong to a core genome form an accessory genome. The genes from pan-, core and accessory genomes are grouped into clusters of homologous genes for further investigations [45,46].

For four sequenced *L. paracasei* strains, a total of 2592 gene clusters were identified in the pan-genome (Figure 7), and the core genome consisted of 2086 gene clusters. While 1975 gene clusters from the core genome contained only genes that presented in a single copy in each genome (i.e., single-copy orthologs), 111 clusters contained genes that presented in multiple copies in at least one of the studied genomes (i.e., paralogs). Importantly, all gene clusters from the complete KEGG pathway modules belonged to the core genome. This suggests that any strain-specific differences in the functional properties described in Sections 3.2.3 and 3.2.4 are most probably the result of differing gene regulation, enzyme activities or both.

Importantly, the major part of the accessory genome of the studied *L. paracasei* strains consisted of gene clusters specific to either kefir- or mahewu-derived strains (Figure 7). Most of these "niche-specific genes" were related to the Mobilome COG category, which is closely related to genome stability. Although limited information regarding the genome stability of different *Lactobacillus* spp. is currently available, there are several pieces of evidences suggesting that genome stability influences trait stability to some extent and, hence, can be linked with a niche adaptation [58–60]. Furthermore, many niche-specific genes in *L. paracasei* genomes were related to transcriptional regulation, which again suggests the regulatory nature of strain-specific differences in the functional properties described in Sections 3.2.3 and 3.2.4. Additionally, the multitude of niche-specific genes related to carbohydrate transport in the genomes of the mahewu-derived strains suggests the presence of regulation not only at the transcriptional level but at the level of the carbohydrate fluxes incoming into cells.

3.3.3. Genome Stability

As most of the niche-specific gene clusters belonged to the Mobilome COG category, the genome stability of the *L. paracasei* strains was compared. The information about the main markers of genome stability, the presence of mobile genetic elements, prophages and plasmids [61], in the studied genomes of *L. paracasei* strains is shown in Table 3.

In terms of mobile genetic elements, the genomes of the kefir-derived strains carried on average 50 insertion sequences (ISs), and 64 such sequences on average were annotated in the genomes of the mahewu-derived strains. Genome analysis with ISFinder showed that, in addition to the IS families IS3, IS5, IS30, IS256 and ISL3 found in the genomes of the kefir-derived strains, the genomes of both mahewu-derived strains contained ISs from the IS6 family, and the *L. paracasei* MA2 genome contained ISs from the IS1182 family. While the majority of ISs detected in the genomes of the kefir-derived strains originated from *Lactobacillus* spp. (mainly *Lactobacillus rhamnosus* and *Lactobacillus casei*), the ISs detected in the genomes of the mahewu-derived strains mainly originated from *Lactococcus lactis*. It should be noted that *L. lactis* is a common strain found in spontaneously fermented mahewu [62–67]. Considering that the starter for the mahewu used in this study most likely originated from some traditionally prepared mahewu, the presence of many ISs from *L. lactis* in the genomes of the mahewu-derived strains can be explained by the close interactions of these LAB species in the original spontaneous fermentation. Although the exact role of ISs in the evolution of bacterial genomes is still debated, their general impact on the architecture of microbial genomes is undeniable [68]. It can be hypothesized that, due to the higher number of ISs, the genomes of the mahewu-derived strains have higher genome instability (plasticity) than those of the kefir-derived strains.

Table 3. Data on the genome stability of the kefir-derived (KF1 and ABK) and mahewu-derived (MA2 and MA3) strains of *L. paracasei*.

		Strains of *L. paracasei*			
		KF1	**ABK**	**MA2**	**MA3**
	Insertion sequences				
IS Family	**Origin**	**BLAST hit**			
IS5	*Lactobacillus rhamnosus*	ISLrh2	ISLrh2	ISLrh2	ISLrh2
IS5	*Lactobacillus rhamnosus*	ISLrh3	ISLrh3	ISLrh3	ISLrh3
IS5	*Lactobacillus casei*	ISLca2	ISLca2	ISLca2	ISLca2
IS3	*Lactobacillus casei*	ISL1	ISL1	ISL1	ISL1
IS30	*Lactobacillus plantarum*	ISLpl1	ISLpl1	ISLpl1	ISLpl1
IS3	*Lactobacillus sanfranciscensis*	IS153	IS153	IS153	IS153
ISL3	*Leuconostoc mesenteroides*	IS1165	IS1165	IS1165	IS1165
IS30	*Pediococcus pentosaceus*	ISPp1	ISPp1	ISPp1	ISPp1
IS256	*Enterococcus hirae*	IS1310	IS1310	IS1310	IS1310
IS6	*Leuconostoc mesenteroides*	-	-	IS1297	IS1297
IS6	*Lactococcus lactis*	-	-	ISS1N	ISS1N
IS6	*Lactococcus lactis*	-	-	ISS1E	ISS1E
IS6	*Lactococcus lactis*	-	-	ISS1M	ISS1M
IS6	*Lactococcus lactis*	-	-	ISS1D	ISS1D
IS6	*Lactococcus lactis*	-	-	ISS1CH	ISS1CH
IS6	*Lactococcus lactis*	-	-	ISS1A	ISS1A
IS6	*Lactococcus lactis*	-	-	IS946V	IS946V
IS6	*Lactococcus lactis*	-	-	ISS1T	ISS1T
IS6	*Lactococcus lactis*	-	-	ISS1S	ISS1S
IS6	*Lactococcus lactis*	-	-	ISS1RS	ISS1RS
IS6	*Lactococcus lactis*	-	-	ISS1B	ISS1B
IS6	*Lactococcus lactis*	-	-	ISS1X	ISS1X
IS6	*Lactococcus lactis*	-	-	ISS1Z	ISS1Z
IS6	*Lactococcus garvieae*	-	-	ISLgar4	ISLgar4
IS5	*Streptococcus thermophilus*	-	-	IS1194	-
IS1182	*Streptococcus agalactiae*	-	-	ISSag8	-
IS1182	*Streptococcus agalactiae*	-	-	IS1563	-
	Prophages				
Most common phage name	**Completeness**	**Number of Total Proteins**			
PHAGE_Lactob_phijl1_NC_006936	Intact	57	59	-	-
PHAGE_Lactob_BH1_NC_048737	Questionable	29	28	-	-
PHAGE_Lactob_iLp84_NC_028783	Incomplete	18	18	-	-
PHAGE_Staphy_phiPV83_NC_002486	Incomplete	9	10	-	-
PHAGE_Staphy_SPbeta_like_NC_029119	Incomplete	22	19	23	19
PHAGE_Lactob_iLp1308_NC_028911	Incomplete	29	26	29	29
PHAGE_Lister_LP_101_NC_024387	Intact	-	-	19	19
PHAGE_Lactob_iA2_NC_028830	Intact	-	-	48	48
PHAGE_Lactob_Lc_Nu_NC_007501	Incomplete	-	-	16	16
	Plasmids				
Best BLAST hit	**Origin**	**Presence**			
pLDW-11	*Companilactobacillus alimentarius* DSM 20249	No	No	Yes	Yes

The search for prophage-containing regions showed that the genomes of all studied strains had only two incomplete prophage regions in common. The genomes of the kefir-derived strains exclusively contained two regions with incomplete prophages, one region with a questionable prophage and one region with an intact prophage. The genomes of the mahewu-derived strains exclusively contained one region with incomplete prophages and two regions with intact prophages. It should be emphasized that none of the studied strains contained CRISPR arrays in their genomes and, consequently, they were equally vulnerable to the incorporation of prophages. However, based on the lower number of intact prophages and the larger number of prophages inactivated by the accumulated mutations, it can be proposed that the kefir-derived strains encountered a lower number of recent prophage-incorporation events compared to the mahewu-derived strains.

In terms of extrachromosomal DNA, no plasmids were detected in the genomes of the kefir-derived strains, while the genomes of the mahewu-derived strains contained one 35.5 kbp plasmid. This plasmid encoded 39 genes, most of which were annotated as unknown. According to the BLAST search, the closest plasmid was plasmid pLDW-11 from *Companilactobacillus alimentarius* DSM 20249, with 96% sequence identity and 83% query coverage. Although the typical habitat (if there is one) of *C. alimentarius* is unknown [23], some of its strains were previously isolated from sourdough [69,70]. It can be proposed that the horizontal transfer of the plasmid from *C. alimentarius* to the mahewu-derived strains of *L. paracasei* occurred due to their interaction during the fermentation of corn inoculated by wheat flour, which is a typical process in mahewu fermentation [64].

3.3.4. Bacteriocin Genome Content

There are several ways by which *Lactobacillus* spp. can inhibit growth of pathogenic microorganisms and each other. While the inhibiting properties of organic acids and peroxide produced by *Lactobacillus* spp. in the process of fermentation have been known for decades [71–73], production of specific antimicrobial proteins—bacteriocins—by these microorganisms is a relatively new discovery [74,75]. Bacteriocins are ribosomal synthesized peptides with antimicrobial activity [76,77]. It is currently believed that Gram-positive bacteria—in particular, LAB—produce bacteriocins with a broader spectrum of antimicrobial activity than Gram-negative bacteria, which produce bacteriocins that inhibit only a number of specific microorganisms typically encountered in their habitat [78–80].

Genome analysis with BAGEL4 showed that all the studied strains of *L. paracasei* contained in their genomes the following identical bacteriocin clusters (Table 4): Butyrivibriocin AR10, ComC/Lactococcin/LSEI_2386, Carnocin CP52 and LSEI 2163. Additionally, the genomes of the kefir-derived strains exclusively contained the Enterolysin A bacteriocin cluster, while the genomes of the mahewu-derived strains exclusively contained the ComC/Acidocin_8912/Acidocin A bacteriocin cluster. Hence, in total, both the kefir-derived strains and the mahewu-derived strains possessed five bacteriocin clusters in their genomes.

Table 4. Data on the bacteriocin genome content of the kefir-derived (KF1 and ABK) and mahewu-derived (MA2 and MA3) strains of *L. paracasei*.

Bacteriocin-Containing Cluster	Strains of *L. paracasei*			
	KF1	ABK	MA2	MA3
Butyrivibriocin AR10	Yes	Yes	Yes	Yes
ComC/Lactococcin/LSEI_2386	Yes	Yes	Yes	Yes
Carnocin CP52	Yes	Yes	Yes	Yes
LSEI 2163	Yes	Yes	Yes	Yes
ComC/Acidocin 8912/Acidocin A	No	No	Yes	Yes
Enterolysin A	Yes	Yes	No	No

Interestingly, in the work of Ghosh et al. [81], who analyzed 75 strains of *L. paracasei*, the genomes of all studied strains contained two to three bacteriocin clusters on average, and the highest number of such clusters (five) was detected only in four strains. Almost all bacteriocin clusters detected in our strains were present in at least 40% of all the genomes studied by Ghosh et al. [81]. The exceptions were the Butyrivibriocin AR10 cluster, which was detected in the genomes of both kefir- and mahewu-derived strains, and the Acidocin 8912 cluster, which was detected only in the genome of the mahewu-derived strains. In the work of Ghosh et al. [81], the Butyrivibriocin AR10 cluster was present in the genomes of 5 out of 75 strains and Acidocin 8912 in the genomes of 4 out of 75 strains. Thus, a distinctive feature of the genomes of both kefir-derived and mahewu-derived strains was the presence of the Butyrivibriocin AR10 cluster, which is rarely observed in other *L. paracasei* strains. The additional peculiarity of the mahewu-derived strains was the absence in their genomes of the Enterolysin A bacteriocin cluster, which is relatively widespread among *L. paracasei* [81].

4. Conclusions

In this work, microbiological, biochemical and genomic analyses were utilized for comparative characterization of *L. paracasei* strains isolated from such non-standard environments as SCOBY—kefir grains (*L. paracasei* strains KF1 and ABK)—and the traditional corn-based nonalcoholic beverage of South Africa mahewu (*L. paracasei* strains MA2 and MA3). It was demonstrated that the biochemical and fermentation characteristics of the strains correlated with their isolation source. Moreover, the genomic analysis demonstrated that both kefir- and mahewu-derived strains possessed a number of gene clusters specific to strains of the same origin. The majority of these niche-specific gene clusters belonged to the Mobilome, Transcription and Carbohydrate Transport and Metabolism COG categories. It was also shown that the mahewu-derived strains possessed more flexible genome content (i.e., more pseudogenes, insertion sequences, intact prophages and plasmids) than the kefir-derived strains. It was proposed that the relative stability of the genomes of the kefir-derived strains reflects their long-term adaptation to the SCOBY environment.

From the technological perspective, all the studied strains demonstrated the ability to produce functional fermented products with antioxidant and antihypertensive properties, and the kefir-derived strains showed promising properties for their use in the recently proposed long-term fermentation processes (which has been proposed to be able to increase their resistance to gastrointestinal digestion and their intestinal adhesion ability). Additionally, all studied strains demonstrated the ability to inhibit growth of common pathogenic bacteria, which highlights their probiotic potential. Further studies of *L. paracasei* strains isolated from non-standard environments and their characterization at several levels, including metabolomic and proteomic, will not only provide a more complete picture of the transitional (nomadic) lifestyle of this LAB species but also help in discovering new strains that have potential health-promoting properties.

Supplementary Materials: The following supporting information can be downloaded at: https://www.mdpi.com/xxx/s1, Table S1: Summary of the genome annotations.

Author Contributions: Conceptualization, K.V.M., O.A.G. and T.V.F.; validation, K.V.M., O.A.G., A.V.B. and T.V.F.; formal analysis, K.V.M. and T.V.F.; investigation, A.V.B., O.S.S. and T.V.F.; writing—original draft preparation, K.V.M., O.A.G. and T.V.F.; writing—review and editing, K.V.M., O.A.G. and T.V.F.; visualization, K.V.M. and O.A.G.; supervision, I.V.R. and T.V.F.; funding acquisition, T.V.F. All authors have read and agreed to the published version of the manuscript.

Funding: This research was funded by the Russian Science Foundation, grant number 22-16-00108.

Data Availability Statement: Data is contained within the article or supplementary material.

Conflicts of Interest: The authors declare no conflict of interest. The funders had no role in the design of the study; in the collection, analyses, or interpretation of data; in the writing of the manuscript; or in the decision to publish the results.

References

1. Bintsis, T. Lactic acid bacteria: Their applications in foods. *J. Bacteriol. Mycol. Open Access* **2018**, *6*, 89–94. [CrossRef]
2. Sharma, R.; Garg, P.; Kumar, P.; Bhatia, S.K.; Kulshrestha, S. Microbial Fermentation and Its Role in Quality Improvement of Fermented Foods. *Fermentation* **2020**, *6*, 106. [CrossRef]
3. Leis, R.; de Castro, M.-J.; de Lamas, C.; Picáns, R.; Couce, M.L. Effects of Prebiotic and Probiotic Supplementation on Lactase Deficiency and Lactose Intolerance: A Systematic Review of Controlled Trials. *Nutrients* **2020**, *12*, 1487. [CrossRef]
4. Nourmohammadi, E.; Mahoonak, A.S. Health Implications of Bioactive Peptides: A Review. *Int. J. Vitam. Nutr. Res.* **2018**, *88*, 319–343. [CrossRef]
5. Ooi, L.-G.; Liong, M.-T. Cholesterol-Lowering Effects of Probiotics and Prebiotics: A Review of in Vivo and in Vitro Findings. *Int. J. Mol. Sci.* **2010**, *11*, 2499–2522. [CrossRef]

6. Adebo, O.A.; Gabriela Medina-Meza, I. Impact of Fermentation on the Phenolic Compounds and Antioxidant Activity of Whole Cereal Grains: A Mini Review. *Molecules* **2020**, *25*, 927. [CrossRef]
7. García-Burgos, M.; Moreno-Fernández, J.; Alférez, M.J.M.; Díaz-Castro, J.; López-Aliaga, I. New perspectives in fermented dairy products and their health relevance. *J. Funct. Foods* **2020**, *72*, 104059. [CrossRef]
8. Natividad, J.M.M.; Verdu, E.F. Modulation of intestinal barrier by intestinal microbiota: Pathological and therapeutic implications. *Pharmacol. Res.* **2013**, *69*, 42–51. [CrossRef]
9. Hiippala, K.; Jouhten, H.; Ronkainen, A.; Hartikainen, A.; Kainulainen, V.; Jalanka, J.; Satokari, R. The Potential of Gut Commensals in Reinforcing Intestinal Barrier Function and Alleviating Inflammation. *Nutrients* **2018**, *10*, 988. [CrossRef]
10. Boricha, A.A.; Shekh, S.L.; Pithva, S.P.; Ambalam, P.S.; Manuel Vyas, B.R. In vitro evaluation of probiotic properties of Lactobacillus species of food and human origin. *LWT* **2019**, *106*, 201–208. [CrossRef]
11. Oleskin, A.V.; Shenderov, B.A. Probiotics and Psychobiotics: The Role of Microbial Neurochemicals. *Probiotics Antimicrob. Proteins* **2019**, *11*, 1071–1085. [CrossRef]
12. Servin, A.L. Antagonistic activities of lactobacilli and bifidobacteria against microbial pathogens. *FEMS Microbiol. Rev.* **2004**, *28*, 405–440. [CrossRef]
13. Maldonado Galdeano, C.; Cazorla, S.I.; Lemme Dumit, J.M.; Vélez, E.; Perdigón, G. Beneficial effects of probiotic consumption on the immune system. *Ann. Nutr. Metab.* **2019**, *74*, 115–124. [CrossRef]
14. Zheng, J.; Wittouck, S.; Salvetti, E.; Franz, C.M.A.P.; Harris, H.M.B.; Mattarelli, P.; O'Toole, P.W.; Pot, B.; Vandamme, P.; Walter, J.; et al. A taxonomic note on the genus Lactobacillus: Description of 23 novel genera, emended description of the genus Lactobacillus Beijerinck 1901, and union of Lactobacillaceae and Leuconostocaceae. *Int. J. Syst. Evol. Microbiol.* **2020**, *70*, 2782–2858. [CrossRef]
15. Agarwala, R.; Barrett, T.; Beck, J.; Benson, D.A.; Bollin, C.; Bolton, E.; Bourexis, D.; Brister, J.R.; Bryant, S.H.; Canese, K.; et al. Database resources of the National Center for Biotechnology Information. *Nucleic Acids Res.* **2018**, *46*, D8–D13. [CrossRef]
16. Sanders, M.E.; Benson, A.; Lebeer, S.; Merenstein, D.J.; Klaenhammer, T.R. Shared mechanisms among probiotic taxa: Implications for general probiotic claims. *Curr. Opin. Biotechnol.* **2018**, *49*, 207–216. [CrossRef]
17. MARSHALL, V.M. Starter cultures for milk fermentation and their characteristics. *Int. J. Dairy Technol.* **1993**, *46*, 49–56. [CrossRef]
18. Bintsis, T. Lactic acid bacteria as starter cultures: An update in their metabolism and genetics. *AIMS Microbiol.* **2018**, *4*, 665–684. [CrossRef]
19. Marco, M.L.; Sanders, M.E.; Gänzle, M.; Arrieta, M.C.; Cotter, P.D.; De Vuyst, L.; Hill, C.; Holzapfel, W.; Lebeer, S.; Merenstein, D.; et al. The International Scientific Association for Probiotics and Prebiotics (ISAPP) consensus statement on fermented foods. *Nat. Rev. Gastroenterol. Hepatol.* **2021**, *18*, 196–208. [CrossRef]
20. Jones, R.M. The Use of Lactobacillus casei and Lactobacillus paracasei in Clinical Trials for the Improvement of Human Health. In *The Microbiota in Gastrointestinal Pathophysiology*; Elsevier: Amsterdam, The Netherlands, 2017; pp. 99–108.
21. Hill, D.; Sugrue, I.; Tobin, C.; Hill, C.; Stanton, C.; Ross, R.P. The Lactobacillus casei Group: History and Health Related Applications. *Front. Microbiol.* **2018**, *9*, 2107. [CrossRef]
22. Huang, C.-H.; Li, S.-W.; Huang, L.; Watanabe, K. Identification and Classification for the Lactobacillus casei Group. *Front. Microbiol.* **2018**, *9*, 1974. [CrossRef]
23. Duar, R.M.; Lin, X.B.; Zheng, J.; Martino, M.E.; Grenier, T.; Pérez-Muñoz, M.E.; Leulier, F.; Gänzle, M.; Walter, J. Lifestyles in transition: Evolution and natural history of the genus Lactobacillus. *FEMS Microbiol. Rev.* **2017**, *41*, S27–S48. [CrossRef]
24. Martino, M.E.; Bayjanov, J.R.; Caffrey, B.E.; Wels, M.; Joncour, P.; Hughes, S.; Gillet, B.; Kleerebezem, M.; van Hijum, S.A.F.T.; Leulier, F. Nomadic lifestyle of Lactobacillus plantarum revealed by comparative genomics of 54 strains isolated from different habitats. *Environ. Microbiol.* **2016**, *18*, 4974–4989. [CrossRef]
25. Kiousi, D.E.; Efstathiou, C.; Tegopoulos, K.; Mantzourani, I.; Alexopoulos, A.; Plessas, S.; Kolovos, P.; Koffa, M.; Galanis, A. Genomic Insight Into Lacticaseibacillus paracasei SP5, Reveals Genes and Gene Clusters of Probiotic Interest and Biotechnological Potential. *Front. Microbiol.* **2022**, *13*, 922689. [CrossRef]
26. Smokvina, T.; Wels, M.; Polka, J.; Chervaux, C.; Brisse, S.; Boekhorst, J.; van Hylckama Vlieg, J.E.T.; Siezen, R.J. Lactobacillus paracasei Comparative Genomics: Towards Species Pan-Genome Definition and Exploitation of Diversity. *PLoS ONE* **2013**, *8*, e68731. [CrossRef]
27. Zhang, Z.-G.; Ye, Z.-Q.; Yu, L.; Shi, P. Phylogenomic reconstruction of lactic acid bacteria: An update. *BMC Evol. Biol.* **2011**, *11*, 1. [CrossRef]
28. Fidanza, M.; Panigrahi, P.; Kollmann, T.R. Lactiplantibacillus plantarum–Nomad and Ideal Probiotic. *Front. Microbiol.* **2021**, *12*, 2911. [CrossRef]
29. Torres-Miranda, A.; Melis-Arcos, F.; Garrido, D. Characterization and Identification of Probiotic Features in Lacticaseibacillus Paracasei Using a Comparative Genomic Analysis Approach. *Probiotics Antimicrob. Proteins* **2022**, *14*, 1211–1224. [CrossRef]
30. Golicz, A.A.; Bayer, P.E.; Bhalla, P.L.; Batley, J.; Edwards, D. Pangenomics Comes of Age: From Bacteria to Plant and Animal Applications. *Trends Genet.* **2020**, *36*, 132–145. [CrossRef]
31. Begunova, A.V.; Savinova, O.S.; Moiseenko, K.V.; Glazunova, O.A.; Rozhkova, I.V.; Fedorova, T.V. Characterization and Functional Properties of Lactobacilli Isolated from Kefir Grains. *Appl. Biochem. Microbiol.* **2021**, *57*, 458–467. [CrossRef]

32. Moiseenko, K.V.; Glazunova, O.A.; Savinova, O.S.; Ajibade, B.O.; Ijabadeniyi, O.A.; Fedorova, T.V. Analytical characterization of the widely consumed commercialized fermented beverages from russia (Kefir and ryazhenka) and south africa (amasi and mahewu): Potential functional properties and profiles of volatile organic compounds. *Foods* **2021**, *10*, 3082. [CrossRef]
33. Dasen, G.; Smutny, J.; Teuber, M.; Meile, L. Classification and Identification of Propionibacteria Based on Ribosomal RNA Genes and PCR. *Syst. Appl. Microbiol.* **1998**, *21*, 251–259. [CrossRef]
34. Stamatakis, A. RAxML version 8: A tool for phylogenetic analysis and post-analysis of large phylogenies. *Bioinformatics* **2014**, *30*, 1312–1313. [CrossRef]
35. Miller, M.A.; Pfeiffer, W.; Schwartz, T. Creating the CIPRES Science Gateway for inference of large phylogenetic trees. In Proceedings of the 2010 Gateway Computing Environments Workshop (GCE), New Orleans, LA, USA, 14 November 2010; IEEE: Piscataway, NJ, USA, 2010; pp. 1–8.
36. Begunova, A.V.; Rozhkova, I.V.; Glazunova, O.A.; Moiseenko, K.V.; Savinova, O.S.; Fedorova, T.V. Fermentation Profile and Probiotic-Related Characteristics of Bifidobacterium longum MC-42. *Fermentation* **2021**, *7*, 101. [CrossRef]
37. Savinova, O.S.; Glazunova, O.A.; Moiseenko, K.V.; Begunova, A.V.; Rozhkova, I.V.; Fedorova, T.V. Exoproteome Analysis of Antagonistic Interactions between the Probiotic Bacteria Limosilactobacillus reuteri LR1 and Lacticaseibacillus rhamnosus F and Multidrug Resistant Strain of Klebsiella pneumonia. *Int. J. Mol. Sci.* **2021**, *22*, 10999. [CrossRef]
38. Adler-Nissen, J. Determination of the degree of hydrolysis of food protein hydrolysates by trinitrobenzenesulfonic acid. *J. Agric. Food Chem.* **1979**, *27*, 1256–1262. [CrossRef]
39. Torkova, A.A.; Ryazantseva, K.A.; Agarkova, E.Y.; Kruchinin, A.G.; Tsentalovich, M.Y.; Fedorova, T.V. Rational design of enzyme compositions for the production of functional hydrolysates of cow milk whey proteins. *Appl. Biochem. Microbiol.* **2017**, *53*, 669–679. [CrossRef]
40. Tatusova, T.; DiCuccio, M.; Badretdin, A.; Chetvernin, V.; Nawrocki, E.P.; Zaslavsky, L.; Lomsadze, A.; Pruitt, K.D.; Borodovsky, M.; Ostell, J. NCBI prokaryotic genome annotation pipeline. *Nucleic Acids Res.* **2016**, *44*, 6614–6624. [CrossRef]
41. Siguier, P. ISfinder: The reference centre for bacterial insertion sequences. *Nucleic Acids Res.* **2006**, *34*, D32–D36. [CrossRef]
42. Arndt, D.; Grant, J.R.; Marcu, A.; Sajed, T.; Pon, A.; Liang, Y.; Wishart, D.S. PHASTER: A better, faster version of the PHAST phage search tool. *Nucleic Acids Res.* **2016**, *44*, W16–W21. [CrossRef] [PubMed]
43. Carattoli, A.; Zankari, E.; García-Fernández, A.; Voldby Larsen, M.; Lund, O.; Villa, L.; Møller Aarestrup, F.; Hasman, H. In Silico Detection and Typing of Plasmids using PlasmidFinder and Plasmid Multilocus Sequence Typing. *Antimicrob. Agents Chemother.* **2014**, *58*, 3895–3903. [CrossRef] [PubMed]
44. van Heel, A.J.; de Jong, A.; Song, C.; Viel, J.H.; Kok, J.; Kuipers, O.P. BAGEL4: A user-friendly web server to thoroughly mine RiPPs and bacteriocins. *Nucleic Acids Res.* **2018**, *46*, W278–W281. [CrossRef] [PubMed]
45. Eren, A.M.; Kiefl, E.; Shaiber, A.; Veseli, I.; Miller, S.E.; Schechter, M.S.; Fink, I.; Pan, J.N.; Yousef, M.; Fogarty, E.C.; et al. Community-led, integrated, reproducible multi-omics with anvi'o. *Nat. Microbiol.* **2020**, *6*, 3–6. [CrossRef] [PubMed]
46. Delmont, T.O.; Eren, A.M. Linking pangenomes and metagenomes: The Prochlorococcus metapangenome. *PeerJ* **2018**, *6*, e4320. [CrossRef] [PubMed]
47. Raak, N.; Rohm, H.; Jaros, D. Enzymatic Cross-Linking of Casein Facilitates Gel Structure Weakening Induced by Overacidification. *Food Biophys.* **2017**, *12*, 261–268. [CrossRef]
48. Jung, S.H.; Hong, D.K.; Bang, S.-J.; Heo, K.; Sim, J.-J.; Lee, J.-L. The Functional Properties of Lactobacillus casei HY2782 Are Affected by the Fermentation Time. *Appl. Sci.* **2021**, *11*, 2481. [CrossRef]
49. Begunova, A.V.; Savinova, O.S.; Glazunova, O.A.; Moiseenko, K.V.; Rozhkova, I.V.; Fedorova, T.V. Development of Antioxidant and Antihypertensive Properties during Growth of Lactobacillus helveticus, Lactobacillus rhamnosus and Lactobacillus reuteri on Cow's Milk: Fermentation and Peptidomics Study. *Foods* **2020**, *10*, 17. [CrossRef]
50. Sarmadi, B.H.; Ismail, A. Antioxidative peptides from food proteins: A review. *Peptides* **2010**, *31*, 1949–1956. [CrossRef]
51. Pihlanto, A. Antioxidative peptides derived from milk proteins. *Int. Dairy J.* **2006**, *16*, 1306–1314. [CrossRef]
52. Udenigwe, C.C.; Aluko, R.E. Food Protein-Derived Bioactive Peptides: Production, Processing, and Potential Health Benefits. *J. Food Sci.* **2012**, *77*, R11–R24. [CrossRef]
53. Donkor, O.N.; Henriksson, A.; Vasiljevic, T.; Shah, N.P. Proteolytic activity of dairy lactic acid bacteria and probiotics as determinant of growth and in vitro angiotensin-converting enzyme inhibitory activity in fermented milk. *Lait* **2007**, *87*, 21–38. [CrossRef]
54. Beltrán-Barrientos, L.M.; Hernández-Mendoza, A.; Torres-Llanez, M.J.; González-Córdova, A.F.; Vallejo-Córdoba, B. Invited review: Fermented milk as antihypertensive functional food. *J. Dairy Sci.* **2016**, *99*, 4099–4110. [CrossRef] [PubMed]
55. Goodhead, I.; Darby, A.C. Taking the pseudo out of pseudogenes. *Curr. Opin. Microbiol.* **2015**, *23*, 102–109. [CrossRef]
56. Mira, A.; Pushker, R. The Silencing of Pseudogenes. *Mol. Biol. Evol.* **2005**, *22*, 2135–2138. [CrossRef]
57. Brockhurst, M.A.; Harrison, E.; Hall, J.P.J.; Richards, T.; McNally, A.; MacLean, C. The Ecology and Evolution of Pangenomes. *Curr. Biol.* **2019**, *29*, R1094–R1103. [CrossRef]
58. Song, Y.; He, Q.; Zhang, J.; Qiao, J.; Xu, H.; Zhong, Z.; Zhang, W.; Sun, Z.; Yang, R.; Cui, Y.; et al. Genomic Variations in Probiotic Lactobacillus plantarum P-8 in the Human and Rat Gut. *Front. Microbiol.* **2018**, *9*, 893. [CrossRef] [PubMed]
59. Sybesma, W.; Molenaar, D.; van IJcken, W.; Venema, K.; Kort, R. Genome Instability in Lactobacillus rhamnosus GG. *Appl. Environ. Microbiol.* **2013**, *79*, 2233–2239. [CrossRef] [PubMed]

60. Feng, C.; Zhang, F.; Wang, B.; Zhang, L.; Dong, Y.; Shao, Y. Genome-wide analysis of fermentation and probiotic trait stability in Lactobacillus plantarum during continuous culture. *J. Dairy Sci.* **2020**, *103*, 117–127. [CrossRef] [PubMed]
61. Darmon, E.; Leach, D.R.F. Bacterial Genome Instability. *Microbiol. Mol. Biol. Rev.* **2014**, *78*, 1–39. [CrossRef] [PubMed]
62. Idowu, O.; Fadahunsi, I.F.; Onabiyi, O.A. Production and Nutritional Evaluation of Mahewu: A Non-Alcoholic Fermented Beaverage of South Africa. *Int. J. Res. Pharm. Biosci.* **2016**, *3*, 27.
63. Fadahunsi, I.; Soremekun, O. Production, Nutritional and Microbiological Evaluation of Mahewu a South African Traditional Fermented Porridge. *J. Adv. Biol. Biotechnol.* **2017**, *14*, 1–10. [CrossRef]
64. Holzapfel, W.; Leonie Taljaard, J. Industrialization of Mageu Fermentation in South Africa. In *Industrialization of Indigenous Fermented Foods*; Steinkraus, K., Ed.; CRC Press: Boca Raton, FL, USA, 2004.
65. Odunfa, S.A.; Oyewole, O.B. African fermented foods. In *Microbiology of Fermented Foods*; Springer: Boston, MA, USA, 1998; pp. 713–752.
66. Simango, C. Lactic acid fermentation of sour porridge and mahewu, a non-alcoholic fermented cereal beverage. *JASSA J. Appl. Sci. S. Afr.* **2005**, *8*, 89–98. [CrossRef]
67. Simatende, P.; Siwela, M.; Gadaga, T.H. Identification of lactic acid bacteria and determination of selected biochemical properties in emasi and emahewu. *S. Afr. J. Sci.* **2019**, *115*, 2–8. [CrossRef] [PubMed]
68. Vandecraen, J.; Chandler, M.; Aertsen, A.; Van Houdt, R. The impact of insertion sequences on bacterial genome plasticity and adaptability. *Crit. Rev. Microbiol.* **2017**, *43*, 709–730. [CrossRef] [PubMed]
69. Gänzle, M.G.; Loponen, J.; Gobbetti, M. Proteolysis in sourdough fermentations: Mechanisms and potential for improved bread quality. *Trends Food Sci. Technol.* **2008**, *19*, 513–521. [CrossRef]
70. Papadimitriou, K.; Zoumpopoulou, G.; Georgalaki, M.; Alexandraki, V.; Kazou, M.; Anastasiou, R.; Tsakalidou, E. Sourdough Bread. In *Innovations in Traditional Foods*; Elsevier: Amsterdam, The Netherlands, 2019; pp. 127–158.
71. Sauer, M.; Russmayer, H.; Grabherr, R.; Peterbauer, C.K.; Marx, H. The Efficient Clade: Lactic Acid Bacteria for Industrial Chemical Production. *Trends Biotechnol.* **2017**, *35*, 756–769. [CrossRef]
72. Ricke, S. Perspectives on the use of organic acids and short chain fatty acids as antimicrobials. *Poult. Sci.* **2003**, *82*, 632–639. [CrossRef]
73. Carr, F.J.; Chill, D.; Maida, N. The Lactic Acid Bacteria: A Literature Survey. *Crit. Rev. Microbiol.* **2002**, *28*, 281–370. [CrossRef]
74. Almeida-Santos, A.C.; Novais, C.; Peixe, L.; Freitas, A.R. Enterococcus spp. as a Producer and Target of Bacteriocins: A Double-Edged Sword in the Antimicrobial Resistance Crisis Context. *Antibiotics* **2021**, *10*, 1215. [CrossRef]
75. Browne, K.; Chakraborty, S.; Chen, R.; Willcox, M.D.; Black, D.S.; Walsh, W.R.; Kumar, N. A New Era of Antibiotics: The Clinical Potential of Antimicrobial Peptides. *Int. J. Mol. Sci.* **2020**, *21*, 7047. [CrossRef]
76. Todorov, S.D.; Popov, I.; Weeks, R.; Chikindas, M.L. Use of Bacteriocins and Bacteriocinogenic Beneficial Organisms in Food Products: Benefits, Challenges, Concerns. *Foods* **2022**, *11*, 3145. [CrossRef] [PubMed]
77. Zimina, M.; Babich, O.; Prosekov, A.; Sukhikh, S.; Ivanova, S.; Shevchenko, M.; Noskova, S. Overview of Global Trends in Classification, Methods of Preparation and Application of Bacteriocins. *Antibiotics* **2020**, *9*, 553. [CrossRef] [PubMed]
78. Perez, R.H.; Perez, M.T.M.; Elegado, F.B. Bacteriocins from Lactic Acid Bacteria: A Review of Biosynthesis, Mode of Action, Fermentative Production, Uses, and Prospects. *Int. J. Philipp. Sci. Technol.* **2015**, *8*, 61–67. [CrossRef]
79. Veskovic-Moracanin, S.; Djukic, D.; Memisi, N. Bacteriocins produced by lactic acid bacteria: A review. *Acta Period. Technol.* **2014**, *283*, 271–283. [CrossRef]
80. Cotter, P.D.; Ross, R.P.; Hill, C. Bacteriocins—A viable alternative to antibiotics ? *Nat. Rev. Microbiol.* **2012**, *11*, 95–105. [CrossRef] [PubMed]
81. Ghosh, S.; Sarangi, A.N.; Mukherjee, M.; Bhowmick, S.; Tripathy, S. Reanalysis of Lactobacillus paracasei Lbs2 Strain and Large-Scale Comparative Genomics Places Many Strains into Their Correct Taxonomic Position. *Microorganisms* **2019**, *7*, 487. [CrossRef]

foods

MDPI

Article

Effect of Probiotic Lactic Acid Bacteria (LAB) on the Quality and Safety of Greek Yogurt

So-Young Yang and Ki-Sun Yoon *

Department of Food and Nutrition, College of Human Ecology, Kyung Hee University, 26 Kyungheedae-ro, Dongdaemun-gu, Seoul 02447, Republic of Korea
* Correspondence: ksyoon@khu.ac.kr; Tel.: +82-2-961-0264

Abstract: Greek yogurt is a strained yogurt with a high protein content that brings nutritional benefits. To enhance the functional benefits of Greek yogurt, Greek yogurt was prepared with various combinations of probiotic lactic acid bacteria (LAB) (*Streptococcus thermophilus*, *Lactobacillus bulgaricus*, *Lactobacillus gasseri* BNR17, and *Lactobacillus plantarum* HY7714). Effects of probiotic LAB on quality, sensory, and microbiological characteristics of Greek yogurt were then compared. Among samples, Greek yogurt fermented by *S. thermophilus* and *L. bulgaricus* showed the highest changes of pH and titratable acidity during 21 d of storage at 4 °C. Greek yogurt fermented with *L. plantarum* HY7714 had a higher viscosity than other samples. Greek yogurt fermented with *S. thermophilus*, *L. bulgaricus*, *L. gasseri* BNR17, and *L. plantarum* HY7714 showed superior physicochemical properties and received the highest preference score from sensory evaluation among samples. Overall, the population of enterohaemorrhagic *Escherichia coli* (EHEC) was more effectively reduced in Greek yogurt fermented with probiotic LAB than in commercial Greek yogurt during storage at 4, 10, and 25 °C. Thus, the addition of *L. gasseri* BNR17 and *L. plantarum* HY7714 as starter cultures could enhance the microbial safety of Greek yogurt and sensory acceptance by consumers.

Keywords: Greek yogurt; *Lactobacillus gasseri* BNR17; *Lactobacillus plantarum* HY7714; enterohaemorrhagic *E. coli*; preference test

Citation: Yang, S.-Y.; Yoon, K.-S. Effect of Probiotic Lactic Acid Bacteria (LAB) on the Quality and Safety of Greek Yogurt. *Foods* **2022**, *11*, 3799. https://doi.org/10.3390/foods11233799

Academic Editor: Xiqing Yue

Received: 20 October 2022
Accepted: 21 November 2022
Published: 25 November 2022

1. Introduction

Greek yogurt is known as a healthy snack that can increase lean muscle mass and decrease body fat [1]. The manufacture of Greek yogurt begins with homogenization of standardized milk. The homogenized milk is then pasteurized and cooled at an incubation temperature up to 40 °C. After a starter culture is inoculated, the yogurt gets a thicker texture through a concentration step [1,2]. This "concentration" step can increase protein content to be around 9–10% and give Greek yogurt a hard texture [3]. Protein in Greek yogurt makes the perception of hunger lower and the time between meals longer [4].

Commercial yogurt is prepared with probiotics to help intestinal function, stabilize gut microflora, and change compositions and numbers of intestinal microflora [5]. *Lactobacillus gasseri* BNR17 originally isolated from human breast milk is known to inhibit weight gain [6]. *L. gasseri* BNR17 can reduce the amount of food intake and 2 h postprandial blood glucose [7]. *L. gasseri* BNR17 can also reduce levels of leptin and insulin [8], waist circumferences, hip circumferences, and visceral adipose tissues [9]. *Lactobacillus plantarum* HY7714 isolated from healthy infant feces is a registered probiotic that can prevent photoaging, restore procollagen, and increase the retention of water in the face and hand [10–13]. Recently, Lee et al. [14] have reported that *L. plantarum* HY7714 can produce exopolysaccharides, which can control tight junctions in intestinal epithelial cells and recover cytotoxicity and hydration capacity in Hs68 cells induced by UVB irradiation. Although *L. gasseri* BNR17 and *L. plantarum* HY7714 are used to manufacture functional capsules or powder, neither *L. gasseri* BNR17 nor *L. plantarum* HY7714 has been applied in the manufacturing of healthy foods such as yogurt.

Enterohaemorrhagic *Escherichia coli* (EHEC) has strong acid resistance and maintains viability at low pH [15]. Acid resistance of EHEC has been clarified more due to the outbreak of EHEC in yogurt, where the risk of foodborne pathogens is very low [16]. EHEC can survive well in yogurt at 4 and 10 °C for 21 d of storage [17]. Recently, the antimicrobial effect of lactic acid bacteria (LAB) on *E. coli* has been reported [18,19]. LAB shows antagonistic activities, expressing higher inhibition effects on *E. coli* and *E. coli* O157:H7 than *Salmonella* typhimurium, *S. Enteritidis*, and *Listeria monocytogenes* [18]. Choi et al. [19] have reported that *Leuconostoc mesenteroides* and *L. plantarum* can inhibit the viability of *E. coli* O157:H7 in kimchi. Although consumers' preference for Greek yogurt is increasing, *L. gasseri* BNR17 and *L. plantarum* HY7714 have not been tested as starter cultures for Greek yogurt manufacturing. Their effects as antimicrobial agents in various functional foods remain unclear.

Therefore, the objective of this study is to manufacture new functional Greek yogurt with *L. gasseri* BNR17 and *L. plantarum* HY7714 known to have various health benefits. How each LAB affected the viscosity, consumer preference, and microbiological safety of Greek yogurt at 4, 10, and 25 °C is also investigated.

2. Materials and Methods

2.1. Starter Culture for Preparation of Greek Yogurt

For Greek yogurt production, probiotic LAB including *Streptococcus thermophilus* (ST, KCTC 3779), *Lactobacillus bulgaricus* (LB, KCTC 3635), *Lactobacillus gasseri* BNR17(KCTC 10902BP), and *Lactobacillus plantarum* HY7714 (KCTC 12400BP) were purchased from Korean Collection for Type Cultures (KCTC). All strains were maintained at −80 °C in De Man, Rogosa, and Sharpe broth (MRS, Difco™, Difco Laboratories, Detroit, MI, USA) with 20% glycerol. Thawed ST was aerobically incubated at 42 °C overnight in an incubator. LB was anaerobically incubated overnight at 36 °C using an anaerobic jar system (Don Whitley Scientific Ltd., Bradford, UK). *L. gasseri* BNR17 and *L. plantarum* HY7714 were aerobically incubated at 36 °C with shaking at 140 rpm in a rotary shaker (VS-8480SR, Vision). *L. gasseri* BNR17 and *L. plantarum* HY7714 were centrifugated at 4500× *g* for 15 min (VS-550, Vision), washed with sterile phosphate-buffered saline (PBS) twice, and resuspended to a final concentration of 10^9 CFU/mL according to Kim et al. [11] with some modifications.

2.2. Preparation of Enterohaemorrhagic E. coli for the Safety Study

Enterohaemorrhagic *E. coli* (EHEC) strains (NCCP 13720, 13721), including *E. coli* O157:H7 (NCTC 12079), were obtained from the Ministry of Food and Drug Safety (MFDS) to investigate the effect of LAB probiotics on the behavior of EHEC in Greek yogurt at various temperatures. Frozen strains were maintained at −80 °C with 20% glycerol in tryptic soy broth (TSB, MB cell, Seoul, Republic of Korea). Then 10 μL of thawed EHEC was inoculated into 10 mL of TSB and incubated at 36 °C in a rotary shaker at 140 rpm overnight. After centrifuging at 4000× *g* for 10 min, the supernatant was removed, and the pellet was washed with 10 mL of 0.1% sterilized peptone water (Difco™ Peptone water, Difco Laboratories). A cocktail of EHEC strains was prepared by resuspending them with 0.1% sterilized peptone water. Then 1 mL aliquot of EHEC was diluted with 0.1% sterilized peptone water for inoculum.

2.3. Manufacturing of Greek Yogurt

Pasteurized whole milk by high-temperature short-time (HTST) was purchased from a local market (Seoul, Republic of Korea) and heated in glass bottles at 42 °C in a water bath (SB-1200, EYELA Co., Ltd., Shanghai, China). Four different Greek yogurts with various combinations of starter culture (5% wt/wt) were prepared as follows: ST and LB as traditional yogurt strains (T1, control), ST, LB, and *L. gasseri* BNR17 (T2), ST, LB, and *L. plantarum* HY7714 (T3), and ST, LB, *L. gasseri* BNR17, and *L. plantarum* HY7714 (T4). All strains were mixed in equal proportions. Fermentation of Greek yogurt was carried out at 42 °C until pH was 4.4. Samples were then distributed into 250 mL polypropylene centrifuge bottles

(Beckman Coulter Avanti™, Indianapolis, IN, USA) and centrifuged at 4500× g for 15 min. After Greek yogurt (60 g) was transferred to sterilized plastic cups, quality evaluation was conducted every three days for 21 days of storage at 4 °C. A flow chart of the manufacturing step for Greek yogurt is shown in Figure 1.

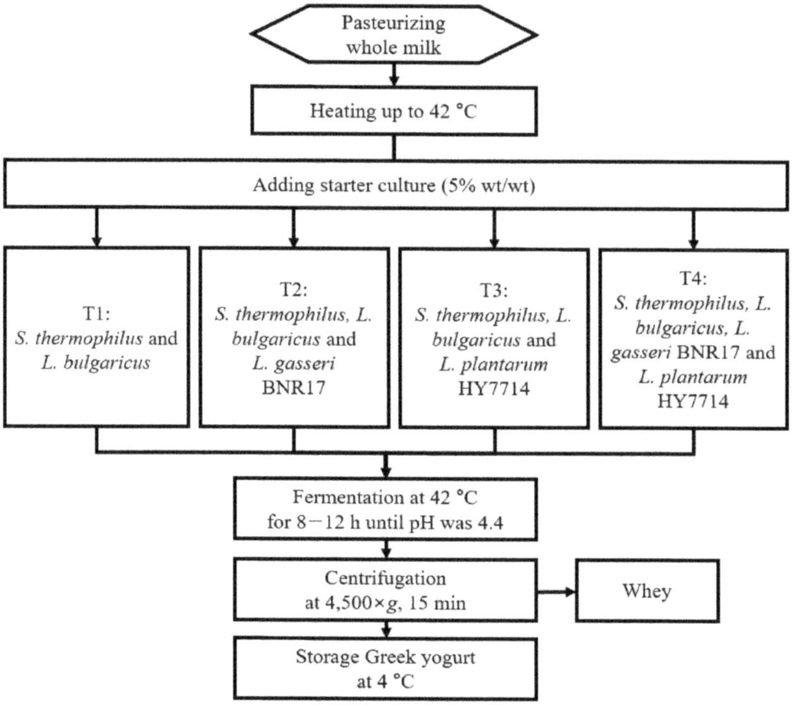

Figure 1. Flow chart of the manufacturing of Greek yogurt with four types of starter culture.

2.4. pH, Titratable Acidity, and Viscosity

The pH and titratable acidity were measured according to the method of AOAC [20]. Greek yogurt (10 g) was homogenized with 90 mL of distilled water using a stomacher (Stomacher, Interscience, Saint-Nom-la-Bretèche, France). The pH was measured with a benchtop pH meter equipped with a glass electrode (OrionTM Star A211, Thermo Fisher Scientific Co., Waltham, MA, USA). Greek yogurt (5 g) was homogenized with 45 mL of distilled water in a sterile filter bag to determine titratable acidity. Then 20 g of the filtrate was titrated with 0.1 N NaOH using 0.5 mL of phenolphthalein indicator until pH reached 8.3. The titratable acidity was expressed as a percentage of lactic acid. It was calculated as follows:

$$\text{Titratable acidity } (\%) = \frac{0.1\ N\ \text{NaOH (mL)} \times 0.1N\ \text{NaOH factor} \times 0.009}{\text{Sample (g)}} \times 100. \quad (1)$$

The viscosity of Greek yogurt was determined according to the method of Ghasempour et al. [21] with modifications. Briefly, all samples were divided into glass bottles in equal volumes. The viscosity of each sample was measured with RV-spindle No. 6 for 20 s at 10 rpm using a Brookfield viscometer (DV1, Brookfield Laboratories, Inc., Middleboro, MA, USA). It was expressed as Pascal-second (Pa·s) and millipascal-second (mPa·s).

2.5. Consumer Prefernce Test

The consumer preference test was performed following the rules of the Institutional Review Board (IRB) to comply with bioethics (KHSIRB-21-354). Four Greek yogurt samples (T1: ST and LB; T2: ST, LB, and *L. gasseri* BNR17; T3: ST, LB, and *L. plantarum* HY7714; T4: ST, LB, *L. gasseri* BNR17, and *L. plantarum* HY7714) were stored at 4 °C before the sensory test. Each sample (30 g) was scooped into paper cups labeled with 3-digit random numbers, which were served monadically to the panelist with spring water and a spoon. The consumer preference test was conducted by 60 panelists (44 women and 16 men) aged between 20 and 68. The purpose of sensory evaluation and the direction of how to score the sensory properties of samples (flavor, sweetness, sourness, viscosity, creaminess, mouthfeel, and overall acceptance) were provided to participants (Table 1). Participants scored each sample with a 7-point Hedonic scale (1 = dislike very much; 2 = dislike moderately; 3 = dislike slightly; 4 = neither like nor dislike; 5 = like slightly; 6 = like moderately, 7 = like very much) [22].

Table 1. Sensory attributes of Greek yogurt.

Attribute	Definition [1]
Flavor	The tangy and dairy-sour flavor
Sweetness	The basic taste associated with sugar
Sourness	The basic taste associated with acid
Viscosity	The force required to move the spoon back and forth
Creaminess	Smooth texture and behave like a fluid product.
Mouthfeel	The physical sensation created by food in the mouth.

[1] Terms were adapted from Desai et al. [23], Cayot et al. [24], and Greis et al. [25].

2.6. Microbiological Analysis

Enumeration of LAB in Greek yogurt was carried out by the standard plate counting method. Each sample (10 g) was diluted 10-fold with 0.9% sterile saline solution (NaCl, Duksan, Ansan-si, Republic of Korea). Then 1 mL of aliquot was inoculated onto MRS agar and incubated at 36 °C for 48 h.

To analyze the behavior of EHEC in Greek yogurt, each sample made with different probiotic LAB was aseptically divided in 10 g into 50 mL conical tubes (SPL Life Science Co., Pocheon-si, Republic of Korea) and compared with a commercial Greek yogurt (Foodis Plain Greek yogurt, ILDONG Foodis Co., Seoul, Republic of Korea) as a control. The commercial Greek yogurt was manufactured with complex lactic acid bacteria ($15 \times 10^{10}/80$ g). Each Greek yogurt was inoculated with a cocktail of diluted EHEC at an initial level of 5~6 log CFU/g and stored at 4, 10, and 25 °C. After an appropriate interval time, samples were homogenized with 0.1% sterilized peptone water, and 1 mL of aliquot was serially diluted. A diluted solution of EHEC was spread on eosin methylene blue agar (EMB agar, Oxoid) and incubated at 36 °C for 24 h to analyze the behavior of EHEC in Greek yogurt. The primary survival model of EHEC in Greek yogurt was applied to the Weibull model [26] (Equation (1)) using the GinaFit V1.7 program [27]. Delta value (time for the first decimal reduction) was then calculated.

$$\text{Weibull equation}: \ \mathrm{Log(N)} = \mathrm{Log}(N_0) - \left(\frac{t}{delta}\right)^{p} \quad (2)$$

N_0: log the initial number of cells.
t: time.
$delta$: time for the first decimal reduction.
p: shape ($p > 1$: concave downward curve; $p < 1$: concave upward curve; $p = 1$: log-linear).

The secondary model for delta value was developed, and the Davey model (Equation (3)) was used to predict delta values as a function of temperature:

$$\text{Davey model}: \ \mathrm{Y} = \mathrm{a} + (\mathrm{b}/\mathrm{T}) + (\mathrm{c}/\mathrm{T}^2) \quad (3)$$

a, b, c: constant.
T: temperature.

2.7. Statistical Analysis

All experiments were conducted three times or more. Results of this study were subjected to ANOVA and Duncan's multiple range test using SAS software ver. 9.4 (SAS Institute, Inc., Cary, NC, USA). The significance was tested at $p < 0.05$ level.

3. Results and Discussion

3.1. Effect of Probiotic LAB on pH and Titratable Acidity

Changes in physicochemical characteristics of Greek yogurt at 4 °C during 21 d of storage are shown in Figures 2 and 3. The pH and titratable acidity of traditional Greek yogurt were 4.06 to 4.64 and 0.919 to 1.579%, respectively [28]. High acidity can negatively affect the water-holding capacity and viscosity of yogurt [29]. After centrifugation, the pH values of all samples decreased from 4.4~4.44 to 4.15~4.17 in this work. The pH of the Greek yogurt made with ST and LB (T1) was further dropped from 4.167 to 3.843 after 21 d of storage, while Greek yogurt made with ST, LB, and *L. gasseri* BNR17 (T2) had relatively constant pH values (Figure 2). Overall, the pH decrease rate was low in Greek yogurt containing *L. gasseri* BNR17 (T2 and T4) during 21 d of storage at 4 °C (Figure 3). The titratable acidity of all Greek yogurt samples increased during 21 d of storage. The increase of titratable acidity of Greek yogurt made with ST and LB (T1) was the highest (+0.780 ± 0.227) among all samples ($p < 0.05$) during storage. The pH dropped while titratable acidity increased in all Greek yogurt samples during storage in this study, similar to the results of previous studies [30,31]. Greek yogurt made with *L. gasseri* BNR17 showed the least changes in pH and titratable acidity during storage, indicating that *L. gasseri* BNR17 did not affect the pH change of yogurt. Increased levels of *L. rhamnosus* GG decrease titratable acidity [32]. *L. casei* AST18 also inhibits the acidogenicity of a commercial yogurt starter [33]. These results confirm that changes in the physicochemical characteristics of Greek yogurt can be controlled by the kind of starter culture used during yogurt manufacturing. Adding various probiotic LAB as starter cultures may contribute to the development of sensory characteristics of a consumer-oriented product.

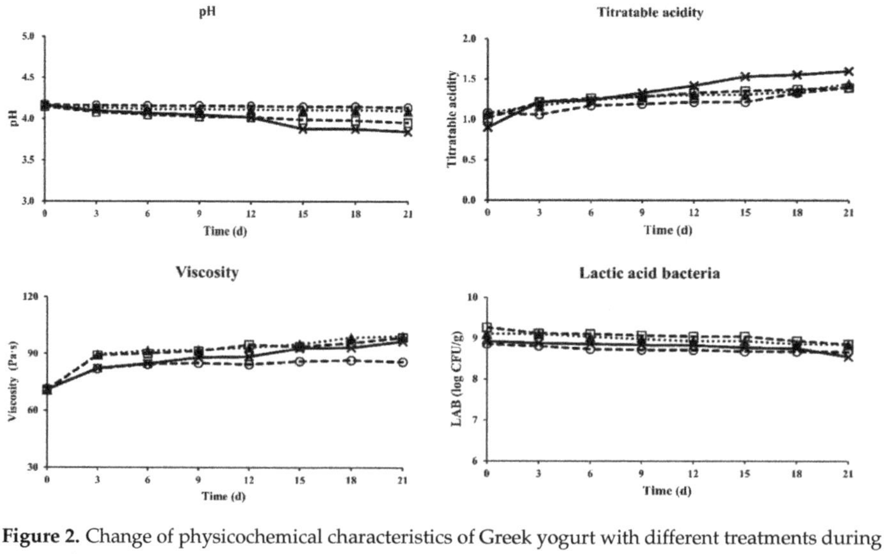

Figure 2. Change of physicochemical characteristics of Greek yogurt with different treatments during storage for 21 days at 4 °C. *S. thermophilus* and *L. bulgaricus* (T1): **x**, *S. thermophilus*, *L. bulgaricus* and *L. gasseri* BNR17 (T2): ○, *S. thermophilus*, *L. bulgaricus* and *L. plantarum* HY7714 (T3): ▲, *S. thermophilus*, *L. bulgaricus*, *L. gasseri* BNR17, and *L. plantarum* HY7714 (T4): □.

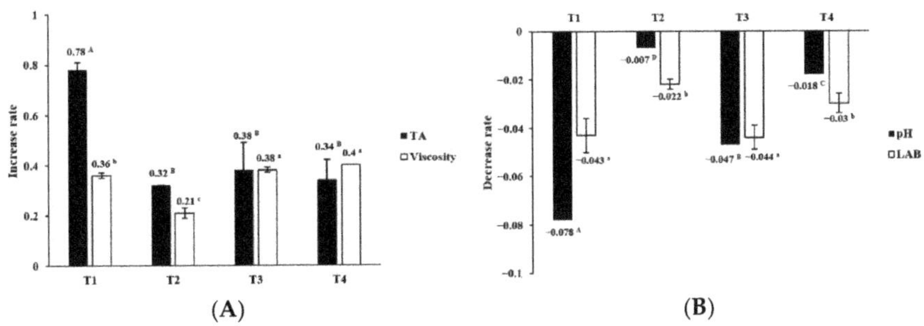

Figure 3. Increase rate of titratable acidity and viscosity (**A**), and decrease rate of pH and LAB (**B**). Decrease rate = (final value − initial value)/initial value, Increase rate = (final value − initial value)/initial value. T1: *S. thermophilus* and *L. bulgaricus*; T2: *S. thermophilus, L. bulgaricus* and *L. gasseri* BNR17; T3: *S. thermophilus, L. bulgaricus* and *L. plantarum* HY7714; T4: *S. thermophilus, L. bulgaricus, L. gasseri* BNR17 and *L. plantarum* HY7714. $^{A–D}$ Means values in the TA and pH categories with different letters are significantly different ($p < 0.05$) $^{a–c}$ Means values in the viscosity and LAB categories with different letters are significantly different ($p < 0.05$).

3.2. Effect of Probiotic LAB on Viscosity and LAB Population

Consumers favor Greek yogurt because of its unique firmness, dense texture, moderate sweet aromatic, milk fat and dairy sour flavors, and moderately sour taste [23]. Thus, viscosity is one of the most important quality characteristics of Greek yogurt. In the present study, the viscosity of Greek yogurt ranged from 70.9 to 71.4 Pa·s (7.09×10^4 to 7.14×10^4 mPa·s) at 0 d after centrifugation. It was increased during 21 d of storage at 4 °C (Figure 2). Greek yogurt containing *L. plantarum* HY 7714 (T3 and T4) had the highest viscosity at 21 d and the highest increase rate of viscosity during 21 d of storage among samples (Figure 3) ($p < 0.05$). *L. plantarum* strains can produce exopolysaccharide (EPS), a macromolecule composed of monosaccharide residues of sugar and sugar derivatives. EPS can act as an important factor in the physicochemical and rheological properties of yogurt due to its role as a natural concentrate agent and stabilizer [34,35]. Nambiar et al. [36] have reported that EPS isolated from *L. plantarum* HM47 has high thermal stability and that it can enhance the texture of yogurt at a low pH (4.0). In this work, *L. plantarum* HY 7714 also increased the viscosity of Greek yogurt.

As shown in Figure 2, Greek yogurt had LAB populations above 10^8 CFU/g after a concentration step. Yogurt should contain a minimum of 10^7 CFU/g of live and active cultures [37]. The LAB population was well maintained above the criteria in Greek yogurt at 4 °C during 21 d of storage in this work. Greek yogurt with ST, LB, and *L. gasseri* BNR17 (T2) showed the lowest decrease in the LAB population (-0.022 ± 0.002) during 21 d of storage at 4 °C. However, no significant difference in the decreased extent of LAB population was observed between T2 and T4 (ST, LB, *L. gasseri* BNR17, and *L. plantarum* HY7714) (Figure 3). These results indicated that populations of probiotic LAB were well maintained in Greek yogurt containing *L. gasseri* BNR17 during storage.

3.3. Antimicrobial Effect of Probiotic LAB on EHEC

Effects of LAB on the survival of EHEC in Greek yogurt stored at 4, 10, and 25 °C are shown in Figure 4. At both 4 °C and 10 °C, the most rapid reduction of EHEC was observed in Greek yogurt made with ST and LB (T1). On the other hand, populations of EHEC were well maintained in all other Greek yogurt stored at 4 °C and 10 °C. At 25 °C, the most rapid reduction of EHEC was observed in Greek yogurt made with ST, LB, and *L. plantarum* HY7714 (T3), in which EHEC was not detected after 4 d of storage. The population of EHEC was maintained in commercial Greek yogurt up to 4 d of storage and then rapidly decreased, indicating that the type of probiotic LAB could affect the behavior of EHEC in Greek yogurt at ambient temperature. Ogwaro et al. [38] have manufactured yogurt with pasteurized full cream milk and found that *E. coli* O157:H7 can survive at 4 °C, while there is no *E. coli* O157:H7 on the fifth day of storage time at 25 °C. Moineau-Jean et al. [39] have also prepared Greek-style yogurt using centrifugation and ultrafiltration methods. Ultrafiltration methods more effectively inhibited the viability of non-pathogenic *E. coli* strains than traditional and centrifugation methods. In addition, *E. coli* had a lower viability at 8 °C than at 4 °C ($p < 0.05$).

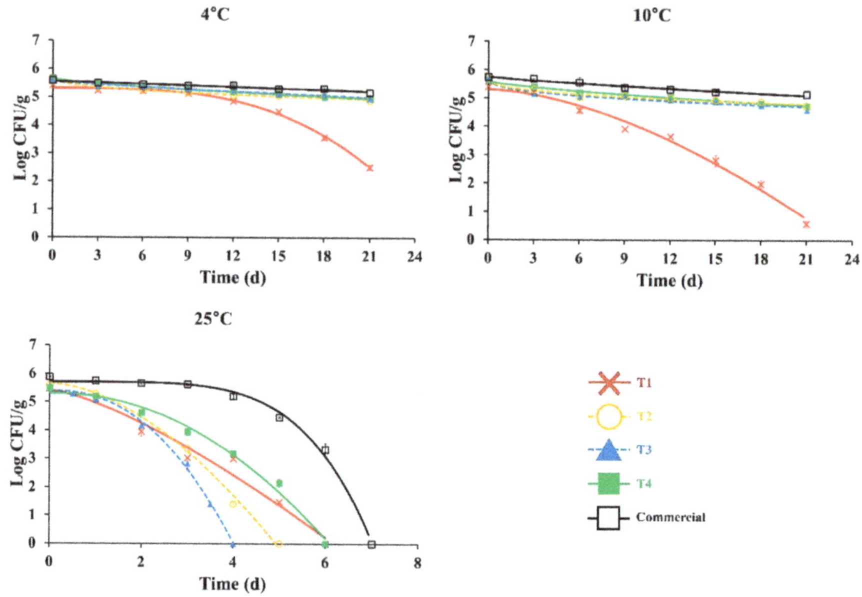

Figure 4. The effect of lactic acid bacteria on the survival of enterohemorrhagic *Escherichia coli* in Greek yogurt at 4, 10, and 25 °C. *S. thermophilus* and *L. bulgaricus* (T1): **X**, *S. thermophilus, L. bulgaricus* and *L. gasseri* BNR17 (T2): ○, *S. thermophilus, L. bulgaricus* and *L. planatarum* HY7714 (T3): ▲, *S. thermophilus, L. bulgaricus, L. gasseri* BNR17 and *L. planatarum* HY7714 (T4): ■, Commercial Greek yogurt (ILDONG Foodis Plain Greek yogurt): □.

Commercial Greek yogurt had higher delta values than other Greek yogurt prepared in this work ($p < 0.05$), indicating that the highest survival ability of EHEC was observed in commercial Greek yogurt at all temperatures. This trend was confirmed with delta values as a function of temperature, which indicated how rapidly pathogens were killed with an increase in temperature (Table 2). The low pH (4.52) and viscosity (59.6 ± 0.26 Pa·s, $5.96 \times 10^4 \pm 0.26$ mPa·s) of commercial Greek yogurt might have affected the survival ability of EHEC. On the contrary, EHEC died more quickly in Greek yogurt prepared with various probiotic LAB in the present study. Especially, Greek yogurt made with ST and LB (T1) had the lowest pH and delta values and the highest titratable acidity among all samples ($p < 0.05$). The presence of lactic acid produced from LAB can control the growth of *E. coli* O157:H7 in yogurt [40]. Hu et al. [41] have observed that organic acid produced from *L. plantarum* exhibits antimicrobial activity against *E. coli*. *E. coli* O157:H7 was also inhibited by reduced pH and lactic acid produced by *L. acidophilus* and *L. casei* [42]. Guraya et al. [43] have reported that a pH below 4.1 can significantly inhibit the growth of EHEC in yogurt. Moreover, cell-free supernatant of LAB strains, including *Leuconostoc mesenteroides* and *L. plantarum* shows antimicrobial activities against EPEC, ETEC, and *E. coli* O157:H7 in kimchi [19].

Table 2. Delta values in Greek yogurt at 4, 10, and 25 °C.

Sample [1]	Temperature		
	4	10	25
T1 [1]	15.47 ± 0.11 [d]	8.30 ± 0.23 [e]	1.63 ± 0.24 [d]
T2	40.93 ± 0.91 [bc]	24.90 ± 1.54 [c]	1.98 ± 0.06 [c]
T3	37.30 ± 3.34 [c]	19.69 ± 2.23 [d]	1.79 ± 0.05 [d]
T4	42.04 ± 1.51 [b]	31.06 ± 5.96 [b]	2.65 ± 0.18 [b]
Commercial [2]	47.76 ± 6.16 [a]	36.77 ± 2.67 [a]	5.00 ± 0.07 [a]

[1] T1: Fermented by *S. thermophilus* and *L. bulgaricus*, T2: Fermented by *S. thermophilus*, *L. bulgaricus* and *L. gasseri* BNR17, T3: Fermented by *S. thermophilus*, *L. bulgaricus* and *L. plantarum* HY7714, T4: Fermented by *S. thermophilus*, *L. bulgaricus*, *L. gasseri* BNR17, and *L. plantarum* HY7714. [2] ILDONG Foodis Plain Greek yogurt. [a–e] Means values in the same column with different letters are significantly different ($p < 0.05$).

3.4. Consumer Preference Test

Consumer preference test results of Greek yogurt are shown in Table 3. There were no significant differences in flavor scores of Greek yogurts among all samples, indicating that the panelist did not recognize the difference in the flavor of Greek yogurts prepared in this work. The average preference score for the flavor of Greek yogurts ranged from 4.82 to 5.08. Overall, Greek yogurt made with ST, LB, *L. gasseri* BNR17, and *L. plantarum* HY7714 (T4) had the highest preference scores of all sensory properties, including flavor (5.08), sweetness (5.05), sourness (5.3), viscosity (5.42), creaminess (5.68), mouthfeel (5.87), and overall acceptability (5.8). Among sensory preference scores, scores for the sourness and overall acceptability of Greek yogurt (T4) were significantly higher than those of other Greek yogurt samples ($p < 0.05$). For sourness, T4 (5.3) had the highest preference score, followed by T2 (4.75), T3 (4.42), and T1 (4.37) ($p < 0.05$). These results are related to the titratable acidity of Greek yogurt. The highest titratable acidity of Greek yogurt made with ST and LB (T1) received the lowest preference score for sourness. Greek yogurt containing *L. gasseri* BNR17 or/and *L. plantarum* HY7714 (T2, T3, T4) received higher scores of sweetness and sourness than Greek yogurt containing only traditional starter culture (T1). Although there were no significant differences in viscosity scores among samples, T4 had the highest viscosity score (5.42) among all samples, followed by T3 (5.28), T2 (5.25), and T1 (5.15). T4 also had the highest scores for creaminess (5.68) and mouthfeel (5.87), followed by T3 (5.55 and 5.57), T2 (5.4 and 5.4), and T1 (5.18 and 5.3). Lastly, Greek yogurt containing both *L. gasseri* BNR17 and *L. plantarum* HY7714 (T4) had the highest score for overall acceptability (5.8), which indicates "like moderately", followed by T2 and T3 with the same score (5.2) and T1 (4.78) ($p < 0.05$). These results show that using *L. gasseri* BNR17 and *L. plantarum* HY7714 as starter cultures in Greek yogurt manufacturing can improve various sensory qualities of yogurt, which are closely related to consumer preference and acceptability. Moreover, the highest preference sensory score for T4 among samples might be attributed to the combination of various probiotic LAB in T4. Coggins et al. [44] have found that taste and texture rather than flavor or appearance make a difference in the preference for yogurt. Aroma, sweetness, sourness, chalky mouthfeel, and viscosity are also significant factors affecting the preference for yogurt drinks [45].

Table 3. Sensory scores of Greek yogurts with different starter cultures by Consumer preference test.

Sample [2]	Sensory Scores [1]						
	Flavor	Sweetness	Sourness	Viscosity	Creaminess	Mouthfeel	Overall Acceptability
T1	5.03 ± 1.35	4.23 ± 1.47 ^{b}	4.37 ± 1.58 ^{b}	5.15 ± 1.30	5.18 ± 1.26 ^{b}	5.30 ± 1.45 ^{b}	4.78 ± 1.56 ^{b}
T2	4.98 ± 1.35	4.77 ± 1.21 ^{a}	4.75 ± 1.27 ^{b}	5.25 ± 1.04	5.40 ± 1.17 ^{ab}	5.40 ± 1.18 ^{b}	5.20 ± 1.33 ^{b}
T3	4.82 ± 1.41	4.77 ± 1.20 ^{a}	4.42 ± 1.36 ^{b}	5.28 ± 1.01	5.55 ± 1.11 ^{ab}	5.57 ± 1.21 ^{ab}	5.20 ± 1.27 ^{b}
T4	5.08 ± 1.33	5.05 ± 1.47 ^{a}	5.30 ± 1.34 ^{a}	5.42 ± 1.12	5.68 ± 1.11 ^{a}	5.87 ± 0.96 ^{a}	5.80 ± 1.13 ^{a}

[1] 1 = dislike very much; 2 = dislike moderately; 3 = dislike slightly; 4 = neither like nor dislike; 5 = like slightly; 6 = like moderately, 7 = like very much. [2] T1: fermented by *S. thermophilus* and *L. bulgaricus*, T2: fermented by *S. thermophilus*, *L. bulgaricus* and *L. gasseri* BNR17, T3: fermented by *S. thermophilus*, *L. bulgaricus* and *L. plantarum HY7714*, T4: fermented by *S. thermophilus*, *L. bulgaricus*, *L. gasseri* BNR17 and *L. plantarum* HY7714. [a,b] Means values in the same column with different letters are significantly different ($p < 0.05$).

LAB can change carbohydrates into lactic acid or other metabolites, caseins into peptides and free amino acids, and milk fat into free fatty acids during fermentation. These mechanisms make the unique flavor of yogurt [46,47]. *Leuconostoc* strains are preferred to increase the butter-like flavor of yogurt due to diacetyl, acetic acid, and ethanol produced during fermentation [48]. Bifidobacteria contribute to the production of acetaldehyde and acetoin with effects on the overall flavor quality of yogurt [49]. When quantitative descriptive analysis and consumer preference evaluation were performed to compare six conventional yogurt samples and three probiotic yogurt samples, probiotic yogurt samples had higher scores of sweet taste, creaminess, and overall sensory quality than conventional yogurt samples. This result appeared to be due to the high preference for probiotic yogurt samples for the degree of uniformity of particles and viscosity in the mouth [50]. Recently, higher preferences for color and overall taste of probiotic yogurt samples containing *L. fermentum* KU200060 than control yogurt have been also reported [51]. Additionally, a combination of *L. rhamnosus* GG, *L. plantarum* NK181, or *L. delbeuckii* KU200171 with a traditional starter culture led to high scores of tastes, texture, flavor, and overall preferences given by trained panelists [52]. In the study of Desai et al. [23], consumer preferences were not significantly different between traditional strained yogurt and fortified Greek yogurt. However, the results of the quality property and consumer preference test in this study confirmed that the use of *L. gasseri* BNR17 and *L. plantarum* HY7714 as starter cultures could enhance consumers' preference for Greek yogurt.

4. Conclusions

Streptococcus thermophilus (ST), *Lactobacillus bulgaricus* (LB), *Lactobacillus gasseri* BNR17, and *Lactobacillus plantarum* HY7714 were used to evaluate the combined effects of various probiotic LAB on the quality and safety aspects of Greek yogurt. The pH and titratable acidity of Greek yogurt made with ST, LB, *L. gasseri* BNR17, and *L. plantarum* HY7714 (T4) was kept relatively constant. In contrast, Greek yogurt made with ST and LB (T1) showed significant changes in pH and titratable acidity ($p < 0.05$), leading to the lowest preference scores for all sensory attributes. Greek yogurt containing *L. plantarum* HY7714 (T3 and T4) had high viscosity, consistent with the results of the viscosity score in the consumer preference test. At 4 °C and 10 °C, the most effective antimicrobial effect against EHEC was observed with T1 due to its low pH and high titratable acidity. At 25 °C, EHEC showed low viability in Greek yogurt containing *L. gasseri* BNR17 and *L. plantarum* HY7714 than in commercial Greek yogurt ($p < 0.05$). Sensory panelists preferred Greek yogurt containing *L. gasseri* BNR17 and *L. plantarum* HY7714 (T4) over other samples. Thus, it is concluded that using probiotic LAB such as *L. gasseri* BNR17 and *L. plantarum* HY7714 as starter cultures for Greek yogurt manufacturing can enhance consumers' preference and functionality.

Author Contributions: Investigation, S.-Y.Y. and K.-S.Y.; methodology, S.-Y.Y. and K.-S.Y.; project administration, K.-S.Y.; writing—original draft preparation, S.-Y.Y.; writing—review and editing, K.-S.Y. All authors have read and agreed to the published version of the manuscript.

Funding: This research received no external funding.

Data Availability Statement: Data is contained within the article.

Conflicts of Interest: The authors declare no conflict of interest.

References

1. Lee, W.J.; Lucey, J.A. Formation and physical properties of yogurt. *Asian Australas. J. Anim. Sci.* **2010**, *23*, 1127–1136. [CrossRef]
2. Li, S.; Ye, A.; Singh, H. Effects of seasonal variations on the quality of set yogurt, stirred yogurt, and Greek-style yogurt. *J. Dairy Sci.* **2021**, *104*, 1424–1432. [CrossRef] [PubMed]
3. Moineau-Jean, A.; Champagne, C.P.; Roy, D.; Raymond, Y.; LaPointe, G. Effect of Greek-style yoghurt manufacturing processes on starter and probiotic bacteria populations during storage. *Int. Dairy J.* **2019**, *93*, 35–44. [CrossRef]
4. Douglas, S.M.; Ortinau, L.C.; Hoertel, H.A.; Leidy, H.J. Low, moderate, or high protein yogurt snacks on appetite control and subsequent eating in healthy women. *Appetite* **2013**, *60*, 117–122. [CrossRef]
5. Daniali, M.; Nikfar, S.; Abdollahi, M. A brief overview on the use of probiotics to treat overweight and obese patients. *Expert Rev. Endocrinol. Metab.* **2020**, *15*, 1–4. [CrossRef]
6. Kang, J.H.; Yun, S.I.; Park, H.O. Composition for Prevention and Treatment of Diabetes Mellitus with *Lactobacillus gasseri* BNR17. Korean Patent KR20080048976A, 29 November 2007.
7. Yun, S.I.; Park, H.O.; Kang, J.H. Effect of *Lactobacillus gasseri* BNR17 on blood glucose levels and body weight in a mouse model of type 2 diabetes. *J. Appl. Microbiol.* **2009**, *107*, 1681–1686. [CrossRef]
8. Kang, J.H.; Yun, S.I.; Park, M.H.; Park, J.H.; Jeong, S.Y.; Park, H.O. Anti-obesity effect of *Lactobacillus gasseri* BNR17 in high-sucrose diet-induced obese mice. *PLoS ONE* **2013**, *8*, e54617. [CrossRef]
9. Kim, J.; Yun, J.M.; Kim, M.K.; Kwon, O.; Cho, B. *Lactobacillus gasseri* BNR17 supplementation reduces the visceral fat accumulation and waist circumference in obese adults: A randomized, double-blind, placebo-controlled trial. *J. Med. Food.* **2018**, *21*, 454–461. [CrossRef]
10. Lee, D.E.; Kim, H.M.; Park, S.D.; Kim, Y.T.; Kim, Y.J.; Jeong, J.W.; Lee, J.H.; Myoung, K.S.; Jang, S.S.; Huh, C.S. Probiotics of *Lactobacillus plantarum* HY7714 for skin wrinkle inhibitory and moisurizing effects and use of thereof as skin anti-wrinkle or moisturizing products. Korean Patent KR101492003B1, 29 April 2013.
11. Kim, H.M.; Lee, D.E.; Park, S.D.; Kim, Y.T.; Kim, Y.J.; Jeong, J.W.; Jang, S.S.; Ahn, Y.T.; Sim, J.H.; Huh, C.S.; et al. Oral administration of *Lactobacillus plantarum* HY7714 protects hairless mouse against ultraviolet B-induced photoaging. *J. Microbiol. Biotechnol.* **2014**, *24*, 1583–1591. [CrossRef]
12. Lee, D.E.; Huh, C.S.; Ra, J.; Choi, I.D.; Jeong, J.W.; Kim, S.H.; Ryu, J.H.; Seo, Y.K.; Koh, J.S.; Lee, J.H.; et al. Clinical evidence of effects of *Lactobacillus plantarum* HY7714 on skin aging: A randomized, double blind, placebo-controlled study. *J. Microbiol. Biotechnol.* **2015**, *25*, 2160–2168. [CrossRef]
13. Ministry of Food and Drug Safety (MFDS). Health Functional Food Code. Available online: https://www.khsa.or.kr/assets/extra/hfood/01.pdf (accessed on 2 July 2021).
14. Lee, K.; Kim, H.J.; Kim, S.A.; Park, S.D.; Shim, J.J.; Lee, J.L. Exopolysaccharide from *Lactobacillus plantarum* HY7714 protects against skin aging through skin–gut axis communication. *Molecules* **2021**, *26*, 1651. [CrossRef] [PubMed]
15. Kim, G.H.; Breidt, F.; Fratamico, P.; Oh, D.H. Acid resistance and molecular characterization of *Escherichia coli* O157: H7 and different non-O157 Shiga toxin-producing *E. coli* serogroups. *J. Food Sci.* **2015**, *80*, M2257–M2264. [CrossRef] [PubMed]
16. Morgan, D.; Newman, C.P.; Hutchinson, D.N.; Walker, A.M.; Rowe, B.; Majid, F. Verotoxin producing *Escherichia coli* O 157 infections associated with the consumption of yoghurt. *Epidemiol. Infect.* **1993**, *111*, 181–188. [CrossRef] [PubMed]
17. Yang, S.Y.; Yoon, K.S. Quantitative Microbial Risk Assessment of *Listeria monocytogenes* and Enterohemorrhagic *Escherichia coli* in Yogurt. *Foods* **2022**, *11*, 971. [CrossRef]
18. Reuben, R.C.; Roy, P.C.; Sarkar, S.L.; Alam, A.R.U.; Jahid, I.K. Characterization and evaluation of lactic acid bacteria from indigenous raw milk for potential probiotic properties. *J. Dairy Sci.* **2020**, *103*, 1223–1237. [CrossRef]
19. Choi, S.J.; Yang, S.Y.; Yoon, K.S. Lactic acid bacteria starter in combination with sodium chloride controls pathogenic *Escherichia coli* (EPEC, ETEC, and EHEC) in kimchi. *Food Microbiol.* **2021**, *100*, 103868. [CrossRef]
20. AOAC International. *Method Official Methods of Analysis*, 16th ed.; AOAC International: Arlington, VA, USA, 1995.
21. Ghasempour, Z.; Javanmard, N.; Langroodi, A.M.; Alizadeh-Sani, M.; Ehsani, A.; Kia, E.M. Development of probiotic yogurt containing red beet extract and basil seed gum; techno-functional, microbial and sensorial characterization. *Biocatal. Agric. Biotechnol.* **2020**, *29*, 101785. [CrossRef]
22. Karagul-Yuceer, Y.; Drake, M. Sensory analysis of yogurt. In *Manufacturing Yogurt and Fermented Milk*, 2nd ed.; Chandan, R.C., Kilara, A., Eds.; John Wiley & Sons, Inc.: New York, NY, USA, 2013.
23. Desai, N.T.; Shepard, L.; Drake, M.A. Sensory properties and drivers of liking for Greek yogurts. *J. Dairy Sci.* **2013**, *96*, 7454–7466. [CrossRef]

24. Cayot, P.; Schenker, F.; Houze, G.; Sulmont-Rosse, C.; Colas, B. Creaminess in relation to consistency and particle size in stirred fat-free yogurt. *Int. Dairy J.* **2008**, *18*, 303–311. [CrossRef]
25. Greis, M.; Sainio, T.; Katina, K.; Kinchla, A.K.; Nolden, A.; Partanen, R.; Seppa, L. Dynamic texture perception in plant-based yogurt alternatives: Identifying temperal drivers of liking by TDS. *Food Qual. Prefer.* **2020**, *86*, 104019. [CrossRef]
26. Mafart, P.; Couvert, O.; Gaillard, S.; Leguérinel, I. On calculating sterility in thermal preservation methods: Application of the Weibull frequency distribution model. *Int. J. Food Microbiol.* **2002**, *72*, 107–113. [CrossRef] [PubMed]
27. Geeraerd, A.H.; Valdramidis, V.P.; Van Impe, J.F. GInaFiT, a freeware tool to assess non-log-linear microbial survivor curves. *Int. J. Food Microbiol.* **2005**, *102*, 95–105. [CrossRef] [PubMed]
28. Serafeimidou, A.; Zlatanos, S.; Laskaridis, K.; Sagredos, A. Chemical characteristics, fatty acid composition, and conjugated linoleic acid (CLA) content of traditional Greek yogurts. *Food Chem.* **2012**, *134*, 1839–1846. [CrossRef]
29. Sodini, I.; Mattas, J.; Tong, P.S. Influence of pH and heat treatment of whey on the functional properties of whey protein concentrates in yoghurt. *Int. Dairy, J.* **2006**, *16*, 1464–1469. [CrossRef]
30. Zhou, T.; Huo, R.; Kwok, L.Y.; Li, C.; Ma, Y.; Mi, Z.; Chen, Y. Effects of applying *Lactobacillus helveticus* H9 as adjunct starter culture in yogurt fermentation and storage. *J. Dairy Sci.* **2019**, *102*, 223–235. [CrossRef] [PubMed]
31. Soni, R.; Jain, N.K.; Shah, V.; Soni, J.; Suthar, D.; Gohel, P. Development of probiotic yogurt: Effect of strain combination on nutritional, rheological, organoleptic and probiotic properties. *J. Food Sci. Technol.* **2020**, *57*, 2038–2050. [CrossRef]
32. Jia, R.; Chen, H.; Chen, H.; Ding, W. Effects of fermentation with *Lactobacillus rhamnosus* GG on product quality and fatty acids of goat milk yogurt. *J. Dairy Sci.* **2016**, *99*, 221–227. [CrossRef]
33. Li, H.; Liu, L.; Zhang, S.; Uluko, H.; Cui, W.; Lv, J. Potential use of *Lactobacillus casei* AST18 as a bioprotective culture in yogurt. *Food Control* **2013**, *34*, 675–680. [CrossRef]
34. Duboc, P.; Mollet, B. Applications of exopolysaccharides in the dairy industry. *Int. Dairy J.* **2001**, *11*, 759–768. [CrossRef]
35. İspirli, H.; Dertli, E. Isolation and identification of exopolysaccharide producer lactic acid bacteria from Turkish yogurt. *J. Food Process. Preserv.* **2018**, *42*, e13351. [CrossRef]
36. Nambiar, R.B.; Sellamuthu, P.S.; Perumal, A.B.; Sadiku, E.R.; Phiri, G.; Jayaramudu, J. Characterization of an exopolysaccharide produced by *Lactobacillus plantarum* HM47 isolated from human breast milk. *Process. Biochem.* **2018**, *73*, 15–22. [CrossRef]
37. Food and Drug Administration (FDA). Milk and Cream Products and Yogurt Products; Final Rule to Revoke the Standards for Lowfat Yogurt and Nonfat Yogurt and to Amend the Standard for Yogurt. Available online: https://www.federalregister.gov/documents/2021/06/11/2021-12220/milk-and-cream-products-and-yogurt-products-final-rule-to-revoke-the-standards-for-lowfat-yogurt-and (accessed on 16 November 2021).
38. Ogwaro, B.A.; Gibson, H.; Whitehead, M.; Hill, D.J. Survival of *Escherichia coli* O157: H7 in traditional African yoghurt fermentation. *Int. J. Food Microbiol.* **2002**, *79*, 105–112. [CrossRef] [PubMed]
39. Moineau-Jean, A.; Guévremont, E.; Champagne, C.P.; Roy, D.; Raymond, Y.; LaPointe, G. Fate of *Escherichia coli* and *Kluyveromyces marxianus* contaminants during storage of Greek-style yogurt produced by centrifugation or ultrafiltration. *Int. Dairy J.* **2017**, *72*, 36–43. [CrossRef]
40. Cutrim, C.S.; de Barros, R.F.; da Costa, M.P.; Franco, R.M.; Conte-Junior, C.A.; Cortez, M.A.S. Survival of *Escherichia coli* O157: H7 during manufacture and storage of traditional and low lactose yogurt. *LWT* **2016**, *70*, 178–184. [CrossRef]
41. Hu, C.H.; Ren, L.Q.; Zhou, Y.; Ye, B.C. Characterization of antimicrobial activity of three *Lactobacillus plantarum* strains isolated from Chinese traditional dairy food. *Food Sci. Nutr.* **2019**, *7*, 1997–2005. [CrossRef] [PubMed]
42. Ogawa, M.; Shimizu, K.; Nomoto, K.; Tanaka, R.; Hamabata, T.; Yamasaki, S.; Takeda, T.; Takeda, Y. Inhibition of in vitro growth of Shiga toxin-producing *Escherichia coli* O157: H7 by probiotic *Lactobacillus* strains due to production of lactic acid. *Int. J. Food Microbiol.* **2001**, *68*, 135–140. [CrossRef]
43. Guraya, R.; Frank, J.F.; Hassan, A.N. Effectiveness of salt, pH, and diacetyl as inhibitors for *Escherichia coli* O157: H7 in dairy foods stored at refrigeration temperatures. *J. Food Prot.* **1998**, *61*, 1098–1102. [CrossRef]
44. Coggins, P.C.; Schilling, M.W.; Kumari, S.; Gerrard, P.D. Development of a sensory lexicon for conventional milk yogurt in the United States. *J. Sens. Stud.* **2008**, *23*, 671–687. [CrossRef]
45. Allgeyer, L.C.; Miller, M.J.; Lee, S.Y. Sensory and microbiological quality of yogurt drinks with prebiotics and probiotics. *J. Dairy Sci.* **2010**, *93*, 4471–4479. [CrossRef]
46. Steele, J.; Broadbent, J.; Kok, J. Perspectives on the contribution of lactic acid bacteria to cheese flavor development. *Curr. Opin. Biotechnol.* **2013**, *24*, 135–141. [CrossRef]
47. Chen, C.; Zhao, S.; Hao, G.; Yu, H.; Tian, H.; Zhao, G. Role of lactic acid bacteria on the yogurt flavour: A review. *Int. J. Food Prop.* **2017**, *20*, S316–S330. [CrossRef]
48. Boumerdassi, H.; Monnet, C.; Desmazeaud, M.; Corrieu, G. Isolation and properties of *Lactococcus lactis* subsp. lactis biovar diacetylactis CNRZ 483 mutants producing diacetyl and acetoin from glucose. *Appl. Environ. Microbiol.* **1997**, *63*, 2293–2299. [CrossRef] [PubMed]
49. Zareba, D.; Ziarno, M.; Obiedzinski, M. Volatile profile of non-fermented milk and milk fermented by *Bifidobacterium animalis* subsp. *lactis*. *Int. J. Food Prop.* **2012**, *15*, 1010–1021. [CrossRef]
50. Majchrzak, D.; Lahm, B.; Duerrschmid, K. Conventional and probiotic yogurts differ in sensory properties but not in consumers' preferences. *J. Sens. Stud.* **2010**, *25*, 431–446. [CrossRef]

51. Lim, S.M.; Lee, N.K.; Kim, K.T.; Paik, H.D. Probiotic *Lactobacillus fermentum* KU200060 isolated from watery kimchi and its application in probiotic yogurt for oral health. *Microb. Pathog.* **2020**, *147*, 104430. [CrossRef] [PubMed]
52. Kim, E.D.; Lee, H.S.; Kim, K.T.; Paik, H.D. Antioxidant and Angiotensin-Converting Enzyme (ACE) Inhibitory Activities of Yogurt Supplemented with *Lactiplantibacillus plantarum* NK181 and *Lactobacillus delbrueckii* KU200171 and Sensory Evaluation. *Foods* **2021**, *10*, 2324. [CrossRef]

foods

MDPI

Article

Vegan Ice Cream Made from Soy Extract, Soy Kefir and Jaboticaba Peel: Antioxidant Capacity and Sensory Profile

Giovana M. N. Mendonça [1], Estela M. D. Oliveira [1], Alessandro O. Rios [2], Carlos H. Pagno [2] and Daniela C. U. Cavallini [1,*]

[1] São Paulo State University (UNESP), School of Pharmaceutical Sciences, Araraquara, São Paulo 01049-010, Brazil
[2] Institute of Science and Food Technology, Federal University of Rio Grande do Sul, UFRGS—Rio Grande do Sul, Porto Alegre 90010-150, Brazil
* Correspondence: d.cavallini@unesp.br

Abstract: Considering the need for functional foods and the use of by-products of the food industry, a potentially functional ice cream was developed, using soy extract, soy kefir and dehydrated jaboticaba peel. Five ice creams were produced using soy kefir (K) and soy extract (S): (1) GS—100% S; (2) GK1-75% S/25% K; (3) GK2-50% S/50% K; (4) GK3-25% S/75% K and (5) GK-100% K; The products were evaluated by physicochemical, microbiological and sensory (check all that apply) analyses. The addition of kefir was found to increase the acidity of the products. The concentrations of total phenolic compounds in the formulations with kefir were approximately ten times higher than the GS formulation. All products presented concentrations of thermotolerant coliforms <3 NMP/g and absence of *Salmonella* ssp. The viability of *Lactobacillus* ssp., *Streptococcus* spp. and *Bifidobacterium* ssp. was higher than 10 log CFU/g during the whole storage period. The GS and GK1 formulations had the lowest scores, while GK ice cream was preferred. The formulations showed distinct sensory profiles in the CATA, and the ice cream with 100% kefir was associated with desirable attributes. The ice creams exhibited microbiological and sensory characteristics that meet the expectations of the product's target audience.

Keywords: probiotic; ice cream; soy extract

Citation: Mendonça, G.M.N.; Oliveira, E.M.D.; Rios, A.O.; Pagno, C.H.; Cavallini, D.C.U. Vegan Ice Cream Made from Soy Extract, Soy Kefir and Jaboticaba Peel: Antioxidant Capacity and Sensory Profile. *Foods* **2022**, *11*, 3148. https://doi.org/10.3390/foods11193148

Academic Editor: Xiqing Yue

Received: 5 August 2022
Accepted: 29 August 2022
Published: 10 October 2022

1. Introduction

Consumer demand for a healthy and balanced diet has driven the food industry to seek alternatives that can meet this demand. Among the products developed focusing on this market are those with reduced trans-fat and sodium content, whole foods and organic, vegan and functional foods, including probiotics [1].

Kefir is defined as a "fermented milk produced by inoculating kefir grains or starter culture, composed of *Lactobacillus kefir*, species of the genera *Leuconostoc* spp., *Lactococcus* spp. and *Acetobacter* spp. and lactose-positive and/or lactose-negative yeasts, which grow synergistically" [2]. This beverage is recognized worldwide as an excellent source of microorganisms, which have potential health benefits. The microorganisms present in kefir and the products of their symbiotic relationship are related to antimicrobial, anti-inflammatory and anti-allergic effects attributed to the beverage [3–6].

Cow's milk kefir is the most popular beverage, but its consumption is limited for lactose-intolerant, dairy-allergic and vegan people, indicating the need to adapt the culture to non-dairy substrates, especially those obtained from cereals and legumes [7].

The water-soluble soy extract can be used as a substitute for cow's milk, due to its good nutritional profile and lactose-free characteristic, for use in products intended for consumers that are lactose-intolerant or allergic to milk protein [8]. The soy extract is mainly consumed as a ready-to-drink beverage or as an ingredient used in various formulations such as flavored drinks, fermented drinks and desserts, including ice cream [9].

Ice cream is a frozen product obtained from an emulsion of fats and proteins or a mixture of water and sugar, allowing the addition of other ingredients, which does not de-characterize the product. This frozen dessert is widely accepted by consumers, with a worldwide consumption per capita of 5.29 L [10,11].

Due to changes in the population's eating habits, the market for vegetable-based ice creams has expanded, with the incorporation of fruits with functional components, such as carotenoids, fibers and phenolic compounds.

Among Brazilian fruits, the jaboticaba stands out for having a sweet and slightly spicy flavor and dark purple peel, rich in minerals, vitamin C, soluble fiber and phenolic compounds (especially anthocyanins). Such compounds can be used as a natural dye and have antioxidant properties, with potential beneficial effects on health, such as modulation of the lipid profile and anti-inflammatory and anti-atherogenic properties [12].

The modification of the raw material and ingredients used in the preparation of ice cream can affect the textural properties, flavor, color and product acceptance. Thus, the aim of this work was to develop a potentially functional ice cream, from soy extract, soy kefir and dehydrated jaboticaba peel, which can be consumed by lactose-intolerant, milk-protein-allergic and vegan individuals.

2. Materials and Methods

2.1. Material

2.1.1. Obtaining Soy Kefir

The soy extract was obtained from the Soy Derivatives Production and Development Unit (Universoja—FCF—UNESP Araraquara) [13]. The soy extract was fermented by a mixed starter culture (CHOOZIT Kefir DC LYO 1000 L), kindly provided by DANISCO (DuPont, Paulinia—SP, Brazil). After the addition of sucrose (10 g per liter), the soy extract was heated to 95 ± 3 °C for 30 min, cooled to 25 ± 2 °C and inoculated with the commercial starter culture (0.005 g/L). The mixture was incubated at 37 ± 2 °C until a pH of 4.80 was reached.

2.1.2. Production of Dehydrated Jaboticaba Peel Flour

The ripe fruits were purchased from local producers of Araraquara-SP in August and October 2018, the jaboticaba's harvest period. The fruits were selected, sanitized in an aqueous sodium chloride solution (100 mg/L) for 30 min and rinsed in running water. The residue was dehydrated in ovens equipped with air circulation at 60 °C to constant moisture content (12.05 ± 0.27 g/100 g). Subsequently, the dehydrated residue was ground in a food processor, placed in glass jars covered with aluminum foil and stored at room temperature until use.

2.1.3. Production of the Ice Cream

Five ice creams were produced, with different concentrations of soy kefir and soy extract Table 1. The following ingredients were used (g/100 g wet basis): 1.3 g of glucose, 1.0 g of emulsifier (Emustab), 0.80 g of dehydrated jaboticaba peel and 5.0 g of condensed soy milk. After homogenizing the ingredients for five minutes, the mixture was left to rest for 24 h, which corresponds to the maturation period. Then it was frozen and air incorporated (overrun) in a ice cream maker (MDG, model MH 80/100—São Carlos- Brazil).

Table 1. Formulations with different concentrations of soy extract (GS) and soy kefir (GK) (g/100 g wet basis).

Ingredients	Formulations (g/100 g Wet Basis)				
	GS	GK1	GK2	GK3	GK
Soy extract	100.0	75.0	50.0	25.0	-
Kefir	-	25.0	50.0	75.0	100.0

2.2. Methods

The formulations were processed on two separate occasions and evaluated during a 90-day storage period at −24 ± 2 °C. The proximate composition of the ice creams was determined in the freshly processed product, and the microbiological safety and viability of the potentially probiotic microorganisms were monitored at 15-day intervals. The other parameters (color, phenolic compounds, antioxidant activity, anthocyanin concentration and sensory profile) were evaluated immediately after preparation (T0) and at the end of the storage period (T90). The physicochemical and microbiological analyses were performed in triplicate.

2.2.1. Microbiological Analyses

Serial decimal dilutions were performed up to 10^{-8} (25 g of ice cream in 225 mL of sterile 0.1% peptone water). Microbiological quality was evaluated by the enumeration of coliforms at 45 °C, using the most probable number (MPN) assay [14] and detection of *Salmonella* spp. [15]. The population of potentially probiotic microorganisms was performed by plating on specific culture media: *Lactobacillus* spp.—Lactobacilli Man Rogosa Sharpe ágar—MRS (Difco, França), *Streptococcus* spp.—M17 ágar and *Bifidobacterium* spp.—BIM-25 (Reinforced Clostridium Ágar—Difco, França—with the addition of nalidixic acid, polymyxin B sulfate, kanamycin sulfate, iodoacetic acid and triphenyl tetrazolium chloride), respectively—using the microdroplet technique [16]. MRS and M17 plates were incubated in aerobiosis at 37 °C/48 h and BIM-25 plates in anaerobiosis at 37 °C/72 h [17]. Yeast enumeration was performed by surface plating on the Yeast Malt medium (YM, Himedia, India) with added chloramphenicol (200 mg/L) and incubation at 22 °C/72–120 h. The results were expressed as colony-forming units per gram (CFU/g).

2.2.2. Physicochemical Analyses

Proximate Composition, Caloric Value, pH and Titratable Acidity

The moisture, ash, protein and lipid contents were determined according to the Analytical Standards of the Adolfo Lutz Institute [18]. Total carbohydrate content was determined by difference [19]. The caloric value of the ice cream was estimated from the carbohydrate (4 kcal/g), protein (4 kcal/g) and fat (9 kcal/g) contents, and the result was expressed in kcal per 100 g. The pH was determined by a potentiometric method using a digital pH meter and the titratable acidity by titration with 0.1 N sodium hydroxide solution and expressed as % lactic acid [18].

Total Phenolic Compounds and Total Antioxidant Activity

The ice cream sample for the determination of phenolic compounds, antioxidant activity and anthocyanin were freeze-dried to preserve their characteristics in the evaluated time periods. The extracts for the determination of total phenolic compounds and antioxidant activity were obtained according to the procedure described by Karaaslan et al. [20]. The quantification of total phenolic compounds was performed by the Folin–Ciocalteu method [21]. Briefly, in a volumetric flask, the ice cream sample (0.1 mL), Folin–Ciocalteau reagent (0.5 mL), 20% sodium carbonate solution (1.5 mL) and distilled water were added to obtain a final volume of 10 mL. The mixture was kept at rest for two hours, and the absorbance was read at room temperature in a spectrophotometer (Shimadzu UV-Vis 1800/08302) at 765 nm. The results were expressed as milligrams of gallic acid equivalents (mgGAE) per 100 g of sample.

The antioxidant capacity was determined by capturing the ABTS free radical reaction according to the application used by Rufino et al. [22]. For the extract preparation, 20 mL of 50% methanol was added to one gram of the sample, followed by homogenization in Ultra-Turrax (IKA®/T25 digital, Staufen, Germany) and incubation for 60 min. The extract was centrifuged (25,400 G/15 min) and the supernatant was stored in an amber flask. Then, the operation was repeated using 70% acetone. In a place without light, the obtained extracts were diluted, and a 100 µL aliquot of each dilution was mixed with 1.0 mL of the

ABTS radical. After homogenization, the samples were left to stand for six minutes, and an absorbance reading was performed at 734 nm in a spectrophotometer (Shimadzu UV-1800). The analyses were performed in triplicate and the results are expressed in μmol Trolox/g.

Anthocyanin's Quantification

For the extraction of anthocyanins, 0.5 g of freeze-dried ice cream samples were diluted (1% HCL methanolic solution), homogenized (Ultra-Turrax®) [23] and vacuum filtered (Millex LCR 0.45 μm, 13 mm filter) [24]. The extractions were repeated until the color of the samples disappeared, and the obtained extracts were stored at 7 ± 1 °C for further analysis. The identification and quantification of anthocyanins were carried out by HPLC [25], using a chromatograph (Agilent®, Series 1100, Santa Clara, CA, USA) equipped with a quaternary pump system and UV-visible detector. A C18 Shim-pak CLC-ODS reverse-phase column (5 μm, 250 × 4.6 mm) was used for the separation of anthocyanins, and the mobile phase consisted of a linear elution gradient of formic acid (5%) and methanol, respecting the ratio of 85:15 (v/v) to 20:80 over 25 min, maintained for 15 min. The flow rate of the mobile phase, the injected volume and the column temperature used were 1.0 mL·min^{-1}, 5 μL and 29 °C, respectively. The chromatograms were processed at 520 nm, and standard curves—constructed from standards of cyanidin 3-glycoside and delphinidin 3-glycoside (Sigma-Aldrich, St. Louis, MI, USA)—were used for the quantification.

Instrumental Color Determination

The color parameters of the samples were determined in a Konica Minolta portable colorimeter (CR-410) using illuminant D65 and 10° visual angle—using the CIEL L * a * b * system—whereas the chroma and hue values were calculated based on parameters a * and b * [22].

Sensory Analysis

A total of 115 consumers (33 men and 82 women, aged between 18 and 59 years old) participated in the acceptance [26] and Check All That Apply tests (CATA). Participants were recruited from students and staff at the School of Pharmaceutical Sciences at UNESP—Araraquara, through questionnaires applied to assess allergies, intolerances or diseases that prevented participation in the tests and to identify aversion to any ingredient of the ice cream. The samples were labeled with 3-digit codes and presented monadically in random order. Data collection took place at the Sensory Laboratory of the School of Pharmaceutical Sciences in sensory booths with controlled temperature (23 ± 2 °C) and white lighting. The sensory analysis was approved by the Research Ethics Committee (REC 003167/2019).

The attributes of appearance, aroma and flavor were evaluated in the acceptance test using a nine-point structured hedonic scale, ranging from "I liked it a lot" to "I disliked it a lot" [27]. In the purchase-intention test, a five-point nominal category scale was used, ranging from "would definitely buy the product" to "would definitely not buy the product" [28].

Consumers were instructed to fill out a CATA questionnaire, containing attributes or phrases describing the ice cream samples [29]. Such attributes or phrases were previously generated using the grid method by ten trained assessors habituated to eating soy-based products and performing descriptive testing. The final list of attributes was defined in consensus with the assessor's team [30]. Consumers were instructed to point out any attributes they considered suitable to describe the ice cream samples.

2.2.3. Statistical Analysis

The data from microbiological and physicochemical tests were evaluated by analysis of variance and Tukey's test of means or a t test, ($p < 0.05$). The acceptance test results were analyzed by analysis of variance, considering the sample and consumers as sources of variation, and Tukey's test of means ($p < 0.05$). For the CATA, the frequency of indication of each term was determined, and the Cochran Q test was applied to identify significant differences among samples for each of the sensory attributes. Furthermore, Multiple Correspondence Analysis (MCA) was performed on the frequency table, containing the samples in rows and the CATA questionnaire terms in columns.

3. Results and Discussion

3.1. Microbiological Analyses

GS, GK1, GK2, GK3 and GK ice creams showed the absence of *Salmonella* spp. and had coliform counts at 45 °C < 3 NMP/g during the storage period (90 days), being considered safe for human consumption [14]. Regarding the potentially probiotic microorganisms, it was not possible to detect the presence of yeasts in the ice creams after 0, 15 and 30 days of storage at −22 °C ± 2. However, after 45 days of storage, the yeast count was above 8 log CFU/g for all the formulations, indicating a possible adaptation to the product conditions. The population of *Bifidobacterium* spp., *Lactobacillus* spp. and *Streptococcus* spp. was higher than 10 log CFU/g in all formulations throughout the storage period. Conventionally, kefir is prepared with cow's milk, an ideal medium for lactic fermentation. In this study, replacing cow's milk with soy extract did not reduce the viability of potentially probiotic bacteria, indicating an adaptation of the starter culture to the ice cream ingredients and processing conditions (Table 2). Santos et al. [31] also observed high counts of *Lactococcus* spp., *Lactobacillus* spp. and yeasts in soymilk kefir with prebiotic addition, after 28 days of storage. Likewise, Walter et al. [32] verified that *Bifidobacterium animalis* subsp. *lactis* BB-12 and *Lactobacillus acidophillus* LA-5 populations remained stable during the storage period using different soy-based raw materials (soybean of the BRS 232 cultivar and commercial soy powder extracts obtained by liquid extraction and solid extraction) to produce kefir.

Table 2. Viability of yeasts, *Bifidobacterium* spp., *Lactobacillus* spp. and *Streptococcus* spp. in ice cream during storage at −22 °C ± 2.

Formulations	Time (Days)	Yeasts Log UFC/g	*Bifidobacterium* spp. LogUFC/g	*Lactobacillus* spp. Log UFC/g	*Streptococcus* spp. Log UFC/g
GK1	0	<1	10.46 ^{abc} ± 0.07	10.60 ^{a} ± 0.07	10.79 ^{a} ± 0.02
	15	<1	10.38 ^{bc} ± 0.04	10.49 ^{ab} ± 0.02	10.76 ^{ab} ± 0.02
	30	<1	10.32 ^{c} ± 0.04	10.32 ^{bc} ± 0.02	10.73 ^{abc} ± 0.01
	45	8.93 ^{a} ± 0.20	10.31 ^{c} ± 0.02	10.35 ^{bc} ± 0.07	10.69 ^{bcd} ± 0.02 ^{b}
	60	8.72 ^{a} ± 0.20	10.48 ^{ab} ± 0.04	10.42 ^{ab} ± 0.12	10.65 ^{cde} ± 0.03
	75	8.30 ^{a} ± 0.00	10.55 ^{a} ± 0.05	10.30 ^{bc} ± 0.07	10.62 ^{de} ± 0.03
	90	8.30 ^{a} ± 0.00	10.49 ^{ab} ± 0.10	10.21 ^{c} ± 0.08	10.57 ^{e} ± 0.04
GK2	0	<1	10.58 ^{a} ± 0.02	10.57 ^{a} ± 0.06	10.77 ^{a} ± 0.01
	15	<1	10.49 ^{ab} ± 0.05	10.53 ^{ab} ± 0.07	10.74 ^{ab} ± 0.01
	30	<1	10.37 ^{bc} ± 0.08	10.41 ^{abc} ± 0.11	10.71 ^{abc} ± 0.02
	45	9.77 ^{a} ± 0.21	10.34 ^{bc} ± 0.07	10.26 ^{bcd} ± 0.10	10.68 ^{bcd} ± 0.03
	60	9.36 ^{b} ± 0.10	10.41 ^{abc} ± 0.10	10.29 ^{abcd} ± 0.01	10.64 ^{cde} ± 0.04
	75	9.60 ^{a} ± 0.00	10.34 ^{bc} ± 0.05	10.15 ^{cd} ± 0.14	10.60 ^{de} ± 0.04
	90	9.30 ^{b} ± 0.00	10.29 ^{c} ± 0.01	10.03 ^{d} ± 0.14	10.56 ^{e} ± 0.04
GK3	0	<1	10.52 ^{a} ± 0.07	10.55 ^{a} ± 0.11	10.75 ^{a} ± 0.02
	15	<1	10.38 ^{a} ± 0.11	10.47 ^{ab} ± 0.02	10.71 ^{ab} ± 0.02
	30	<1	10.32 ^{a} ± 0.07	10.31 ^{bc} ± 0.02	10.68 ^{abc} ± 0.03
	45	9.92 ^{a} ± 0.08	10.40 ^{a} ± 0.09	10.31 ^{bc} ± 0.02	10.64 ^{bcd} ± 0.03
	60	9.70 ^{ab} ± 0.17	10.48 ^{a} ± 0.05	10.27 ^{bc} ± 0.04	10.60 ^{cd} ± 0.03
	75	9.36 ^{b} ± 0.10	10.41 ^{a} ± 0.05	10.14 ^{cd} ± 0.12	10.56 ^{de} ± 0.03
	90	9.40 ^{b} ± 0.17	10.35 ^{a} ± 0.06	10.06 ^{d} ± 0.10	10.50 ^{e} ± 0.04

Table 2. *Cont.*

Formulations	Time (Days)	Yeasts Log UFC/g	*Bifidobacterium* spp. LogUFC/g	*Lactobacillus* spp. Log UFC/g	*Streptococcus* spp. Log UFC/g
GK	0	<1	10.48 ^{a} ± 0.09	10.57 ^{a} ± 0.08	10.79 ^{a} ± 0.03
	15	<1	10.35 ^{ab} ± 0.08	10.46 ^{ab} ± 0.09	10.77 ^{ab} ± 0.03
	30	<1	10.29 ^{b} ± 0.01	10.38 ^{ab} ± 0.11	10.74 ^{ab} ± 0.03
	45	9.77 ^{a} ± 0.21	10.34 ^{ab} ± 0.07	10.37 ^{ab} ± 0.08	10.70 ^{abc} ± 0.03
	60	9.46 ^{a} ± 0.15	10.40 ^{ab} ± 0.07	10.43 ^{ab} ± 0.13	10.66 ^{bcd} ± 0.04
	75	9.60 ^{a} ± 0.00	10.48 ^{ab} ± 0.07	10.34 ^{ab} ± 0.04	10.62 ^{cd} ± 0.04
	90	9.56 ^{a} ± 0.07	10.34 ^{ab} ± 0.03	10.29 ^{b} ± 0.01	10.58 ^{d} ± 0.04

Means (±standard deviation) followed by different lowercase letters in the same column differ statistically from each other, according to the Tukey test ($p \leq 0.05$). GS (100% S), GK1 (75% S, 25% kefir), GK2 (50% S, 50% kefir), GK3 (25% S, 75% kefir), GK (100% kefir). Average of the triplicate of two different processes.

Ice cream's processing steps, such as homogenization, churning and the consequent incorporation of oxygen, as well as low storage temperatures, may result in decreasing the population of beneficial microorganisms. Temperature fluctuations during storage, resulting in the formation of ice crystals, can also reduce the survival of strains [32]. However, in the present work, the bacteria and yeast from the kefir starter culture were resistant to the stress conditions of the process, requiring no additional strategies to protect them.

There is no consensus about an appropriate dose of probiotic organisms required to achieve beneficial effects. According to the International Scientific Association for Probiotics and Prebiotics (ISAPP), all developed ice creams should be designated as a product "containing live and active cultures". These products could not be considered a probiotic since they have a diverse microbial community that is not completely defined, in terms of the composition and stability of the strains, and without proof of beneficial effects. All ice cream also showed a minimum of 9 log CFU per serving, as suggested by ISAPP [33].

3.2. Physicochemical Analyses

The results of the physicochemical analyses are presented in Table 3. Overall, kefir caused a decrease in the pH values and an increase in the titratable acidity of the ice creams. This effect was expected since kefir is a product fermented by lactic and acetic bacteria, which metabolizes carbohydrates and produces acids, mainly lactic, acetic, citric, propionic and butyric acids [34,35].

Table 3. Physicochemical analysis and caloric value of the five ice cream formulations (g/100 g).

Formulations	pH	Acidity %	Ashes	Protein %	Lipid %	Moisture %	Carbohydrates %	Kcal/100 g
GS	6.15 ^{a} ± 0.05	6.36 ^{b} ± 0.36	0.64 ^{d} ± 0.00	4.55 ^{a} ± 0.24	3.82 ^{a} ± 0.51	80.25 ^{a} ± 0.14	11.36 ^{c} ± 0.14	98.02 ^{c} ± 3.17
GK1	4.92 ^{c} ± 0.02	10.06 ^{a} ± 0.32	0.72 ^{b} ± 0.01	5.33 ^{a} ± 0.04	4.24 ^{a} ± 0.39	74.91 ^{b} ± 1.52	15.49 ^{b} ± 1.35	121.44 ^{b} ± 7.91
GK2	5.04 ^{b} ± 0.02	10.61 ^{a} ± 0.09	0.77 ^{a} ± 0.01	5.11 ^{a} ± 1.25	3.32 ^{a} ± 0.42	72.70 ^{c} ± 0.18	18.16 ^{b} ± 1.47	122.96 ^{b} ± 5.77
GK3	4.92 ^{c} ± 0.01	10.86 ^{a} ± 0.15	0.76 ^{a} ± 0.01	5.74 ^{a} ± 0.10	3.67 ^{a} ± 0.89	73.36 ^{bc} ± 0.36	17.20 ^{b} ± 1.08	124.79 ^{b} ± 3.72
GK	4.76 ^{d} ± 0.06	10.92 ^{a} ± 0.11	0.68 ^{c} ± 0.01	5.37 ^{a} ± 0.02	3.52 ^{a} ± 0.45	69.62 ^{d} ± 0.33	21.59 ^{a} ± 0.12	139.52 ^{a} ± 0.38

Means (±standard deviation) followed by different lowercase letters in the same column differ statistically from each other, according to the Tukey test ($p \leq 0.05$). GS (100% S), GK1 (75% S, 25% kefir), GK2 (50% S, 50% kefir), GK3 (25% S, 75% kefir), GK (100% kefir). S = soy extract. Average of the triplicate of two different processes.

Kefir also caused a significant decrease in the moisture content of the ice cream (moisture GK = 69.61 ± 0.33 g/100 g and GS = 80.25% ± 0.14 g/100 g), and this result was expected because kefir has a lower moisture content than soy extract (Kefir: 84.47 ± 0.18 g/100 g; soy extract 94.25 ± 0.08 g/100 g). The GK formulation (100% kefir) showed the highest carbohydrate content (21.59 ± 0.12 g/100 g), differing from the other formulations produced with a mixture of kefir and soy extract. This behavior is explained by the addition of sugar (10% *w*/*v*) to the soy extract used to obtain the soy kefir. The GK and GS formulations have the highest and lowest caloric values, with 139.52 and 98.02 Kcal/100 g, respectively, because of their carbohydrate contents. The protein and lipid contents were similar among all formulations.

3.3. Total Phenolic Compounds and Antioxidant Capacity

In T0, formulation GS presented the lowest content of total phenolic compounds (567.65 ± 35.60 mgGAE/100 g, $p < 0.05$) while GK had the highest average of such compounds (7631.69 ± 47.73 mgGAE/100 g, $p < 0.05$). Higher stability in total phenolic content was observed for GK3, GK and GK2 formulations with a reduction of only 6.94%, 7.73% and 8.29%, respectively, after 90 days of storage. However, the opposite effect was observed in the GS formulation with a reduction of 52.26% (Table 4).

Table 4. Total phenolic compounds and antioxidant capacity of ice cream after 0 and 90 days of storage.

Formulations	Total Phenolics (mgEAG/100 g)		Antioxidant Capacity (µmol Trolox/g)	
	T0	T90	T0	T90
GS	567.65 ^{aD} ± 35.60	271.00 ^{bD} ± 18.82	2.14 ^{aB} ± 0.46	1.80 ^{aA} ± 0.10
GK1	5970.93 ^{aC} ± 595.43	4755.51 ^{bC} ± 58.59	3.32 ^{aA} ± 0.05	1.84 ^{bA} ± 0.04
GK2	6783.94 ^{aB} ± 78.10	6221.41 ^{bB} ± 23.27	3.35 ^{aA} ± 0.02	1.95 ^{bA} ± 0.07
GK3	6670.40 ^{aBC} ± 32.63	6207.21 ^{aB} ± 19.74	2.81 ^{aAB} ± 0.20	1.93 ^{bA} ± 0.06
GK	7631.69 ^{aA} ± 47.73	7042.00 ^{bA} ± 48.34	2.99 ^{aA} ± 0.27	1.89 ^{bA} ± 0.05

Means (±standard deviation) followed by different lowercase letters on the same line differ statistically from each other, according to Test T (times—$p \leq 0.05$). Means (±standard deviation) followed by different capital letters in the same column differ statistically from each other (formulations—$p \leq 0.05$), according to the Tukey test ($p \leq 0.05$). GS (100% S), GK1 (75% S, 25% kefir), GK2 (50% S, 50% kefir), GK3 (25% S, 75% kefir), GK (100% kefir). S = soy extract.

The concentrations of total phenolics in each ice cream formulation were higher than those found by other authors who used jaboticaba in the processing of symbiotic concentrated yogurt with 1% of jaboticaba bark flour (292.5 ± 2.5 mg GAE/100 g) [36] and jaboticaba juice (150.4 ± 0.6 mg/L) [37]. The bark of jaboticaba is rich in phenolic compounds, such as gallic and ellagic acid, rutin and quercetin [38,39], and soy extract is a source of these same compounds and isoflavones, ferulic, gallic and vanillic acids [40], which justifies the high levels of total phenolics in ice cream. The higher concentration of total phenolic compounds in the samples containing kefir is likely due to the metabolic activity of the starter culture. Enzymes derived from microorganisms—such as ß-glycosidase—hydrolyze complex phenolic compounds into simpler ones, leading to an increase in the total phenolic content [41,42].

The ice creams processed with kefir also exhibited the highest antioxidant capacity, without differing from each other. However, after 90 days of storage, only the sample without kefir (GS) did not show a significant reduction in this parameter. This result cannot be attributed to the concentration of total phenolic compounds because the formulations with kefir showed a lower reduction in these compounds at the end of the storage period. The antioxidant activity of a food is related to the concentration and chemical structure of bioactive compounds (phenolic compounds, vitamins and enzymes superoxide dismutase, catalase and peroxidase), which may change during storage. In addition, synergy among compounds may result in increased antioxidant activity [43,44] and explain the observed difference.

3.4. Anthocyanin Content

The concentration of anthocyanins was expressed as cyanidin 3-glucoside and delphinidin 3-glucoside (Table 5), since these compounds are found in higher concentrations in jaboticaba bark and are related to its purple color.

Table 5. Quantification of anthocyanins cyanidin 3-glycoside and delphinidin 3-glycoside in ice cream after 0 and 90 days of storage.

Formulations	Cyanidin 3-Glycoside (mg/100 g) T0	Cyanidin 3-Glycoside (mg/100 g) T90	Delphinidin 3-Glycoside (mg/100 g) T0	Delphinidin 3-Glycoside (mg/100 g) T90
GS	$0.29^{aA} \pm 0.00$	$0.26^{bA} \pm 0.00$	$11.11^{aA} \pm 0.07$	$9.89^{bA} \pm 0.00$
GK1	$0.23^{aB} \pm 0.01$	$0.22^{bB} \pm 0.01$	$8.77^{aC} \pm 0.30$	$8.26^{aB} \pm 0.00$
GK2	$0.22^{aBC} \pm 0.00$	$0.18^{bC} \pm 0.00$	$8.36^{aC} \pm 0.07$	$7.07^{bC} \pm 0.00$
GK3	$0.21^{aC} \pm 0.00$	$0.16^{bD} \pm 0.00$	$8.15^{aC} \pm 0.10$	$6.95^{bC} \pm 0.00$
GK	$0.19^{aD} \pm 0.00$	$0.17^{bD} \pm 0.00$	$10.23^{aB} \pm 0.02$	$6.72^{bC} \pm 0.00$

Means (±standard deviation) followed by different lowercase letters on the same line differ statistically from each other, according to Test T (times—$p \leq 0.05$). Means (±standard deviation) followed by different capital letters in the same column differ statistically from each other (formulations—$p \leq 0.05$), according to the Tukey test ($p \leq 0.05$). GK1 (75% S, 25% kefir), GK2 (50% S, 50% kefir), GK3 (25% S, 75% kefir), GK (100% kefir). S = soy extract.

The concentration of anthocyanins in ice cream formulations is related to the amount of peel added (0.02%), which was identical for all formulations. Delphinidin was the most abundant anthocyanin, contrary to the data obtained by Reynertson et al. [39] and Inada et al. [37] who reported the predominance of cyanidin in jaboticaba and jaboticaba juice, respectively. The formulation without kefir (GS) presented the highest values of cyanidin 3-glycoside and delphinidin 3-glycoside, differing significantly from the other formulations at the two evaluated times. The lower values of anthocyanins found in samples GK, GK1 and GK2 may be related to the activity of kefir strains. Some microorganisms can produce enzymes that hydrolyze anthocyanins into less stable aglycone forms and/or can produce hydrogen peroxide that facilitates their degradation [45]. A significant reduction in the concentration of anthocyanins was observed during storage for all formulations. Such behavior was expected, since anthocyanins are unstable to variations of temperature, oxygen, light, pH and acidity of the medium [46,47]. The concentration of anthocyanins can also be affected by a mixture of copper, iron and manganese, which act as catalysts for the oxidation reaction and are found in the bark of jaboticaba [48].

One of the challenges of using anthocyanins as natural dyes in foods is their instability in the face of pH variations, which affect the color conferred by these compounds. While a low pH is associated with greater stability of the compound, an increase in the pH results in a reduction in intensity and a change in the color pattern conferred [49]. The pH of the ice cream (Table 3) did not interfere positively with the concentration and stability of anthocyanins (Table 5), and we can infer that the different processing steps, which lead to the incorporation of oxygen in the samples, may have influenced the results obtained.

3.5. Color Determination

Table 6 shows the results for the color parameters (L *, a *, b *, C * and H°) during storage (T0 and T90). Briefly, during the storage period, the samples produced with kefir became darker (L * reduction) and with lower color intensity (C * reduction). The red color was predominant in all samples, except for GK1.

The GK and GK3 formulations were the lightest samples showing the highest averages for parameter L * at T0 and T90. The values of a * were positive for all formulations, indicating a predominance of red color, and only GK1 showed a significant reduction in this parameter at the end of the storage time. The b * value was also positive for all formulations, with a tendency of a yellow color.

Chroma (C *) is a quantitative parameter related to the color intensity perceived by the human eye. The GK1, GK3 and GK samples have the highest averages for C * at T0, and the formulation with 50% kefir (GK2) exhibited the greatest reduction in this parameter at the end of 90 days. The results of the hue angle (h°—qualitative attribute of the color) indicate that GK (T0 and T90), GK3 (T0) and GK2 (T90) are characterized by a reddish color, while in the other samples, there was a predominance of yellow.

Table 6. Instrumental color parameters (L *, a *, b *, C * and H°) of ice creams after 0 and 90 days of storage.

		Formulations				
		GS	**GK1**	**GK2**	**GK3**	**GK**
L *	T 0	0.60 ^{cA} ± 0.12	0.84 ^{bA} ± 0.07	0.47 ^{cA} ± 0.08	0.91 ^{bA} ± 0.07	1.21 ^{aA} ± 0.03
	T 90	0.59 ^{bA} ± 0.10	0.26 ^{cB} ± 0.22	0.32 ^{bcB} ± 0.06	1.02 ^{aA} ± 0.05	0.94 ^{aB} ± 0.16
a *	T 0	0.15 ^{bA} ± 0.05	0.22 ^{abA} ± 0.07	0.15 ^{bA} ± 0.05	0.29 ^{abA} ± 0.04	0.36 ^{aA} ± 0.10
	T 90	0.16 ^{abA} ± 0.06	0.05 ^{bB} ± 0.01	0.09 ^{bA} ± 0.05	0.20 ^{abA} ± 0.14	0.27 ^{aA} ± 0.04
b *	T 0	0.27 ^{abA} ± 0.07	0.33 ^{abA} ± 0.02	0.33 ^{abA} ± 0.02	0.25 ^{bA} ± 0.08	0.40 ^{aA} ± 0.02
	T 90	0.22 ^{aA} ± 0.04	0.25 ^{aB} ± 0.02	0.11 ^{bB} ± 0.03	0.23 ^{aA} ± 0.08	0.25 ^{aB} ± 0.04
C *	T 0	0.31 ^{bA} ± 0.07	0.40 ^{abA} ± 0.04	0.36 ^{bA} ± 0.01	0.38 ^{abA} ± 0.08	0.40 ^{aA} ± 0.02
	T 90	0.28 ^{aA} ± 0.05	0.26 ^{abB} ± 0.02	0.15 ^{bB} ± 0.03	0.25 ^{abA} ± 0.11	0.25 ^{aB} ± 0.04
h°	T 0	60.78 ^{abA} ± 6.99	56.50 ^{abB} ± 7.97	65.47 ^{aA} ± 8.67	40.00 ^{bA} ± 7.50	40.38 ^{bA} ± 7.79
	T 90	54.32 ^{abA} ± 11.38	78.71 ^{aA} ± 3.18	41.35 ^{bB} ± 8.06	53.50 ^{abA} ± 14.02	40.50 ^{bA} ± 3.79

Means (±standard deviation) followed by different lowercase letters in the same row (sample comparison) and different uppercase letters in the same column (time comparison) differ statistically according to the Tukey's test ($p \leq 0.05$). GK1 (75% S, 25% kefir), GK2 (50% S, 50% kefir), GK3 (25% S, 75% kefir), GK (100% kefir). S = soy extract.

The characteristic coloration of the anthocyanin pigment is variable and can give the product a pink, red or blue color, depending on the source. Although equal concentrations of dehydrated jaboticaba peel were used, the addition of different concentrations of kefir may interfere with the color of the product, as well as with its stability. In summary, the results indicate that the ice creams are characterized as dark, with a color ranging from red to yellow and with low saturation. GK1 and GK2 exhibited the lowest color stability during storage, and the one prepared with 100% kefir (GK) showed a tendency toward the most pronounced red color.3.6. Sensory Analysis.

In T0, GK and GK2 showed the highest acceptance averages for the flavor attribute, without differing from GK3 and GS. There were no significant differences ($p > 0.05$) among the ice creams in appearance, color and aroma. After 90 days of storage, GK showed the highest hedonic averages for the sensory attributes evaluated, differing from all formulations regarding flavor. It is also noteworthy that the sample with 100% kefir (GK) was the only one that did not show a reduction in color acceptance during the storage time (Table 7).

Table 7. Ice cream acceptance at T0 and T90 days of storage.

Time	Formulations				
	GS	**GK1**	**GK2**	**GK3**	**GK**
T0					
Appearance	6.73 ^{aA} ± 1.66	6.88 ^{aA} ± 1.67	7.09 ^{aA} ± 1.60	7.21 ^{aA} ± 1.37	7.17 ^{aA} ± 1.58
Color	6.75 ^{aA} ± 1.64	6.79 ^{aA} ± 1.56	7.02 ^{aA} ± 1.50	7.23 ^{aA} ± 1.36	7.18 ^{aA} ± 1.48
Aroma	5.73 ^{aA} ± 1.62	5.63 ^{aA} ± 1.56	5.98 ^{aA} ± 1.44	5.90 ^{aA} ± 1.40	5.95 ^{aA} ± 1.67
Flavor	5.57 ^{abA} ± 2.05	5.18 ^{bA} ± 1.98	5.92 ^{aA} ± 1.85	5.84 ^{abA} ± 1.91	6.06 ^{aB} ± 2.12
T90					
Appearance	6.31 ^{bA} ± 1.86	6.28 ^{bB} ± 1.94	6.56 ^{abB} ± 1.67	6.69 ^{abB} ± 1.64	7.15 ^{aA} ± 1.59
Color	6.22 ^{bB} ± 1.87	6.24 ^{bB} ± 1.83	6.55 ^{abB} ± 1.71	6.59 ^{abB} ± 1.71	7.03 ^{aA} ± 1.62
Aroma	5.90 ^{aA} ± 1.71	5.85 ^{aA} ± 1.51	5.84 ^{aA} ± 1.47	6.10 ^{aA} ± 1.54	6.29 ^{aA} ± 1.65
Flavor	5.87 ^{bA} ± 1.98	5.02 ^{cA} ± 1.88	5.47 ^{bcA} ± 1.83	5.95 ^{bA} ± 1.70	6.86 ^{aA} ± 1.94

Means (±standard deviation) followed by different lowercase letters on the same line differ statistically from each other, and different uppercase letters on the same column statistically according to the Tukey test ($p \leq 0.05$). GK1 (75% S, 25% kefir), GK2 (50% S, 50% kefir), GK3 (25% S, 75% kefir), GK (100% kefir). S = soy extract.

Formulations GK, GK3 and GK2 showed the highest frequencies of positive purchase intention ("would certainly or probably buy the products") at T0 (44.2%, 38.4% and 34.2%) and T90 (59.8%, 34.5% and 31.1%) (Table 7). On the other hand, the highest frequencies of negative purchase intention ("probably or certainly would not buy") were obtained by GS (35.8% at T0 and 35.2% at T90) and GK1 samples (42.5% at T0 and 47.5% at T90). These results agree with those obtained in the hedonic test for storage time T0, where formulations GS and GK1 exhibited the lowest acceptance averages for all the attributes evaluated and GK was the most accepted.

A total of 24 attributes were previously selected by the trained assessors to characterize the ice cream by the CATA method. The frequencies of the attributes and the results of Cochran's Q-test for the freshly processed product (T0) are reported in Table 8. The most selected attributes were soy flavor, aftertaste, refreshing flavor, natural jaboticaba flavor and sweetness. Cochran's Q test showed significant differences in 6 of the 24 attributes analyzed, and all of them were related to flavor (sweet, jaboticaba flavor, acid taste, refreshing flavor, soy flavor and yogurt flavor), indicating its importance for detecting differences and characterizing the ice cream.

Table 8. Frequency of sensory attributes and Cochran's Q test associated with each ice cream at the beginning of the storage period (T0).

	Formulations					
Attributes	**GS**	**GK1**	**GK2**	**GK3**	**GK**	***p***
Sweet	24	32	32	30	53	<0.0001
Milk flavor	22	19	30	26	16	0.065
Artificial jaboticaba flavor	20	19	19	23	23	0.822
Natural jaboticaba flavor	24	25	25	33	38	0.032
Creamy texture	48	51	49	43	58	0.299
Soft texture	48	39	44	41	40	0.680
Mild flavor	42	36	49	41	40	0.413
Watered down	23	31	29	25	19	0.243
Acid taste	04	05	14	18	26	<0.0001
Refreshing flavor	31	23	31	38	44	0.005
Vanilla flavor	17	19	19	20	18	0.970
Fruit scent	22	18	13	22	23	0.240
Aftertaste	47	46	32	38	35	0.062
Fermented flavor	26	28	24	29	27	0.890
Artificial scent	10	14	12	09	10	0.695
Soy flavor	77	69	57	58	55	0.001
Roughness	37	38	31	40	35	0.649
Acid scent	03	03	05	08	05	0.380
Fermented scent	24	21	21	25	26	0.808
Yogurt flavor	21	20	30	25	36	0.033
Astringency	15	18	20	17	20	0.791
Purple fruit color (berries)	44	42	42	39	45	0.805
Natural jaboticaba color	39	38	39	48	42	0.218
Attractive color	42	37	47	47	50	0.138

n = 24. $p \leq 0.05$ indicates a significant difference in Cochran's Q test. GK1 (75% S, 25% kefir), GK2 (50% S, 50% kefir), GK3 (25% S, 75% kefir), GK (100% kefir) S = soy extract.

After 90 days of storage, the CATA test was applied again to identify possible sensory changes as a function of time. The results of Cochran's Q test showed significant differences ($p < 0.05$) in 14 of the attributes analyzed (Table 9). The attributes sweet, creamy texture, soy flavor, natural jaboticaba flavor, roughness, purple fruit color (berries), refreshing flavor and attractive color were mentioned more frequently, showing a change in the sensory profile of the formulations during the storage period.

Table 9. Frequency of sensory attributes and Cochran's Q test associated with each ice cream at the end of the storage period (T90).

Attributes	Formulations					
	GS	GK1	GK2	GK3	GK	p
Sweet	50	26	28	40	75	<0.0001
Milk flavor	29	25	21	19	22	0.369
Artificial jaboticaba flavor	22	22	25	26	19	0.702
Natural jaboticaba flavor	32	25	31	43	50	<0.0001
Creamy texture	76	51	43	51	72	<0.0001
Soft texture	53	40	43	38	47	0.182
Mild flavor	56	39	47	45	49	0.186
Watered down	15	39	36	38	16	<0.0001
Acid taste	03	10	10	22	24	<0.0001
Refreshing flavor	28	19	27	38	52	<0.0001
Vanilla flavor	28	27	27	21	24	0.694
Fruit scent	23	16	22	30	38	0.001
Aftertaste	39	44	45	40	30	0.115
Fermented flavor	20	27	23	35	29	0.081
Artificial scent	18	19	16	12	11	0.238
Soy flavor	57	67	65	43	34	<0.0001
Roughness	31	41	44	49	37	0.044
Acid scent	03	06	05	09	09	0.225
Fermented scent	15	10	17	19	20	0.192
Yogurt flavor	16	17	27	29	39	0.000
Astringency	17	32	28	32	18	0.001
Purple fruit color (berries)	47	41	57	48	56	0.002
Natural jaboticaba color	39	37	42	41	45	0.590
Attractive color	26	29	37	44	53	<0.0001

n = 24. $p \leq 0.05$ indicates a significant difference in Cochran's Q test.GK1 (75% S, 25% kefir), GK2 (50% S, 50% kefir), GK3 (25% S, 75% kefir), GK (100% kefir). S = soy extract.

Multiple correspondence analyses showed that the first and second dimensions explain 92.86% of the variance of the experiment at T0 (Figure 1a). The statistical test uses only the most frequently cited attributes since those with a low frequency can lead to false results [50]. The formulations were positioned in three different groups and obtained good attribute separation among the three quadrants. GS and GK1 were characterized by an aftertaste and a soy flavor, GK3 by a refreshing flavor and yogurt flavor, and GK2 did not obtain any relationship with the descriptor terms. The GK formulation, which exhibited the highest absolute mean for the flavor attribute, was characterized by a sweet and natural jaboticaba flavor, evidencing the impact of these characteristics on the product consumer. At T90 (Figure 1b), the first and second dimensions explained 94.62% of the variation between samples. The ice creams were positioned in four different quadrants and characterized by different attributes. The attribute that best described the GS formulation was a creamy texture; GK1 and GK2 were characterized by a purple fruit color (berries) and GK3 by a fermented flavor. Once again, the formulation with 100% kefir (GK) presented a distinct sensory profile, being associated with pleasant attributes such as a fruit aroma, refreshing flavor, natural jaboticaba flavor, attractive color and yogurt flavor.

The change in the descriptive profile generated by CATA after 90 days of storage may be associated with physicochemical characteristics of color and anthocyanin content, since color-related sensory attributes were not considered important in the freshly processed product. The color of foods and beverages has been found to play important roles in consumer perception of other sensory attributes (taste and aroma), product acceptance and emotional responses [51]. In this study, the results of instrumental color showed that the ice cream produced with 100% kefir (GK) had the highest intensity of red color, which likely contributed to the better sensory profile of this formulation.

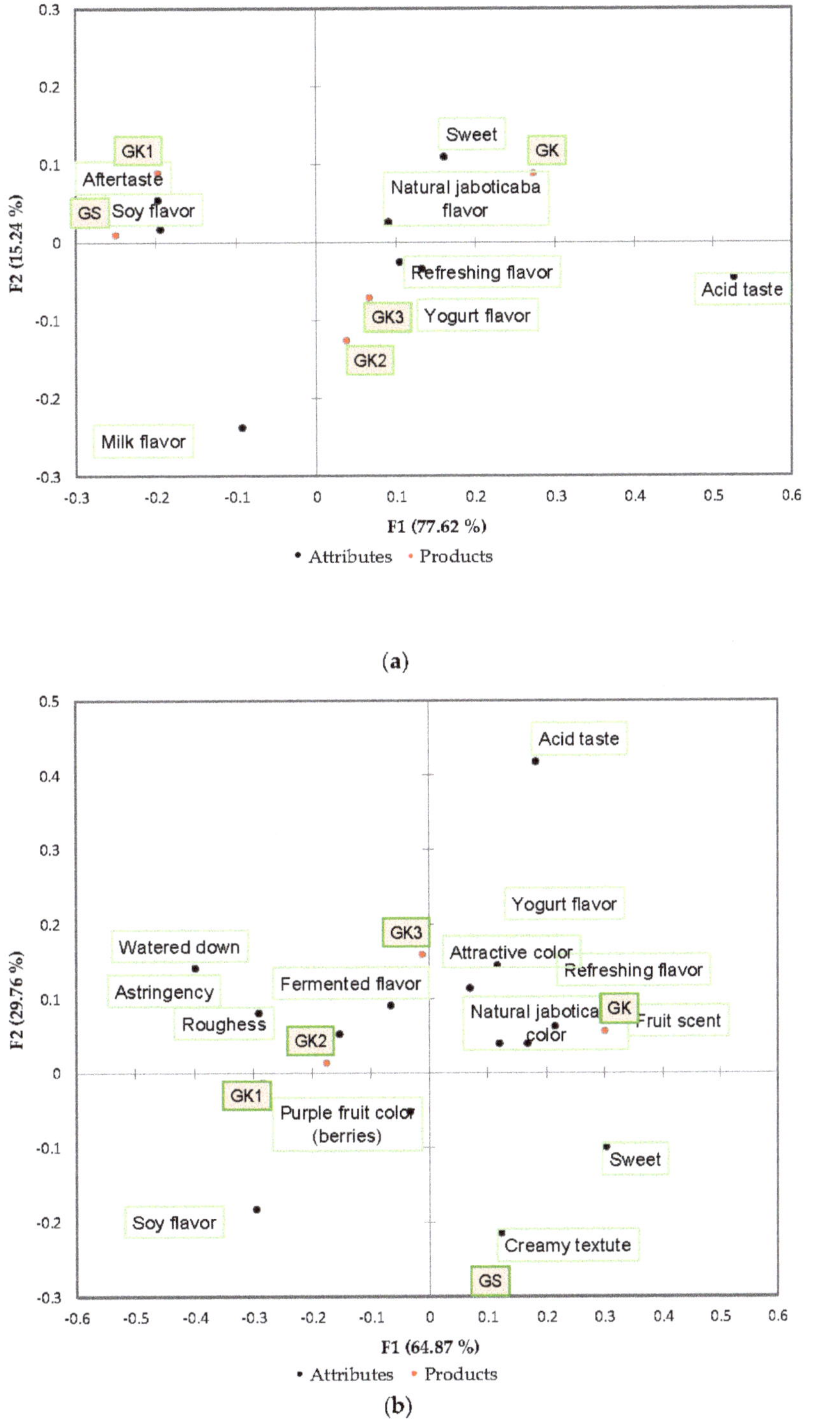

Figure 1. Correspondence analysis of the sensory descriptors in the CATA analysis for the initial time (T0) (**a**) and final time (T90) (**b**).

4. Conclusions

The results indicated that these ice creams, especially the one prepared with 100% kefir (GK), are suitable vehicles for potential probiotic strains. The jaboticaba peels imparted a purple color and improved the antioxidant capacity and anthocyanin profile of the products. The use of alternative raw materials and ingredients made it possible to obtain a product with functional potential that can be ingested by individuals who adopt a diet with the restriction of lactose, milk protein or foods of animal origin.

Author Contributions: Conceptualization, D.C.U.C.; methodology, D.C.U.C. and G.M.N.M.; validation, G.M.N.M.; formal analysis, G.M.N.M., E.M.D.O., A.O.R. and C.H.P.; investigation, G.M.N.M.; writing—original draft preparation, G.M.N.M. and D.C.U.C.; writing—review and editing, G.M.N.M. and D.C.U.C.; supervision, D.C.U.C.; funding acquisition, G.M.N.M. and D.C.U.C. All authors have read and agreed to the published version of the manuscript.

Funding: The authors thank the Coordination of Improvement of Higher Education Personnel—Brazil (CAPES)—Financial Code 001 and the School of Pharmaceutical Sciences (UNESP-Campus ARARAQUARA-SP) for their financial support.

Institutional Review Board Statement: Ethics approval was required for this research number (caae: 05932918.1.0000.5426).

Informed Consent Statement: Informed consent was obtained from all subjects involved in the study.

Data Availability Statement: The datasets generated during analyzed during the current study are available from the corresponding author on request.

Acknowledgments: The authors would like to thank Eliane Vale da Silva, Mariela Zucolo and Maria Jara Montibeller for supporting the execution of sensory analysis.

Conflicts of Interest: The authors declare no conflict of interest.

References

1. Avelar, A.E.D.; Rezende, D.C.D. *Hábitos Alimentares Fora Do Lar: Um Estudo De Caso Em Lavras—MG;* Eating away from Home: A Case Study in Lavras—MG; Federal University of Lavras: Lavras, Brazil, 2013; pp. 137–152.
2. FAO. Codex Standards for Fermented Milks. Immun Inflamm Heal Dis Emerg Roles Nutraceuticals Funct Foods Immune Support. Available online: https://www.fao.org/fao-who-codexalimentarius/sh-proxy/en/?lnk=1&url=https%253A%252F%252Fworkspace.fao.org%252Fsites%252Fcodex%252FStandards%252FCXS%2B243-2003%252FCXS_243e.pdf (accessed on 1 April 2018).
3. Romanin, D.; Serradell, M.; González Maciel, D.; Lausada, N.; Garrote, G.L.; Rumbo, M. Down-regulation of intestinal epithelial innate response by probiotic yeasts isolated from kefir. *Int. J. Food Microbiol.* **2010**, *140*, 102–108. [CrossRef] [PubMed]
4. Lee, M.Y.; Ahn, K.S.; Kwon, O.K.; Kim, M.J.; Kim, M.K.; Lee, I.Y.; Oh, S.R.; Lee, H.K. Anti-inflammatory and anti-allergic effects of kefir in a mouse asthma model. *Immunobiology* **2007**, *212*, 647–654. [CrossRef] [PubMed]
5. Rodrigues, K.L.; Gaudino Caputo, L.R.; Tavares Carvalho, J.C.; Evangelista, J.; Schneedorf, J.M. Antimicrobial and healing activity of kefir and kefiran extract. *Int. J. Antimicrob. Agents* **2005**, *25*, 404–408. [CrossRef]
6. Otles, S.; Çagindi, O. Kefir: A Probiotic Dairy-Composition, Nutritional and Therapeutic Aspects. *Pak. J. Nutr.* **2003**, *2*, 54–59.
7. Rosa, D.D.; Dias, M.M.S.; Grześkowiak, Ł.M.; Reis, S.A.; Conceição, L.L.; Peluzio, M.D.C.G. Milk kefir: Nutritional, microbiological and health benefits. *Nutr. Res. Rev.* **2017**, *30*, 82–96. [CrossRef]
8. de Góes-Favoni, S.P.; Beléia, A.D.P.; Carrão-Panizzi, M.C.; Mandarino, J.M.G. Isoflavonas em produtos comerciais de soja. *Ciênc. Tecnol. Aliment.* **2004**, *24*, 582–586. [CrossRef]
9. Dos Santos, P.; Leite, N.; Martins, L.; Lodete, A.; Motta, R. Bebida fermentada a base de soja com sabor de ameixa e suplementada com inulina em substituição ao iogurte tradicional. *Vet. Zootec.* **2017**, *24*, 724–733.
10. Brasil. Regulamento Técnico Para Gelados Comestíveis E Preparados Para Gelados Comestíveis. Available online: https://www.in.gov.br/materia/-/asset_publisher/Kujrw0TZC2Mb/content/id/62789384/do1-2019-02-11-resolucao-da-diretoria-colegiada-rdc-n-266-de-8-de-fevereiro-de-2019-62789329 (accessed on 1 April 2018).
11. Associação Brasileira das Indústrias de Sorvetes. Available online: http://www.abis.com.br/mercado/ (accessed on 1 October 2019).
12. Cipriano, P.d.A. *Antocianinas de Açaí (Euterpe Oleracea Mart.) e Casca de Jabuticaba (Myrciaria jaboticaba) na Formulação de Bebidas Isotônicas;* Federal University of Viçosa: Viçosa, Brazil, 2011; p. 150.
13. Maia, M.J.L.; Rossi, E.A.; Carvalho, M.R. Qualidade e Rendimento do "Leite" de Soja da Unidade de Produção de Derivados da Soja—Unisoja—FCF/UNESP. *Alim. Nutr.* **2006**, *17*, 65–72.

14. Anvisa. Rdc No 12, De 2001—Regulamento Técnico Sobre Os Padrões Microbiológicos Para Alimentos. 2001. Available online: https://www.gov.br/agricultura/pt-br/assuntos/inspecao/produtos-vegetal/legislacao-1/biblioteca-de-normas-vinhos-e-bebidas/resolucao-rdc-no-12-de-2-de-janeiro-de-2001.pdf (accessed on 4 August 2022).
15. *ISO 6579:2002*; Microbiology of Food and Animal Feeding Stuffs—Horizontal Method for the Detection of Salmonella spp. International Organization for Standardization: Geneva, Switzerland, 2002.
16. Herigstad, B.; Hamilton, M.; Heersink, J. How to optimize the drop plate method for enumerating bacteria. *J. Microbiol. Methods* **2001**, *44*, 121–129. [CrossRef]
17. Silva, N.; Junqueira, V.C.A.; Silveira, N.F.A. *Manual de Métodos de Análise Microbiológica de Alimentos*, 3rd ed.; Livraria Varela: São Paulo, Brazil, 2001; 317p.
18. Instituto Adolfo Lutz. Normas Analíticas do Instituto Adolfo Lutz. In *Métodos Químicos e Físicos para Análises de Alimentos*, 4th ed.; Ministério da Saúde: Brasilia, Brazil, 2005.
19. Fuchs, R.H.B.; Borsato, D.; Bona, E.; Hauly, M.C.d.O. "Iogurte" de soja suplementado com oligofrutose e inulina. *Ciênc. Tecnol. Aliment.* **2005**, *25*, 175–181. [CrossRef]
20. Karaaslan, M.; Ozden, M.; Vardin, H.; Turkoglu, H. Phenolic fortification of yogurt using grape and callus extracts. *LWT-Food Sci. Technol.* **2011**, *44*, 1065–1072. [CrossRef]
21. Singleton, V.L.; Orthofer, R.; Lamuela-Raventós, R. Analysis of total phenols and other oxidation substrates and antioxidants by means of folin-ciocalteu reagent. *Methods Enzymol.* **1999**, *299*, 152–178.
22. Rufino, M.d.S.M.; Alves, R.E.; Brito, E.S.d.; Morais, S.M.d.; Sampaio, C.d.G.; Pérez-Jiménez, J.; Saura-Calixto, F.D. *Metodologia Científica: Determinação da Atividade Antioxidante Total em Frutas Pela Captura do Radical Livre DPPH*; Embrapa Agroindústria Tropical-Comunidade Técnico (INFOTECA-E); Embrapa: Brasilia, Brazil, 2007.
23. Zanatta, C.F.; Cuevas, E.; Bobbio, F.O.; Winterhalter, P.; Mercadante, A.Z. Determination of anthocyanins from Camu-camu (*Myrciaria dubia*) by HPLC-PDA, HPLC-MS, and NMR. *J. Agric. Food Chem.* **2005**, *53*, 9531–9535. [CrossRef] [PubMed]
24. Vargas, E.F.; Jablonsky, A.; Flores, S.H.; Rios, A.O. Pelargonidin 3-Glucoside Extraction from the Residue from Strawberry Processing (*Fragaria* × *ananassa*). *Curr. Bioact. Compd.* **2016**, *4*, 269–275. [CrossRef]
25. Schmidt, H. *Caracterização Físico-Química, Nutricional e de Compostos Bioativos de Sete Espécies da Família Myrtaceae Nativas da Região Sul do Brasil*; Universidade Federal do Rio Grande do Sul: Porto Alegre, Brazil, 2018.
26. Hough, G.; Wakeling, I.; Mucci, A.; Chambers, I.V.E.; Gallardo, I.M.; Alves, L.R. Number of consumers necessary for sensory acceptability tests. *Food Qual. Prefer.* **2006**, *17*, 522–526. [CrossRef]
27. Stone, H.; Sidel, J. *Sensory Evaluation Practices*, 3rd ed.; Food Science and Technology; Elsevier: Amsterdam, The Netherlands, 2004; 408p.
28. Meilgaard, M.; Civille, G.; Carr, B. *Sensory Evaluation Techniques*, 4th ed.; CRC Press: Boca Raton, FL, USA; New York, NY, USA, 2006; 276p.
29. Ares, G.; Varela, P.; Rado, G.; Giménez, A. Identifying ideal products using three different consumer profiling methodologies. Comparison with external preference mapping. *Food Qual. Prefer.* **2011**, *22*, 581–591. [CrossRef]
30. Moskowitz, H.R. *Product Testing and Sensory Evaluation of Foods*; Food and Nutrition Press: Westport, CT, USA, 1983; p. 605.
31. dos Santos, D.C.; de Oliveira Filho, J.G.; Santana, A.C.A.; de Freitas, B.S.M.; Silva, F.G.; Takeuchi, K.P.; Egea, M.B. Optimization of soymilk fermentation with kefir and the addition of inulin: Physicochemical, sensory and technological characteristics. *LWT* **2019**, *104*, 30–37. [CrossRef]
32. Walter, E.H.M.; Carneiro, M.S.; Felberg, I.; Oliveira, D.R.; Costa, S.D.O.; Conte, C. *Obtenção de Bebidas Fermentadas Por Probióticos a Partir de Diferentes Matérias-Primas da Soja*; Embrapa Agroindústria de Alimentos: Rio de Janeiro, Brazil, 2016.
33. Hill, C.; Guarner, F.; Reid, G.; Gibson, G.R.; Merenstein, D.J.; Pot, B.; Morelli, L.; Canani, R.B.; Flint, H.J.; Salminen, S.; et al. Expert consensus document: The international scientific association for probiotics and prebiotics consensus statement on the scope and appropriate use of the term probiotic. *Nat. Rev. Gastroenterol. Hepatol.* **2014**, *11*, 506–514. [CrossRef]
34. Leite, A.M.O.; Leite, D.C.A.; Del Aguila, E.M.; Alvares, T.S.; Peixoto, R.S.; Miguel, M.A.L.; Silva, J.T.; Paschoalin, V.M.F. Microbiological and chemical characteristics of Brazilian kefir during fermentation and storage processes. *J. Dairy Sci.* **2013**, *96*, 4149–4159. [CrossRef]
35. Peres, M. Microbiota e benefícios a saúde, quanto ao keffir como alimento funcional. *Rev. FAROL* **2019**, *8*, 451–470.
36. Bartnikowsky, S.; Pelegrini, J.; Olivo, I.; Ribeiro, D.; Chaves, A.C.S.D. Desenvolvimento e caracterização de iogurte concentrado simbiótico de jabuticaba com farinha de casca de jabuticaba (*Myrciaria cauliflora*). *J. Chem. Inf. Model.* **2014**, *24*, 19–22. Available online: http://www.elsevier.com/locate/scp (accessed on 25 May 2018).
37. Inada, K.O.P.; Duarte, P.A.; Lapa, J.; Miguel, M.A.; Monteiro, M. Suco de jabuticaba (*Myrciaria jaboticaba*) obtido por extração a vapor: Perfil de compostos fenólicos, capacidade antioxidante, estabilidade microbiológica e aceitabilidade sensorial. *J. Ciênc. Tecnol. Aliment.* **2018**, *1*, 52–61.
38. Plaza, M.; Rodriguez-Meizoso, I. Advanced Extraction Processes to Obtain Bioactives from Marine Foods. *Bioact. Compd.* **2013**, *16*, 343–371.
39. Reynertson, K.A.; Wallace, A.M.; Adachi, S.; Gil, R.R.; Yang, H.; Basile, M.J. Bioactive depsides and anthocyanins from jaboticaba (*Myrciaria cauliflora*). *J. Nat. Prod.* **2006**, *8*, 1228–1230. [CrossRef]
40. Reynerstson, K.A.; Yang, H.; Jiang, B.; Basile, M.J.; Kennelly, E.J. Quantitative analysis of antiradical phenolic constituents from fourteen edible *Myrtaceae fruits*. *Food Chem.* **2008**, *4*, 883–890. [CrossRef] [PubMed]

41. Coda, R.; Lanera, A.; Trani, A.; Gobbetti, M.; Cagno, R. Yogurt-like beverages made of a mixture of cereals, soy and grape must: Microbiology, texture, nutritional and sensory properties. *Int. J. Food Microbiol.* **2012**, *155*, 120–127. [CrossRef] [PubMed]
42. Sabokbar, N.; Khodaiyan, F. Total phenolic content and antioxidant activities of pomegranate juice and whey based novel beverage fermented by kefir grains. *Int. J. Food Sci. Technol.* **2016**, *53*, 739–747. [CrossRef]
43. Kirakosyan, A.; Seymour, E.M.; Noon, K.R.; Llanes, D.E.U.; Kaufman, P.B.; Warber, S.L. Interactions of antioxidants isolated from tart cherry (*Prunus cerasus*) fruits. *Food Chem.* **2010**, *122*, 78–83. [CrossRef]
44. Saito, T.; Martins-Madalão, M.C.; Bernardes, P.C.; Bosi, M.G.; Della Lucia, S.M.; Saraiva, S.H. *Jabuticaba* (*Myrciaria cauliflora*) peel extract increases bioactive compounds in petit-suisse cheese. *Int. Food Res. J.* **2019**, *26*, 277–285.
45. Ścibisz, I.; Ziarno, M.; Mitek, M.; Zaręba, D. Effect of probiotic cultures on the stability of anthocyanins in blueberry yoghurts. *LWT-Food Sci. Technol.* **2012**, *49*, 208–212. [CrossRef]
46. Shimamoto, G.G.; Rossi, A.V. Antocianinas e gelo seco para visualizar equilíbrios ácido/base numa abordagem contextualizada. *Educ. Quím.* **2010**, *3*, 31–36.
47. Paula, A.G.P.; Heemann, A.C.W.; Heemann, R.; Lima, C.P. Avaliação da estabilidade das antocianinas do açaí no período de 28 dias em diferentes condições. *Braz. J. Heal. Rev.* **2019**, *2*, 4811–4823. [CrossRef]
48. Inada, K.O.P.; oliveira, A.A.; Revorêdo, T.B.; Martins, A.B.N.; Lacerda, E.C.Q.; Freire, A.S. Screening of the chemical composition and occurring antioxidants in jabuticaba (*Myrciaria jaboticaba*) and jussara (*Euterpe edulis*) fruits and their fractions. *J. Funct. Foods* **2015**, *17*, 422–433. [CrossRef]
49. Araújo, A.C. *Obtenção e Estabilidade de Corante de Antocianinas Extraídas do Repolho Roxo (Brassica oleracea)*; Universidade Federal de Campina Grande: Campina Grande, Brazil, 2019.
50. Popper, R.; Abdi, H.; Williams, A.; Kroll, B.J. Multi-Block Hellinger Analysis for Creating Perceptual Maps from Check-All-That-Apply Questions. *Rose-Marie Sens. Sci. Symp.* **2011**, *9*, 1–33.
51. Jeesan, S.A.; Seo, H.S. Color-induced aroma illusion: Color cues can modulate consumer perception, acceptance, and emotional responses toward cooked rice. *Foods* **2020**, *9*, 1845. [CrossRef]

foods

MDPI

Review

Research Progress for Probiotics Regulating Intestinal Flora to Improve Functional Dyspepsia: A Review

Xinyu Shen [1,†], Aijun Xie [2,†], Zijing Li [1], Chengxi Jiang [1], Jiaqi Wu [1], Mohan Li [1,*] and Xiqing Yue [3,*]

1 College of Food Science, Shenyang Agricultural University, Shenyang 110866, China; shenxinyu98218@163.com (X.S.); 2022149037@stu.syau.edu.cn (Z.L.); 2022149044@stu.syau.edu.cn (C.J.); 2022240300@stu.syau.edu.cn (J.W.)

2 Department of Chemical and Biomolecular Engineering, National University of Singapore, Singapore 119077, Singapore; axie26@wisc.edu

3 Shenyang Key Laboratory of Animal Product Processing, Shenyang Agricultural University, Shenyang 110866, China

* Correspondence: limohan@syau.edu.cn (M.L.); yuexq@syau.edu.cn (X.Y.)

† These authors contributed equally to this work.

Abstract: Functional dyspepsia (FD) is a common functional gastrointestinal disorder. The pathophysiology remains poorly understood; however, alterations in the small intestinal microbiome have been observed. Current treatments for FD with drugs are limited, and there are certain safety problems. A class of active probiotic bacteria can control gastrointestinal homeostasis, nutritional digestion and absorption, and the energy balance when taken in certain dosages. Probiotics play many roles in maintaining intestinal microecological balance, improving the intestinal barrier function, and regulating the immune response. The presence and composition of intestinal microorganisms play a vital role in the onset and progression of FD and serve as a critical factor for both regulation and potential intervention regarding the management of this condition. Thus, there are potential advantages to alleviating FD by regulating the intestinal flora using probiotics, targeting intestinal microorganisms. This review summarizes the research progress of probiotics regarding improving FD by regulating intestinal flora and provides a reference basis for probiotics to improve FD.

Keywords: probiotics; gastrointestinal tract; intestinal flora; function dyspepsia

Citation: Shen, X.; Xie, A.; Li, Z.; Jiang, C.; Wu, J.; Li, M.; Yue, X. Research Progress for Probiotics Regulating Intestinal Flora to Improve Functional Dyspepsia: A Review. *Foods* **2024**, *13*, 151. https://doi.org/10.3390/foods13010151

Academic Editor: Andrea Gomez-Zavaglia

Received: 14 November 2023
Revised: 5 December 2023
Accepted: 8 December 2023
Published: 2 January 2024

1. Introduction

Functional dyspepsia (FD) is a chronic non-organic gastrointestinal disease that is one of the most common diseases of the digestive system and one of the most common diseases of the gut–brain interaction worldwide [1,2]. Approximately 16% of individuals within the general population are affected by epigastric pain syndrome (EPS) and postprandial distress syndrome (PDS), which are two primary forms of FD [3,4]. In countries outside of Asia, the incidence is approximately 10–40%, and the incidence in Asian countries is approximately 5–30%. Of the people who experience FD, approximately 40% choose to seek medical treatment due to discomfort [5]. In approximately 80% of patients suffering from dyspepsia there is no structural explanation for their symptoms, which is then termed FD. The Rome IV criteria serve as the established diagnostic standard for FD. It is defined as a symptom of dyspepsia originating in the stomach and duodenum in the absence of evidence of organic, systemic, or metabolic disease that explains the symptom with its defining symptoms comprising postprandial satiety, epigastric pain, early satiety, or epigastric burning that persists for a minimum duration of six months [6,7]. As far as the pathophysiology of the disease is concerned, the Rome IV criteria as well as a recent multinational consensus of European experts support the role for impaired gastric accommodation, gastric distention hypersensitivity, disturbances in gastric emptying, and altered central nervous system signal processing [8]. The presence of lower gastrointestinal symptoms like diarrhea and constipation enhance the capacity of physicians to differentiate between individuals with functional digestive disorders and those experiencing non-FD [9,10]. FD is a chronic condition without a known cure, and as a result, it profoundly influences both the physical and mental health and the overall quality of life of patients [11].

The etiology and pathogenesis of FD remain unclear, but it is typically believed that it may be related to factors that include (1) gastric motility disturbance, (2) visceral hypersensitivity, (3) decreased gastric fundus receptivity diastolic function, (4) *Helicobacter pylori* infection, (5) gut–brain axis disturbance, (6) mental and social factors, and (7) increased eosinophilic cells, epithelial barrier disruption, and mucosal inflammation in the duodenum accompanied by elevated mast cell levels [12–16]. Additionally, an imbalance in intestinal flora is one of the important pathogenesis processes of FD. The imbalance in intestinal flora will lead to the disturbance of the intestinal environment, ultimately resulting in the reduction in the total amount of probiotics and a series of acute and chronic diseases [17,18]. Thus, the correlation between the gut microbiome and the development of human diseases is of paramount importance [19]. There are several lines of evidence suggesting that both locoregional duodenal and systemic changes may also be present in FD. Duodenal eosinophilia, epithelial barrier defect, and subtle mucosal inflammation, along with higher levels of mast cells, have been reported in FD, whereas the role of local and systemic inflammatory changes and increased small-bowel-homing T cells were not highlighted until very recently [20,21]. Intestinal flora may improve the clinical symptoms of FD by improving intestinal barrier function and visceral hypersensitivity, and regulating gastrointestinal motility. Changing the type and composition of intestinal flora may provide a safe and effective treatment to relieve FD symptoms [22].

Probiotics are active microorganisms that are beneficial to the host. Furthermore, strains that can meet the basic conditions of probiotics can become probiotics. Therefore, probiotics cover numerous types of microorganisms, and the main probiotics in this study include *Bifidobacterium*, *Lactobacillus*, *Saccharomycetes*, *Bacillus*, and others [23,24]. Concurrently, the health effects of probiotics have also become a research hotspot (Table 1). Presently, probiotics have been demonstrated to maintain the normal structure of intestinal flora, resist pathogen infection, improve constipation and diarrhea, relieve lactose intolerance, reduce serum cholesterol levels, and promote immune system development [25]. The mechanism of action of probiotics primarily includes three aspects: (1) enhancing the host defense capacity, (2) directly fighting microorganisms, and (3) metabolites playing an important function [26] (Figure 1). Probiotics can effectively relieve functional gastrointestinal disorders such as irritable bowel syndrome [27]. Several intestinal flora colonize the gastrointestinal tract, an organ involved in the pathogenesis of FD due to mucosal damage and inflammation [15]. Therefore, improving FD symptoms by regulating the gut microbiota may be beneficial.

Table 1. The health effects of common probiotics.

Probiotic Species	Bacterial Strain	Mechanism of Action	Health Benefits	References
Lactobacillus	*Lactobacillus acidophilus*	Adjuvant antigen-specific immune response;	Immunity enhancement; lower inflammatory factors; restoration of nervous system function	[28,29]
	Lactobacillus casei	Through antagonism, colonization competition, increasing antibody production, enhancing the systemic immune effect, and the antibacterial action of metabolites, it can resist the invasion of pathogens	Inhibition of pathogenic bacteria in the gut; protecting the internal environment; maintaining intestinal microecology	[30,31]

Table 1. *Cont.*

Probiotic Species	Bacterial Strain	Mechanism of Action	Health Benefits	References
	Lactobacillus paracasei	Regulating the production of anti-inflammatory cytokines by Th1/Th2 cells and reducing the release of toxic nitrogen-containing metabolites	Relieving inflammation; relieving chronic metabolic diseases	[32–34]
	Lactobacillus rhamnosus	Competitive colonization, inhibition of *H. pylori* growth, and adhesion to mucosal cells	Maintaining intestinal barrier integrity; inhibiting inflammation and oxidative stress; regulating gut–brain communication	[35]
	Lactobacillus reuteri	Increasing the number of *Bifidobacteria* in the gut, and transforming the intestinal dominant flora suitable for breaking down proteins into the flora suitable for sugar metabolism, thereby reducing the production of toxic and spoilage metabolites	Secreting antimicrobial compounds; regulating the host immune system; preventing diarrhea and colitis; reducing the prevalence of acute abdominal pain in infants	[36–38]
Bifidobacterium	*Bifidobacterium longum*	Increasing the intestinal flora and inhibiting pathogenic bacteria	Maintaining intestinal health in early life; promoting the establishment of intestinal microbiota	[39,40]
	Bifidobacterium animals	Increasing antibody production	Resisting foreign pathogenic microorganisms	[41]
	Bifidobacterium infantis	Preventing excessive intestinal immune responses	Regulating intestinal flora; anti-hepatic fibrosis; anti-infection	[42,43]
Others	*Bacillus coagulans*	Enhancing the specific defense of non-specific antigens	Pathogen suppression; improved immune ability and growth performance of the body; improved intestinal digestion and absorption of nutrients; improved utilization of mineral elements in the body	[44–46]
	Streptococcus thermophilus	Producing vitamins and cofactors; activating immune function; affecting bile salt concentration	Improved body immunity; improved lactose intolerance	[47,48]

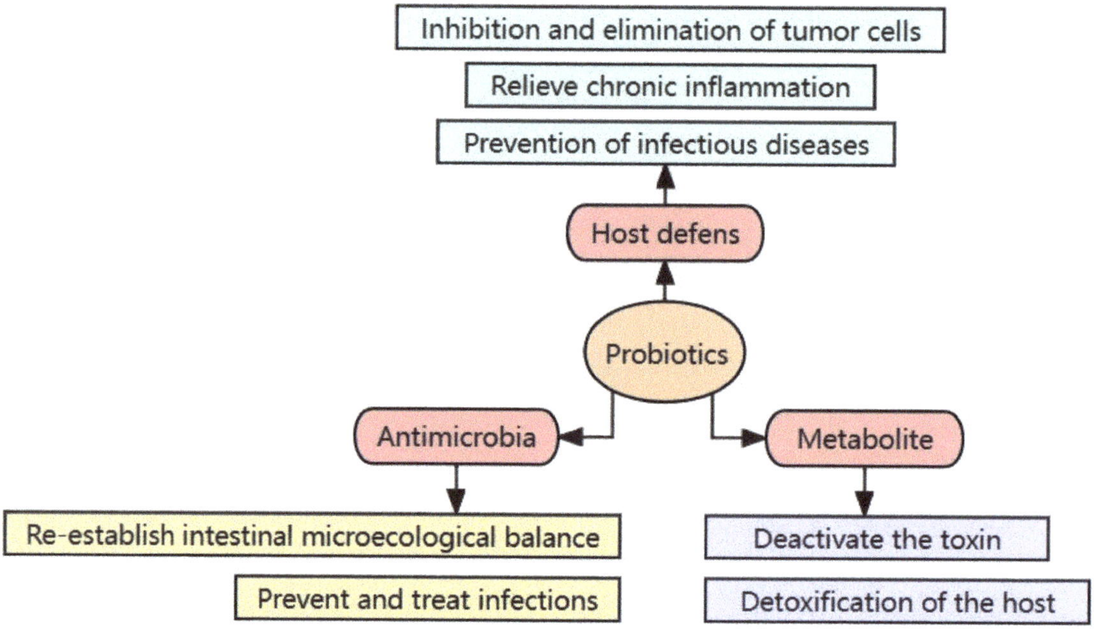

Figure 1. Mechanism of probiotic actions.

2. Relationship between Probiotics and Intestinal Flora

There is a close relationship between probiotics and gut flora. As a safe and effective intervention method, probiotics can affect the occurrence and development of various diseases by regulating the intestinal flora, and this also makes probiotics more attractive in the fields of basic and clinical research [49]. Humans acquire their initial gut flora from their mothers from birth, and the gut flora in infants is dominated by *Bifidobacterium*. With the establishment of human gut microbiota symbiosis, the gut microbiota changes with factors such as the environment, diet, and age, and the gut microbiota possesses a wide range of personalization [50]. Probiotics and intestinal flora interact through nutritional competition, antagonism, and symbiosis [51]. There is further proof that probiotics in this situation lessen the negative effects of antibiotic treatment, and certain probiotics possess the potential to completely eradicate *H. pylori* through an inhibitory impact on the bacterium. [52]. The direct regulation of probiotics on the intestinal flora depends upon the composition of individual flora, and the probiotic effects of different probiotic strains vary. Additionally, not all probiotic strains can exert beneficial effects on specific diseases [53].

2.1. *Effects of Probiotics on Gastrointestinal Health*

The notion of probiotics traces back to a hypothesis initially advanced by Russian researcher Elie Metchnikoff. He linked the extended lifespan of Bulgarian farmers to their consumption of fermented dairy products, and this idea was subsequently expanded upon [23,54–56]. In recent years, the research and product development of probiotics have both attracted greater attention. Probiotics can improve and regulate intestinal health, effectively relieve intestinal discomfort such as constipation, diarrhea, inflammatory gastrointestinal diseases, irritable bowel syndrome, and FD, and protect intestinal health to the greatest extent [57–59]. Through their colonization of the intestinal mucosa and interactions with the mucus layer, probiotics regulate the immune response, thus enhancing the ability to ward off external threats [60]. Following their establishment in the intestinal tract, the bioactive compounds responsible for their beneficial effects can help to maintain the equilibrium of the gut microbiota; this includes boosting the population of beneficial bacteria in the intestinal ecosystem while decreasing the abundance of harmful bacteria, ultimately bolstering intestinal immunity [61]. Changes in the composition of the intestinal flora can lead to gastrointestinal diseases, and probiotics can be applied to the intestinal environment to alleviate pathological conditions [51]. Bioactive molecules produced by probiotic strains, such as bacteriocins, vitamins, short-chain fatty acids, enzymes, and amino acids, exert certain beneficial effects on the host (Table 2). Studies have demonstrated that lactic acid bacteria can produce bioactive sequences of different compounds, such as peptides, sugar polymers, and fatty acids, that promote human health [62]. Hence, consuming fermented foods rich in probiotics can enhance the immune system and lower the risk of developing diseases. This occurs through continuous interaction between the bacteria and the host immune system, ultimately leading to alterations in the gut microbial composition that favor the growth of beneficial microbiota while regulating the presence of pathogenic flora [63,64].

Table 2. The health role of active substances in probiotics in the gut.

Biologically Active Compound	Probiotics	Mechanism of Action	Health Benefits in the Gut	References
Bacteriocins	*Lactococcus lactis*	Located in the cytoplasmic membrane region of receptor binding on bacterial surfaces	Inhibition of pathogenic bacteria in the gut; helps gastrointestinal bacteria to survive	[65,66]
Vitamins	*Lactobacillus reuteri* JCM1112; *Lactobacillus fermenti* CECT 5716	Promotes *Bifidobacteria* to produce B-complex vitamins and maintain the health of the host intestine	Promotes energy metabolism; reduces intestinal inflammation	[67,68]
SCFAs	*Butyricicoccus pullicaecorum*; *Bifdobacterium sps.*	Receptors maintain homeostasis in the host cells by controlling the energy utilization of the host	The energy source of the colon cells	[69]
Enzymes	*Lactobacillus sp.* G3_4_1TO2; *Lactobacillus fermentum* E-3	Amylases and peptidases produced by probiotic organisms play a role in biochemical reactions of the host metabolism	Scavenges free radicals; hydrolyzes *β-galactoside*	[70,71]
Amino acids	*Fusobacterium varium*; *Clostridium sps.*	Regulates carbohydrate and lipid metabolism and alters host physiology	Provides host with essential amino acids	[72,73]

2.2. *Effect of Probiotics on Patients with FD*

The pathogenesis of FD is complex. The strains of probiotics are specific; there are certain differences in the physiology and metabolism of the strains of different species, and the influence of different growth states of probiotics on intestinal flora and host metabolism is also different [74]. Studies have demonstrated that after the intervention of probiotics or their fermented products, the clinical symptoms in FD patients have been relieved to varying degrees, and their quality of life has been significantly improved (Table 3).

Wauters et al. [75] conducted a 16-week randomized, double-blind, placebo-controlled study of 68 FD subjects (as defined by Rome IV criteria) aged ≥18 years, in which 32 participants received probiotics and 36 received the placebo. The experimental group were provided with probiotic capsules (*Bacillus coagulans* MY01 and *Bacillus subtilis* MY02 in a 1:1 mixture with a bacterial count of 2.5 × 109 CFU) twice each day, and the placebo group were provided with maltodextrin capsules (without any symbiotic bacteria). The findings indicated a substantial increase in the quality of life and short-chain fatty acid (SCFAs) levels in the experimental group of FD patients ($p < 0.05$) coupled with a significant reduction in clinical symptom scores ($p < 0.05$). In contrast, the placebo group did not exhibit significant changes. The abundance of the intestinal flora *Rhodotella* and *Leuconostoidea* increased. The results revealed that *Bacillus coagulans* MY01 and *Bacillus subtilis* MY02 exerted some alleviating effects on FD. Navarro-Rodriguez et al. [76] conducted an 8-week randomized, double-blind, placebo-controlled clinical trial on 107 subjects diagnosed with *H. pylori* infection with FD. The results indicated that the bacterial eradication rate was 89.8% in the active probiotic group and 85.1% in the placebo group ($p = 0.49$). There was no significant change in the efficacy of bacterial eradication or the adverse effects of *H. pylori* eradication in the placebo and probiotic groups. This may be due to the low concentration of probiotics. Sun et al. [77] conducted a randomized clinical trial to determine the effect of beverages containing *Lactobacillus paracei* LC-37 and its ability to relieve FD symptoms. The results demonstrated that after 14 and 28 days of treatment with a beverage containing LC-37, the symptoms of 26 FD patients were relieved, clinical symptom scores were significantly reduced, abdominal pain and hiccups were significantly reduced after 14 days, and the symptoms almost completely disappeared after 28 days. There was a significant increase in *Lactobacillus, Lactococcus,* and *Weikerella,* while there was a significant decrease in the abundance of harmful bacteria such as *Bacillus folLiculiformis.* In general, the beverage containing LC-37 attenuates symptoms of FD, in part through strain-specific effects by increasing probiotics, such as *Lactobacillus* and *Weissella,* and decreasing harmful bacteria including *Lachnocliostridium.* Rahmani et al. [78] conducted a 4-week randomized, double-blind, placebo-controlled clinical trial in 125 FD patients according to Rome III criteria, in which 65 subjects received probiotics (*Lactobacillus reuteri*) and the rest received placebo. The results revealed that all FD-related variables, such as frequency, severity, and duration of pain, were significantly lower at the end of week 4 compared to the baseline. In this study, there was a relative success of probiotic-based treatment without significant improvement in recovery. Differences between the strains of bacteria used is a possible explanation of the discrepancies. In a study conducted by Nakea et al. [79], a comparison was made between the fundamental physiological properties of gastric fluid (GF) and the microflora structure within the GF among 44 healthy individuals and 44 FD patients. Subsequently, the FD patients were administered yogurt infused with the probiotic strain *Lactobacillus gasseri* OLL2716 (LG21 yogurt). The investigation assessed its impact on bacteriological parameters and symptoms while seeking to elucidate their interconnection. The results indicated that the volume of GF recovered in the stomach of FD patients after overnight fasting was greater than that of healthy subjects, while the volume of GF in the stomach of FD patients with improved symptoms after treatment with LG21 yogurt was reduced. Additionally, the overall structure of the bacterial community in the GF of FD patients and the abundance of *Prevotella* was significantly different from that of healthy subjects. In patients with FD, treatment with LG21 yogurt reversed this ecological imbalance. In a 12-week clinical trial, Ohtsu et al. [80] enrolled 116 FD patients with an average age of 42.8 ± 9.0 years, and all were Helicobacter-pylori-infected. The trial followed a double-blind, parallel, placebo-controlled, and randomized design. Participants were randomly allocated to either the group consuming yogurt enriched with *Lactobacillus gasseri* OLL2716 (the *Lactobacillus gasseri* OLL2716 group) or the group consuming yogurt without *Lactobacillus gasseri* OLL2716 (the placebo group). The findings indicated that *Lactobacillus gasseri* OLL2716 exerted a notably more favorable impact on gastrointestinal symptoms compared to that of the placebo group and exhibited statistical trends ($p = 0.073$). Specifically, the elimination rates of the primary FD symptoms were 17.3% for the placebo group and 35.3% for the *Lactobacillus gasseri* OLL2716 group ($p = 0.048$), thus underscoring the beneficial effect of *Lactobacillus gasseri* OLL2716 in regard to alleviating FD symptoms.

Table 3. The effect of probiotics on functional dyspepsia.

Probiotics	Experimental Subject	Intervention Time	Primary Outcome	Reference
Bacillus coagulans MY01, *Bacillus subtilis* MY02	68 patients	16 weeks	Quality of life and SCFA content↑	Wauters et al. [75]
Lactobacillus acidophilus, *Lactobacillus rhamnosus*, *Bifidobacterium bifidum* and *Streptococcus faecium*	107 patients	8 weeks	H pylori and adverse reaction↓	Navarro-Rodriguez et al. [76]
Lactobacillus paracei LC-37	26 patients	14 days, 28 days	14 days: Abdominal pain and hiccups↓ 28 days: Abdominal pain and hiccups disappear, *Lactobacillus*, *Lactococcus*, and *Weikerella*↑, *Bacillus folliculiformis*↓	Sun et al. [77]
Lactobacillus reuteri	125 patients	4 weeks	The frequency, severity, and duration of pain↓	Rahmani et al. [78]
Lactobacillus OLL2716	44 healthy control participants, 44 patients	3 mouths	GF volume in stomach↓	Nakea et al. [79]
Lactobacillus gasseri OLL2716	116 patients	12 weeks	Rate of elimination of major symptoms↑	Ohtsu et al. [80]

3. Relationship between Intestinal Flora and FD

Compared with healthy people, the species composition of the gut microbiota in FD patients showed significant changes. Intestinal flora may improve the clinical symptoms of FD by improving the intestinal barrier function, visceral hypersensitivity and regulating gastrointestinal motility.

3.1. Research Progress on FD

FD is often attributed to disorders of gastric physiology, such as slow gastric emptying, an inability to relax the gastric fundus after meals, or gastric hypersensitivity caused by gastric dilation [81]. The relaxation of the lower esophageal sphincter in gastroesophageal reflux is related to FD to some extent [82]. FD can also be induced by *Salmonella*, *Escherichia coli* O157, *Campylobacter jejuni*, *Giardia lamblia*, and norovirus [83]. FD can develop due to inflammation in the proximal small intestine or stomach following an intestinal infection, while irritable bowel syndrome (IBS) may be triggered by inflammation in the distal small intestine or colon. When inflammation affects both the proximal and distal segments of the small intestine, it can lead to a syndrome that combines features of both IBS and FD [84].

In clinical practice, drug therapy is still the main means of treating FD, including Helicobacter pylori eradication drugs, proton pump inhibitors (PPI), H2 receptor antagonists, antidepressants, and anti-anxiety drugs [85,86]. An analysis of 10 PPI randomized trials involving 3347 patients reported that histamine H2-receptor antagonists exerted a more pronounced effect than PPIs [87]. A PPI was effective in patients reporting regurgitant or ulcer-like FD but not in patients reporting dyspraxia like functional gastroenteritis [88]. For FD patients with abnormal gastric motility and fundus regulation, studies have demonstrated that motility-promoting drugs such as cisapride, domperidone, and itopride are more effective than placebo. They promote gastric motility and gastrointestinal peristalsis by stimulating smooth muscle contraction and 5-HT 4 receptors of neurotransmitter-regulating neurons. This is more suitable for patients with postprandial discomfort syndrome (PDS). Although the drug treatment exerts a certain effect, long-term use of drugs will cause diarrhea, dizziness, vomiting, rash, and other adverse reactions and may even be accompanied by headache, atrophic gastritis, stomach polyps, and other side effects [89–91]. In addition to drug treatment, there are placebo treatments and psychological treatments. A randomized clinical trial comparing the use of placebos to untreated IBS patients revealed that the use of placebos was significantly more likely to provide sufficient relief from symptoms [92]. For those who experience emotional disorders, psychological treatment can be considered. Although placebo treatment and psychotherapy resulted in fewer side effects, they provided less relief from FD symptoms [93]. Probiotics modulate intestinal dysbiosis and are in their infancy as a potential treatment for FD. Currently, the probiotics used in FD clinical trials include *Bifidobacterium bifidum* tablets, *Bacillus subtilis bifidum* capsules, *Clostridium butyricum* live bacteria capsules, and others [94,95]. Probiotics are considered to be a safe dietary supplement and offer potential benefits in managing the development and progression of symptoms in patients with IBS. This is potentially attributed to their anti-inflammatory effects or regulation of visceral hypersensitivity. Similarly (as with IBS), a mounting body of evidence suggests that addressing FD by implementing strategies aimed at restoring gastrointestinal flora represents a promising and evolving approach [96].

3.2. The Mechanism of Intestinal Flora on FD

3.2.1. Regulation of Intestinal Barrier Function

Probiotics can regulate the intestinal flora, maintain the integrity of the intestinal barrier structure, and protect the intestinal tissue from damage by pathogens. The integrity of the intestinal barrier is a sign of intestinal health and a reflection of the ability of the microflora to adapt to the dynamic balance of mucosal function. The integrity of the mucus barrier is the first line of defense to protect the gastrointestinal tract, and the microbiome plays a major role in the factors affecting the mucus barrier by driving mucus changes [97]. The impaired intestinal barrier function caused by the imbalance in intestinal flora is associated with mild intestinal mucosal inflammation and immune activation [98]. FD patients exhibited lower mucin expression levels and higher mast cell (MC) counts compared with healthy individuals, indicating impaired mucosal integrity, i.e., increased intestinal permeability and mild inflammation [99]. FD patients with duodenal epithelial cells exhibited an abnormal expression of adhesion protein and increased paracellular channels, and there was a significant correlation between the expression level of intercellular adhesion protein and the degree of increased permeability and the severity of low-grade inflammation [100]. Intestinal microorganisms play an important role in maintaining the mucosal barrier function. Conversely, by synthesizing and secreting a variety of metabolites (such as SCFAs, indole derivatives, etc.), microorganisms promote the production and secretion of gastrointestinal mucus while increasing the expression of tight junction protein and cytoskeleton-related proteins. This plays a crucial role in regulating epithelial permeability [101,102]. Mediators such as histamine, rennet, and prostaglandin D2 released during MC activation regulate the secretion and permeability of chloride ions and water in epithelial cells. MC-mediated intestinal barrier changes are also associated with neuropeptides, neurotransmitters, hormones (vasoactive intestinal peptides, SP, NGF, estrogens, and estradiol), and inflammatory mediators (tumor necrosis factor-alpha, interferon-gamma, and cytokines) released by other immune cells [103]. Intestinal epithelial cell barrier function is impaired, ultimately resulting in the increased penetration of pathogenic bacteria and the activation of T lymphocytes to release inflammatory cytokines mediating inflammation and further aggravating the inflammatory response.

3.2.2. Effects on Visceral Sensitivity

Visceral hypersensitivity indicates that the gastrointestinal mucosa of FD patients is strongly responsive to normal stimulation, and 35% to 50% of FD patients also have visceral hypersensitivity [104]. Visceral hypersensitivity is a crucial factor in functional gastrointestinal disorder syndrome, greatly influencing the occurrence and severity of symptoms. This is due to an abnormal mucosal immune response activation in relation to the level of inflammatory response and the close proximity of immune cells and sensory neurons [105]. There are numerous factors related to visceral hypersensitivity and signal transduction, such as the expression of cytokines, the secretion of corticosterone, the release of SCFAs, and microbial metabolites [106]. Research has revealed that a symbiotic intestinal microbiota is an essential requirement for the proper stimulation of intestinal sensory neurons, thereby offering a plausible mechanism for the transmission of information between the microbiota and the nervous system [107]. Increased visceral sensitivity can lead to abnormal pain or pain sensitivity in patients, and this is the main pathophysiological change in EPS. The sensitivity of the stomach and duodenum to mechanical and chemical stimulation was increased in patients with FD, and the degree of visceral sensitivity was consistent with the severity of clinical symptoms [108]. The hypersensitive state can be improved by downregulating the expression level of TRPV1 in the duodenum. TRPV1 can be selectively activated to induce the release of neuropeptides such as calcitonin gene-related peptide (CGRP), substance P, and others [109]. Evaluation of the physical injury response induced by inflammatory stimuli in both sterile and conventional mice suggests that certain symbiotic bacteria are necessary for the development of inflammatory hypersensitivity in mice. This suggests an important role of the interactions between symbiotic microbiota and the host in facilitating adaptation to environmental stress [110]. The intestinal microbiota significantly affects visceral hypersensitivity, and this is a new target for the treatment of FD-related visceral hypersensitivity.

3.2.3. Regulating Gastrointestinal Motility

Gastrointestinal motility dysfunction is one of the main pathological mechanisms of FD. The generation of gastrointestinal motility is related to the slow-wave potential induced by interstitial cells of Cajal (ICCs). The change in the number or structure of ICCs is the key to gastrointestinal motility disorder, and the reduction in the c-kit gene may reflect the decrease in the number of ICCs [111]. It has been established that the decrease in plasma motilin content in FD patients is related to the decrease in gastric emptying capacity, gastric electrical rhythm disturbance, and antrum–pyloric–duodenal motor coordination disorder. Studies have reported that metabolites fermented by intestinal microorganisms such as short-chain fatty acids (SCFAs) or peptides, in addition to being an important energy source for gastrointestinal motility, can also directly activate the intestinal nervous system to regulate the synthesis and secretion of certain gastrointestinal hormones by intestinal endocrine cells. These include intestinal hormone peptide YY, cholecystokinin, glucagon-like peptide 1, gastric inhibitory peptide, and motion-related peptide, and they can also regulate gastrointestinal motility and gastric emptying [112,113]. Additionally, SCFAs stimulate the central nervous system (CNS) and enteric nervous system (ENS) to release the key neurotransmitter 5-HT, which can promote intestinal motor disorders through intestinal smooth muscle contraction and is one of the important factors affecting gastrointestinal motility [114]. Changes in the biosynthesis, release, or reuptake of 5-HT exert an important impact on the control of the central nervous system and gastrointestinal tract and are involved in the regulation of emotions, psychological states, and sensorimotor functions of the gastrointestinal tract [115,116]. The intestinal microbiota and its metabolites affect gastrointestinal peristalsis through various pathways, including intestinal neurons, glia, and intestinal macrophages [117]. During digestion, the complex migratory movement of the stomach and small intestine can mechanically migrate gastrointestinal contents and pathogenic bacteria to the distal intestine. Meanwhile, FD often exhibits non-transmissive and retrogressive MMC stage III activity that may induce or aggravate the migration of proximal intestinal bacteria to the stomach along with the regurgitant duodenal fluid. Delayed gastric emptying results in the retention of regurgitant bacteria in the stomach and duodenum for a longer period of time. Concurrently, the proliferation of foreign bacteria can produce lipopolysaccharides to stimulate the immune response and then inhibit gastrointestinal peristalsis, ultimately leading to the aggravation of FD symptoms [118,119]. Abnormal gastrointestinal motility may lead to a flora imbalance that, in turn, may further affect gastrointestinal motility and mediate the development of FD.

4. Mechanism of Probiotics to Improve FD Symptoms

Probiotics improve FD symptoms by regulating immune function and regulating the probiotics of metabolites such as short-chain fatty acids, bile acids, and neurotransmitters.

4.1. Regulation of Immune Function

Probiotics exert a significant impact on shaping the gut microbiota composition, as they can hinder the establishment of pathogenic bacteria in the intestinal environment. They assist the host in developing a robust protective mucosal layer in the intestines, fortify the host's immune system, and generate beneficial metabolites that contribute to maintaining overall health [120,121]. Local intestinal immune response is induced by the interaction of probiotics with intestinal epithelial cells and mucosal lamina propria immune cells (Figure 2). The gastrointestinal tract microbiota modulates the movement and role of neutrophils and influences the division of populations of T cells into various forms of T helper cells (Th), namely Th1, Th2, and Th17, or into regulatory T cells [122]. Mast cells (MCs) are membrane cells connected to the epithelial tissue; epithelial cells (ECs) and dendritic cells (DCs) are treated with probiotics and enter the intestinal lumen in different internalization manners. Following their interaction with epithelial cells, probiotics or their components are internalized and initially engage with antigen-presenting cells (APC), including macrophages and dendritic cells in the tunica propria of the digestive tract. The interaction between probiotics and epithelial cells triggers the release of IL-6 that, in turn, stimulates the clonal expansion of IgA-producing B lymphocytes, thereby elevating their population. These IgA-producing B lymphocytes then migrate into the protoplasmic cells of the intestinal tunica propria through these antigen-presenting cells [123,124]. Macrophages and dendritic cells engulf probiotics or fragments, thereby inducing the production of cytokines such as TNF-α and IFN-γ to enhance epithelial excitation and mutual interference among immune-related cells [125]. IL-6 and TNF-α are two important cytokines that play an important role in inflammation, anti-tumor activity, and the regulation of immune function. In the intestinal mucosal immune system, probiotic bacteria or their metabolites can be acquired and recognized by mucosal M cells as antigens. Thus, this promotes the development of the intestinal mucosal immune system, activating macrophages and B lymphocytes, forming germinal centers in the intestinal mucosal lymphoid tissue, and finally, transforming B lymphocytes into plasma cells to secrete mucosal antibody IgA to mediate mucosal immunity [126,127]. Probiotic strains can increase the levels of anti-inflammatory cytokines such as IL-10, reduce the levels of inflammatory cytokines such as TNF-α, IL-1β, and IL-8, and exert significant effects on reducing intestinal inflammation and improving colitis [128]. Additionally, different probiotics maintain gut health in different ways. *Lactobacillus* and *Bifidobacterium* primarily compete with pathogenic microorganisms for favorable adhesion sites, improve intestinal barrier function, regulate intestinal microflora, enhance intestinal immunity, and inhibit or kill harmful bacteria. However, *Bacillus* primarily enhances intestinal immunity through its metabolites and resists the invasion of pathogenic bacteria. Additionally, the probiotic effect of probiotics is strain-specific in regard to antibodies, and different strains utilize different modes of action and exert different effects on different parts of the intestine. For example, *Bacillus subtilis* MY02 is more effective at increasing SCFAs levels and *Lactobacillus paracei* LC-37 is more effective at reducing abdominal pain. Based on this, it is necessary to further precisely regulate the probiotics acting on different parts to provide full play to their probiotic effect on FD and other intestinal diseases [129,130].

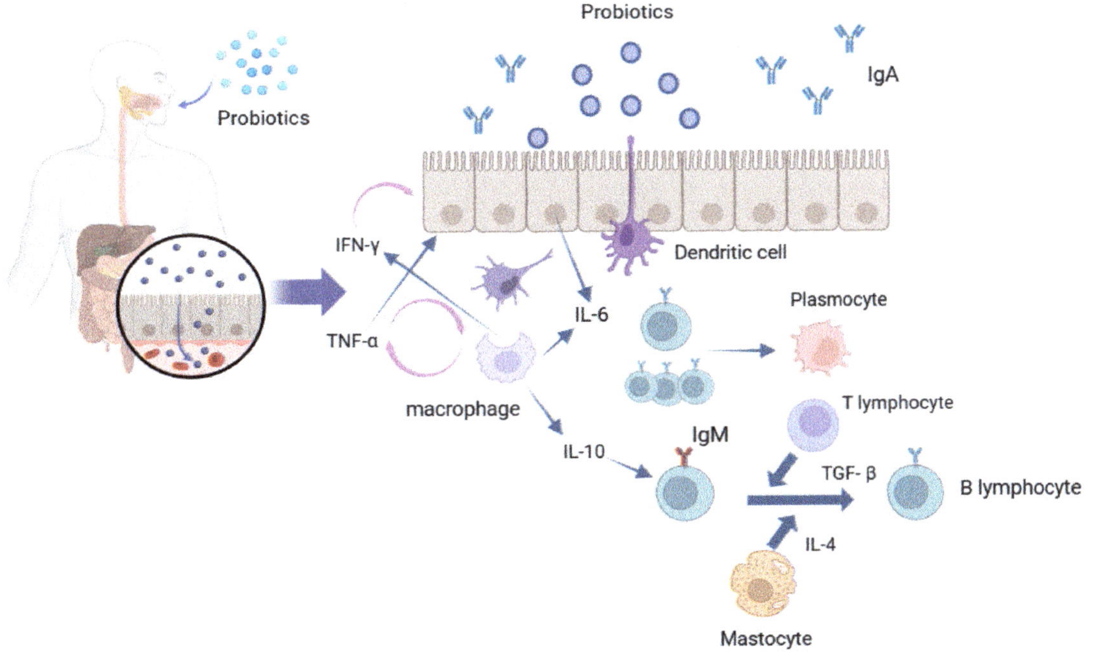

Figure 2. Local intestinal immune response induced by the interaction of probiotics and intestinal epithelial cells.

4.2. Probiotics Work by Regulating the Metabolites of the Flora

In addition to directly acting on the intestinal flora, probiotics can also play a beneficial role by indirectly regulating metabolites of the intestinal flora such as short-chain fatty acids (SCFAs), bile acids, and neurotransmitters, relieving FD symptoms to improve body health (Figure 3).

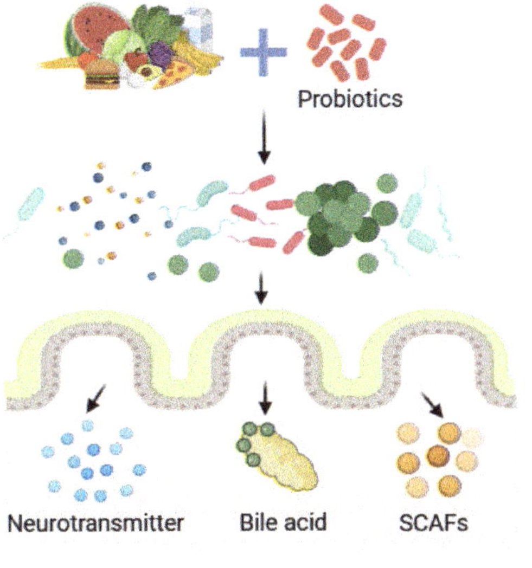

Figure 3. Probiotics regulate biota metabolites.

4.2.1. Promotion of the Production of Short-Chain Fatty Acids

As significant byproducts of gut flora, SCFAs have the ability to strengthen the immunity and function as chemical messengers for brain–gut interactions [131,132]. Research has revealed that administering probiotics to mice led to a considerable increase in SCFAs and beneficial bacteria (including *Oscillibacter* and *Prevotella*) in their gut flora as compared to levels in the control group [133]. The gut is the primary site where SCFAs mediate intestinal epithelial integrity or mucosal immune response. Intestinal flora disturbances that lead to reduced SCFAs are associated with colon disease. SCFAs exert an anti-inflammatory function by regulating immune cell chemotaxis, reactive oxygen species release, and cytokine release [134]. Lactobacillus and *Bifidobacterium* produce lactic acid and acetic acid, which are major end products of carbohydrate metabolism. These organic acids, when produced in situ, can reduce intracavitary pH and inhibit the growth of pathogenic bacteria [135,136]. *Bifidobacterium* mainly produces SCFAs through fermentation, where the oligosaccharides of Chinese yams can be used as carbon sources by *Lactobacillus plantarum*, *Bifidobacterium*, and other intestinal probiotics in the simulated colon environment. The content of acetic acid in the fermentation broth of *Bifidobacterium* after 48 h was as high as 1.85 mg/mL, and after 8 h fermentation, was as high as 0.082 mg/mL [137]. SCFAs are produced by various pathways, where the most common is through glycolysis, and certain bacterial groups such as *Bifidobacterium* can also use the pentose phosphate pathway to produce the same metabolite [138].

4.2.2. Promotion of Neurotransmitter Production

There is bidirectional regulation between the gastrointestinal tract, enteric nervous system, and central nervous system. Certain probiotic strains produce small molecules that exert different effects on the host and its gut microbes [139]. Neurotransmitters are important signaling molecules between neurons, and between neurons and effector cells and include dopamine, gamma-aminobutyric acid (GABA), and 5-HT [140]. Both dopamine and GABA are important neurotransmitters that regulate various functions in the central nervous system. Studies have demonstrated that a partially specific gut microbiota produces neurotransmitters such as dopamine and GABA. 5-HT, also known as serotonin, is an important metabolite of tryptophan, and the majority of 5-HT is produced by enterochromaffin cells that play an important role as an important neurotransmitter and signaling molecule in the two-way communication system between the brain and the gut [141,142]. The gut microbiota can promote the production of some neurotransmitters, which are often associated with central nervous system diseases such as the brain. Therefore, targeting the regulation of microbial metabolites, such as neurotransmitters, may be a potential way to improve neurological-system-related diseases. It has also been shown that *Candida*, *Streptococcus*, and *Enterococcus* can produce neurotransmitters such as serotonin; *Bacillus* and *Saccharomyces* species can produce noradrenaline; while *Lactobacillus* and *Bifidobacterium* species can synthesize and release GABA. These microbially synthesized neurotransmitters can act locally and also cross the intestinal mucosa to act locally, but potentially also the central nervous system via nerval signaling [143]. Through the regulation of neurotransmitters, the brain–gut axis can improve the symptoms of intestinal peristalsis, upper abdominal discomfort, loss of appetite, and constipation in FD patients [144].

4.2.3. Promotion of Bile Acid Production

Under the action of gut bacteria, the primary bile acids formed in the liver are modified into secondary bile acids. The circulation process between the liver and the intestine is called enterohepatic circulation, which is also an important means to regulate the composition of bile acids [145,146]. Based on the important role of intestinal flora in the modification of bile acids, probiotics, as an intervention targeting intestinal flora, may indirectly regulate the metabolism of intestinal flora by bile acids, thereby playing a probiotic role in alleviating FD symptoms.

5. Probiotics as a Potential Treatment for FD

Interactions between the microbiota and host crosstalk are plausible underlying mechanisms, which will help to establish probiotics as a novel, tailored therapeutic approach for FD. Probiotics play a multifaceted role, and they contribute to the amelioration of FD symptoms through several mechanisms, such as eliminating pathogenic bacteria to reestablish microbial balance [98], modulating epithelial barrier permeability, influencing visceral hypersensitivity, exerting both local and systemic anti-inflammatory effects, and regulating intestinal motility [147,148]. These factors are very beneficial to intestinal health. Probiotics play a certain role in maintaining the integrity of the duodenal mucosa. FD is related to a defect in the duodenal barrier that is caused by the immune response of food and microorganisms in the local area and the whole body, thus producing FD symptoms [149]. *E. coli*/*Shigella* bacteria represent a significant origin of toxic lipopolysaccharides that may impede gastric emptying. The intake of probiotics, particularly *Bifidobacteria*, can efficiently lower their concentrations and reinstate normal motor function in the small intestine [150]. Patients with FD exhibit both local and systemic immune activation. Probiotics exert their influence by modulating toll-like receptors (specifically TLR2 and TLR4) and generating pro-inflammatory cytokines through the metabolites produced by the gut microbiota. The interaction between intestinal neurons and microorganisms increases neuronal survival and gastrointestinal motility. The TLR4 agonist lipopolysaccharide promotes the survival of intestinal neurons by activating TLR4 and NF-B. Factors that regulate neuronal TLR4 signaling may alter gastrointestinal motility [151]. Thus, TLR4 and its downstream signaling molecules could be potential therapeutic targets for the treatment of gastrointestinal motility disorders.

6. Conclusions and Future Trends

Probiotics, by regulating intestinal flora, offer a potential advantage in alleviating FD, addressing the disorder's onset and progression through their roles in maintaining microecological balance, improving barrier function, and regulating the immune response. Currently, the diversity, structure, abundance, and distribution of FD-associated intestinal microbiota are abnormal. However, there is no unified conclusion as to which specific microbiota are closely related to the occurrence and development of FD. Regarding improving FD symptoms, there is a tendency for probiotics to exert beneficial effects, but there is still confusion regarding which strains or species may be beneficial or most beneficial. Furthermore, for all the randomized controlled trials that we identified, the longest treatment duration was limited to 16 weeks, thus leaving the long-term effectiveness of probiotics or prebiotics in FD undetermined. Variability in the clinical outcomes of probiotics for FD exists due to factors such as individual living conditions, medication use, and the underlying causes of the condition. Therefore, future research should place greater emphasis on examining the influence of probiotics in diverse populations and various FD subtypes. Moreover, the interaction between microbiota and host crosstalk is a possible underlying mechanism that may be useful to explore in order to further elucidate the mechanisms by which probiotics enhance FD.

Author Contributions: Conceptualization, X.S., A.X., M.L. and X.Y.; writing—original draft preparation, X.S., A.X., Z.L., C.J., J.W., M.L. and X.Y.; writing—review and editing, M.L. and X.Y.; supervision, M.L. and X.Y.; project administration, M.L. and X.Y. All authors have read and agreed to the published version of the manuscript.

Funding: This research received no external funding.

Institutional Review Board Statement: Not applicable.

Informed Consent Statement: Not applicable.

Data Availability Statement: Data are contained within the article.

Conflicts of Interest: The authors declare no conflict of interest.

References

1. Zhou, G.; Qin, W.; Zeng, F.; Liu, P.; Yang, X.; von Deneen, K.M.; Gong, Q.; Liang, F.; Tian, J. White-Matter Microstructural Changes in Functional Dyspepsia: A Diffusion Tensor Imaging Study. *Am. J. Gastroenterol.* **2013**, *108*, 260–269. [CrossRef] [PubMed]
2. Stanghellini, V.; Tack, J.; Chan, F.K.L.; Malagelada, J.R.; Suzuki, H.; Hasler, W.L.; Talley, N.J. Gastroduodenal Disorders. *Gastroenterology* **2013**, *150*, 1380–1392. [CrossRef] [PubMed]
3. Drago, L.; Meroni, G.; Pistone, D.; Pasquale, L.; Milazzo, G.; Monica, F.; Aragona, S.; Ficano, L.; Vassallo, R. Evaluation of main functional dyspepsia symptoms after probiotic administration in patients receiving conventional pharmacological therapies. *J. Int. Med. Res.* **2021**, *49*, 0300060520982657. [CrossRef] [PubMed]
4. Ford, A.C.; Mahadeva, S.; Carbone, M.F.; Lacy, B.E.; Talley, N.J. Functional dyspepsia. *Lancet* **2020**, *396*, 1689–1702. [CrossRef]
5. Ford, A.C.; Forman, D.; Bailey, A.G.; Cook, M.B.; Axon, A.T.R.; Moayyedi, P. Who Consults with Dyspepsia? Results from a Longitudinal 10-Yr Follow-Up Study. *Am. J. Gastroenterol.* **2007**, *102*, 957–965. [CrossRef] [PubMed]
6. Wang, C.; Fang, X. Inflammation and Overlap of Irritable Bowel Syndrome and Functional Dyspepsia. *J. Neurogastroenterol. Motil.* **2021**, *27*, 153–164. [CrossRef]
7. Tack, J.; Carbone, F. Functional dyspepsia and gastroparesis. *Curr. Opin. Gastroenterol.* **2017**, *33*, 446–454. [CrossRef]
8. Tziatzios, G.; Gkolfakis, P.; Leite, G.; Mathur, R.; Damoraki, G.; Giamarellos-Bourboulis, E.J.; Triantafyllou, K. Probiotics in Functional Dyspepsia. *Microorganisms* **2023**, *11*, 351. [CrossRef]
9. Matsuzaki, J.; Suzuki, H.; Asakura, K.; Fukushima, Y.; Inadomi, J.M.; Takebayashi, T.; Hibi, T. Classification of functional dyspepsia based on concomitant bowel symptoms. *Neurogastroenterol. Motil.* **2012**, *24*, 325-e164. [CrossRef]
10. Bousdouni, P.; Kandyliari, A.; Koutelidakis, A.E. Probiotics and Phytochemicals: Role on Gut Microbiota and Efficacy on Irritable Bowel Syndrome, Functional Dyspepsia, and Functional Constipation. *Gastrointest. Disord.* **2022**, *4*, 30–48. [CrossRef]
11. Sander, G.B.; Mazzoleni, L.E.; Francesconi, C.F.d.M.; Balbinotto, G.; Mazzoleni, F.; Wortmann, A.C.; Cardoso, I.d.Q.; Klamt, A.L.; Milbradt, T.C. Influence of Organic and Functional Dyspepsia on Work Productivity: The HEROES-DIP Study. *Value Health* **2011**, *14*, S126–S129. [CrossRef] [PubMed]
12. Luo, C.-F.; Yuan, M.; Chen, M.-S.; Liu, S.-M.; Zhu, L.; Huang, B.-Y.; Liu, X.-W.; Xiong, W. Pharmacokinetics, tissue distribution and relative bioavailability of puerarin solid lipid nanoparticles following oral administration. *Int. J. Pharm.* **2011**, *410*, 138–144. [CrossRef] [PubMed]
13. Gong, H.-L.; Tang, W.-F.; Yu, Q.; Xiang, J.; Xia, Q.; Chen, G.-Y.; Huang, X.; Liang, M.-Z. Effect of severe acute pancreatitis on pharmacokinetics ofDa-Cheng-Qi Decoction components. *World J. Gastroenterol.* **2009**, *15*, 5992–5999. [CrossRef] [PubMed]
14. Nakagawa, K.; Hara, K.; Fikree, A.; Siddiqi, S.; Woodland, P.; Masamune, A.; Aziz, Q.; Sifrim, D.; Yazaki, E. Patients with dyspepsia have impaired mucosal integrity both in the duodenum and jejunum: In vivo assessment of small bowel mucosal integrity using baseline impedance. *J. Gastroenterol.* **2020**, *55*, 273–280. [CrossRef] [PubMed]
15. Wang, X.; Li, X.P.; Ge, W.Q.; Huang, J.; Li, G.Q.; Cong, Y.Q.; Li, F.K.; Liu, Z.; Liu, Z.Y.; Li, Y.Q.; et al. Quantitative evaluation of duodenal eosinophils and mast cells in adult patients with functional dyspepsia. *Ann. Diagn. Pathol.* **2015**, *19*, 50–56. [CrossRef] [PubMed]
16. Van Oudenhove, L.; Aziz, Q. The role of psychosocial factors and psychiatric disorders in functional dyspepsia. *Nat. Rev. Gastroenterol. Hepatol.* **2013**, *10*, 158–167. [CrossRef]
17. John, G.K.; Wang, L.; Nanavati, J.; Twose, C.; Singh, R.; Mullin, G. Dietary Alteration of the Gut Microbiome and Its Impact on Weight and Fat Mass: A Systematic Review and Meta-Analysis. *Genes* **2018**, *9*, 167. [CrossRef]
18. Hatton, G.B.; Madla, C.M.; Rabbie, S.C.; Basit, A.W. All disease begins in the gut: Influence of gastrointestinal disorders and surgery on oral drug performance. *Int. J. Pharm.* **2018**, *548*, 408–422. [CrossRef]
19. Zmora, N.; Suez, J.; Elinav, E. You are what you eat: Diet, health and the gut microbiota. *Nat. Rev. Gastroenterol. Hepatol.* **2019**, *16*, 35–56. [CrossRef]
20. Hadizadeh, F.; Bonfiglio, F.; Belheouane, M.; Vallier, M.; Sauer, S.; Bang, C.; Bujanda, L.; Andreasson, A.; Agreus, L.; Engstrand, L.; et al. Faecal microbiota composition associates with abdominal pain in the general population. *Gut* **2018**, *67*, 778. [CrossRef]
21. Liebregts, T.; Adam, B.; Bredack, C.; Gururatsakul, M.; Pilkington, K.R.; Brierley, S.M.; Blackshaw, L.A.; Gerken, G.; Talley, N.J.; Holtmann, G. Small Bowel Homing T Cells Are Associated with Symptoms and Delayed Gastric Emptying in Functional Dyspepsia. *Am. J. Gastroenterol.* **2011**, *106*, 1089–1098. [CrossRef] [PubMed]
22. Tziatzios, G.; Giamarellos-Bourboulis, E.J.; Papanikolaou, I.S.; Pimentel, M.; Dimitriadis, G.D.; Triantafyllou, K. Is small intestinal bacterial overgrowth involved in the pathogenesis of functional dyspepsia? *Med. Hypotheses* **2017**, *106*, 26. [CrossRef] [PubMed]
23. Gao, J.; Li, X.Y.; Zhang, G.H.; Sadiq, F.A.; Simal-Gandara, J.; Xiao, J.B.; Sang, Y.X. Probiotics in the dairy industry-Advances and opportunities. *Compr. Rev. Food Sci. Food Saf.* **2021**, *20*, 3937–3982. [CrossRef] [PubMed]
24. Xie, A.; Dong, Y.; Liu, Z.; Li, Z.; Shao, J.; Li, M.; Yue, X. A Review of Plant-Based Drinks Addressing Nutrients, Flavor, and Processing Technologies. *Foods* **2023**, *12*, 3952. [CrossRef]
25. Behnsen, J.; Deriu, E.; Sassone-Corsi, M.; Raffatellu, M. Probiotics: Properties, Examples, and Specific Applications. *Cold Spring Harb. Perspect. Med.* **2013**, *3*, a010074. [CrossRef]
26. Oelschlaeger, T.A. Mechanisms of probiotic actions—A review. *Int. J. Med. Microbiol.* **2010**, *300*, 57–62. [CrossRef]
27. Didari, T.; Mozaffari, S.; Nikfar, S.; Abdollahi, M. Effectiveness of probiotics in irritable bowel syndrome: Updated systematic review with meta-analysis. *World J. Gastroenterol.* **2015**, *21*, 3072–3084. [CrossRef]

28. Cheng, H.X.; Ma, Y.; Liu, X.H.; Tian, C.; Zhong, X.L.; Zhao, L.B. A Systematic Review and Meta-Analysis: *Lactobacillus acidophilu* for Treating Acute Gastroenteritis in Children. *Nutrients* **2022**, *14*, 682. [CrossRef]
29. Asl, Z.R.; Sepehri, G.; Salami, M. Probiotic treatment improves the impaired spatial cognitive performance and restores synaptic plasticity in an animal model of Alzheimer's disease. *Behav. Brain Res.* **2019**, *376*, 112183.
30. Chen, H.L.; Lai, Y.W.; Chen, C.S.; Chu, T.W.; Lin, W.; Yen, C.C.; Lin, M.F.; Tu, M.Y.; Chen, C.M. Probiotic Lactobacillus casei Expressing Human Lactoferrin Elevates Antibacterial Activity in the Gastrointestinal Tract. *Biometals* **2010**, *23*, 543–554. [CrossRef]
31. Sakai, T.; Kubota, H.; Gawad, A.; Gheyle, L.; Oishi, K. Effect of fermented milk containing *Lactobacillus casei* strain Shirota on constipation-related symptoms and haemorrhoids in women during puerperium. *Benef. Microbes* **2015**, *6*, 253–262. [CrossRef]
32. Verhaar, B.J.H.; Prodan, A.; Nieuwdorp, M.; Muller, M. Gut Microbiota in Hypertension and Atherosclerosis: A Review. *Nutrients* **2020**, *12*, 1082–1095. [CrossRef] [PubMed]
33. Kao, L.; Liu, T.H.; Tsai, T.Y.; Pan, T.M. Beneficial effects of the commercial lactic acid bacteria product, Vigiis 101, on gastric mucosa and intestinal bacterial flora in rats. *J. Microbiol. Immunol. Infect.* **2020**, *53*, 266–273. [CrossRef] [PubMed]
34. Ke, X.X.; Walker, A.; Haange, S.B.; Lagkouvardos, I.; Liu, Y.W.; Schmitt-Kopplin, P.; Von Bergen, M.; Jehmlich, N.; He, X.; Clavel, T.; et al. Synbiotic-driven improvement of metabolic disturbances is associated with changes in the gut microbiome in diet-induced obese mice. *Mol. Metab.* **2019**, *22*, 96–109. [CrossRef] [PubMed]
35. Segers, M.E.; Lebeer, S. Towards a better understanding of *Lactobacillus rhamnosus* GG-host interactions. *Microb. Cell Factories* **2014**, *13*, S7. [CrossRef] [PubMed]
36. Liu, Y.Y.; Tian, X.J.; He, B.K.; Hoang, T.K.; Taylor, C.M.; Blanchard, E.; Freeborn, J.; Park, S.; Luo, M.; Couturier, J.; et al. *Lactobacillus reuteri* DSM 17938 feeding of healthy newborn mice regulates immune responses while modulating gut microbiota and boosting beneficial metabolites. *Am. J. Physiol. Gastrointest. Liver Physiol.* **2019**, *317*, G824–G838. [CrossRef] [PubMed]
37. Talarico, T.L.; Dobrogosz, W.J. Chemical characterization of an antimicrobial chemical characterization of an antimicrobial substance produced by *Lactobacillus reuteri*. *Antimicrob. Agents Chemother.* **1989**, *33*, 674–679. [CrossRef] [PubMed]
38. Indrio, F.; Riezzo, G.; Raimondi, F.; Bisceglia, M.; Cavallo, L.; Francavilla, R. The effects of probiotics on feeding tolerance, bowel habits, and gastrointestinal motility in preterm newborns. *J. Pediatr.* **2008**, *152*, 801–806. [CrossRef]
39. Colston, J.M.; Taniuchi, M.; Ahmed, T.; Ferdousi, T.; Kabir, F.; Mduma, E.; Nshama, R.; Iqbal, N.T.; Haque, R.; Ahmed, T.; et al. Intestinal Colonization with *Bifidobacterium longum* Subspecies Is Associated with Length at Birth, Exclusive Breastfeeding, and Decreased Risk of Enteric Virus Infections, but Not with Histo-Blood Group Antigens, Oral Vaccine Response or Later Growth in Three Birth Cohorts. *Front. Pediatr.* **2022**, *10*, 804798.
40. Chichlowski, M.; Shah, N.; Wampler, J.L.; Wu, S.S.; Vanderhoof, J.A. *Bifidobacterium longum* Subspecies *infantis* (*B. infantis*) in Pediatric Nutrition: Current State of Knowledge. *Nutrients* **2020**, *12*, 1581. [CrossRef]
41. Feng, X.B.; Su, Y.; Jiang, J.; Li, N.; Ding, W.W.; Wang, Z.M.; Hu, X.H.; Zhu, W.Y.; Li, J.S. Changes in Fecal and Colonic Mucosal Microbiota of Patients with Refractory Constipation after a Subtotal Colectomy. *Am. Surg.* **2015**, *81*, 198–206. [CrossRef] [PubMed]
42. Mormone, E.; George, J.; Nieto, N. Molecular pathogenesis of hepatic fibrosis and current therapeutic approaches. *Chem. Biol. Interact.* **2011**, *193*, 225–231. [CrossRef] [PubMed]
43. Zuo, L.; Yuan, K.T.; Yu, L.; Meng, Q.H.; Chung, P.C.K.; Yang, D.H. *Bifidobacterium infantis* attenuates colitis by regulating T cell subset responses. *World J. Gastroenterol.* **2014**, *20*, 18316–18329. [CrossRef] [PubMed]
44. Adami, A.; Cavazzoni, V. Occurrence of selected bacterial groups in the faeces of piglets fed with Bacillus coagulans as probiotic. *J. Basic Microbiol.* **1999**, *39*, 3–10. [CrossRef]
45. Giang, H.H.; Viet, T.Q.; Ogle, B.; Lindberg, J.E. Growth performance, digestibility, gut environment and health status in weaned piglets fed a diet supplemented with potentially probiotic complexes of lactic acid bacteria. *Livest. Sci.* **2010**, *129*, 95–103. [CrossRef]
46. Vercammen, A.; Vivijs, B.; Lurquin, I.; Michiels, C.W. Germination and inactivation of *Bacillus coagulans* and *Alicyclobacillus acidoterrestris* spores by high hydrostatic pressure treatment in buffer and tomato sauce. *Int. J. Food Microbiol.* **2012**, *152*, 162–167. [CrossRef] [PubMed]
47. Adolfsson, O.; Meydani, S.N.; Russell, R.M. Yogurt and gut function. *Am. J. Clin. Nutr.* **2004**, *80*, 245–256. [CrossRef]
48. Iyer, R.; Tomar, S.K.; Maheswari, T.U.; Singh, R. *Streptococcus thermophilus* strains: Multifunctional lactic acid bacteria. *Int. Dairy J.* **2010**, *20*, 133–141. [CrossRef]
49. Holzapfel, W.H.; Schillinger, U. Introduction to pre-and probiotics. *Food Res. Int.* **2002**, *35*, 109–116. [CrossRef]
50. Ouwehand, A.C.; Forssten, S.; Hibberd, A.A.; Lyra, A.; Stahl, B. Probiotic approach to prevent antibiotic resistance. *Ann. Med.* **2016**, *48*, 246–255. [CrossRef]
51. Van Baarlen, P.; Wells, J.M.; Kleerebezem, M. Regulation of intestinal homeostasis and immunity with probiotic lactobacilli. *Trends Immunol.* **2013**, *34*, 208–215. [CrossRef]
52. Fang, H.R.; Zhang, G.Q.; Cheng, J.Y.; Li, Z.Y. Efficacy of *Lactobacillus*-supplemented triple therapy for *Helicobacter pylori* infection in children: A meta-analysis of randomized controlled trials. *Eur. J. Pediatr.* **2019**, *178*, 7–16. [CrossRef]
53. Hou, Q.C.; Zhao, F.Y.; Liu, W.J.; Lv, R.R.; Khine, W.W.T.; Han, J.; Sun, Z.H.; Lee, Y.K.; Zhang, H.P. Probiotic-directed modulation of gut microbiota is basal microbiome dependent. *Gut Microbes* **2020**, *12*, 1736974. [CrossRef] [PubMed]
54. Carrel, A. The problem of the prolongation of life. *J. Insur. Med.* **1946**, *1*, 33–35.
55. Ding, R.X.; Li, M.H.; Zou, Y.T.; Wang, Y.R.; Yan, C.Y.; Zhang, H.B.; Wu, R.N.; Wu, J.R. Effect of normal and strict anaerobic fermentation on physicochemical quality and metabolomics of yogurt. *Food Biosci.* **2022**, *46*, 101368. [CrossRef]

56. Li, M.H.; Yu, H.K.; Chen, J.L.; Abdlla, R.; Liu, A.C.; Song, W.Y.; Zhang, J.; Zhang, X.M.; Yue, X.Q.; Li, Q.L. Novel insights into whey protein differences between donkey and bovine milk. *Food Chem.* **2021**, *365*, 130397. [CrossRef] [PubMed]
57. Lin, T.L.; Shu, C.C.; Lai, W.F.; Tzeng, C.M.; Lu, C.C. Investiture of next generation probiotics on amelioration of diseases–Strains do matter. *Med. Microecol.* **2019**, *1–2*, 100002. [CrossRef]
58. Sivan, A.; Corrales, L.; Hubert, N.; Williams, J.B.; Aquino-Michaels, K.; Earley, Z.M.; Benyamin, F.W.; Lei, Y.M.; Jabri, B.; Alegre, M.L.; et al. Commensal *Bifidobacterium* promotes antitumor immunity and facilitates anti-PD-L1 efficacy. *Science* **2015**, *350*, 1084–1089. [CrossRef]
59. Xie, A.J.; Zhao, S.S.; Liu, Z.F.; Yue, X.Q.; Shao, J.H.; Li, M.H.; Li, Z.W. Polysaccharides, proteins, and their complex as microencapsulation carriers for delivery of probiotics: A review on carrier types and encapsulation techniques. *Int. J. Biol. Macromol.* **2023**, *242*, 124784. [CrossRef]
60. Liu, X.J.; Wang, M.M.; Zhu, C.; Gou, M.X.; Yan, X.H. Research progress of functional lactic acid bacteria. In Proceedings of the 6th International Conference on Energy, Environment and Sustainable Development (ICEESD), Zhuhai, China, 11–12 March 2017; pp. 643–646.
61. Saarela, M.H. Safety aspects of next generation probiotics. *Curr. Opin. Food Sci.* **2019**, *30*, 8–13. [CrossRef]
62. Martin, R.; Langella, P. Emerging Health Concepts in the Probiotics Field: Streamlining the Definitions. *Front. Microbiol.* **2019**, *10*, 1047. [CrossRef]
63. Dimidi, E.; Cox, S.R.; Rossi, M.; Whelan, K. Fermented Foods: Definitions and Characteristics, Impact on the Gut Microbiota and Effects on Gastrointestinal Health and Disease. *Nutrients* **2019**, *11*, 1806. [CrossRef] [PubMed]
64. La Fata, G.; Weber, P.; Mohajeri, M.H. Probiotics and the Gut Immune System: Indirect Regulation. *Probiotics Antimicrob. Proteins* **2018**, *10*, 11–21. [CrossRef] [PubMed]
65. Sankar, N.R.; Priyanka, V.D.; Reddy, P.S.; Rajanikanth, P.; Kumar, V.K.; Indira, M. Purification and characterization of bacteriocin produced by Lactobacillus plantarum isolated from cow milk. *Int. J. Microbiol. Res.* **2012**, *3*, 133–137.
66. Sturme, M.H.J.; Kleerebezem, M.; Nakayama, J.; Akkermans, A.D.L.; Vaugha, E.E.; de Vos, W.M. Cell to cell communication by autoinducing peptides in gram-positive bacteria. *Antonie Van Leeuwenhoek* **2002**, *81*, 233–243. [CrossRef]
67. Wu, Z.R.; Liu, D.L.; Deng, F.H. The Role of Vitamin D in Immune System and Inflammatory Bowel Disease. *J. Inflamm. Res.* **2022**, *15*, 3167–3185. [CrossRef]
68. Cárdenas, N.; Laiño, J.E.; Delgado, S.; Jiménez, E.; del Valle, M.J.; de Giori, G.S.; Sesma, F.; Mayo, B.; Fernández, L.; LeBlanc, J.G.; et al. Relationships between the genome and some phenotypical properties of *Lactobacillus fermentum* CECT 5716, a probiotic strain isolated from human milk. *Appl. Microbiol. Biotechnol.* **2015**, *99*, 4343–4353. [CrossRef]
69. Geirnaert, A.; Calatayud, M.; Grootaert, C.; Laukens, D.; Wiele, T.V.D. Butyrate-producing bacteria supplemented in vitro to Crohn's disease patient microbiota increased butyrate production and enhanced intestinal epithelial barrier integrity. *Sci. Rep.* **2017**, *7*, 11450. [CrossRef]
70. Tallapragada, P.; Rayavarapu, B.; Rao, P.P.; Ranganath, N.N.; Veerabhadrappa, P.P. Screening of potential probiotic lactic acid bacteria and production of amylase and its partial purification. *J. Genet. Eng. Biotechnol.* **2018**, *16*, 357–362.
71. Amaretti, A.; Nunzio, M.; Pompei, A.; Raimondi, S.; Rossi, M.; Bordoni, A. Antioxidant properties of potentially probiotic bacteria: In vitro and in vivo activities. *Appl. Microbiol. Biotechnol.* **2013**, *97*, 809–817. [CrossRef]
72. Dai, Z.L.; Wu, Z.L.; Hang, S.Q.; Zhu, W.Y.; Wu, G.Y. Amino acid metabolism in intestinal bacteria and its potential implications for mammalian reproduction. *Mol. Hum. Reprod.* **2015**, *21*, 389–409. [CrossRef] [PubMed]
73. Metges, C.C.; Eberhard, M.; Petzke, K.J. Synthesis and absorption of intestinal microbial lysine in humans and non-ruminant animals and impact on human estimated average requirement of dietary lysine. *Curr. Opin. Clin. Nutr. Metab. Care* **2006**, *9*, 37–41. [CrossRef] [PubMed]
74. Li, C.; Nie, S.P.; Zhu, K.X.; Xiong, T.; Xie, M.Y. *Lactobacillus plantarum* NCU116 fermented carrot juice evokes changes of metabolites in serum from type 2 diabetic rats. *Food Res. Int.* **2016**, *80*, 36–40. [CrossRef]
75. Wauters, L.; Slaets, H.; De Paepe, K. Efficacy and safety of spore-forming probiotics in the treatment of functional dyspepsia: A pilot randomised, double-blind, placebo-controlled trial. *Lancet Gastroenterol. Hepatol.* **2021**, *6*, 784–792. [CrossRef] [PubMed]
76. Navarro Rodriguez, T.; Silva, F.M.; Barbuti, R.C.; Mattar, R.; Moraesfilho, J.P.; Oliveira, M.N.d.; Bogsan, C.S.; Chinzon, D.; Eisig, J.N. Association of a probiotic to a Helicobacter pylori eradication regimen does not increase efficacy or decreases the adverse effects of the treatment: A prospective, randomized, double-blind, placebo-controlled study. *BMC Gastroenterol.* **2013**, *13*, 56. [CrossRef] [PubMed]
77. Sun, E.N.; Zhang, X.M.; Zhao, Y.F.; Li, J.Y.; Sun, J.; Mu, Z.S.; Wang, R. everages containing *Lactobacillus paracasei* LC-37 improved functional dyspepsia through regulation of the intestinal microbiota and their metabolites. *J. Dairy Sci.* **2021**, *104*, 6389–6398. [CrossRef]
78. Rahmani, P.; Ghouran-Orimi, A.; Motamed, F.; Moradzadeh, A. Evaluating the effects of probiotics in pediatrics with recurrent abdominal pain. *Clin. Exp. Pediatr.* **2020**, *63*, 485–490. [CrossRef]
79. Nakae, H.; Tsuda, A.; Matsuoka, T.; Mine, T.; Koga, Y. Gastric microbiota in the functional dyspepsia patients treated with probiotic yogurt. *BMJ Open Gastroenterol.* **2016**, *3*, e000109. [CrossRef]
80. Ohtsu, T.; Takagi, A.; Uemura, N.; Inoue, K.; Sekino, H.; Kawashima, A.; Uchida, M.; Koga, Y. The Ameliorating Effect of *Lactobacillus gasseri* OLL2716 on Functional Dyspepsia in *Helicobacter pylori*-Uninfected Individuals: A Randomized Controlled Study. *Digestion* **2017**, *96*, 92–102. [CrossRef]

81. Carbone, F.; Tack, J. Gastroduodenal Mechanisms Underlying Functional Gastric Disorders. *Dig. Dis.* **2014**, *32*, 222–229. [CrossRef]
82. Samsom, M.; Verhagen, M.A.; van Berge Henegouwen, G.P.; Smout, A.J. Abnormal clearance of exogenous acid and increased acid sensitivity of the proximal duodenum in dyspeptic patients. *Gastroenterology* **1999**, *116*, 515–520. [CrossRef] [PubMed]
83. Futagami, S.; Itoh, T.; Sakamoto, C. Systematic review: Post-infectious functional dyspepsia. *Aliment. Pharmacol. Ther.* **2014**, *41*, 177–188. [CrossRef] [PubMed]
84. Spiller, R. Postinfectious Functional Dyspepsia and Postinfectious Irritable Bowel Syndrome: Different Symptoms but Similar Risk Factors. *Gastroenterology* **2010**, *138*, 1660–1663. [CrossRef] [PubMed]
85. Miwa, H.; Ghoshal, U.C.; Gonlachanvit, S.; Gwee, K.A.; Ang, T.L.; Chang, F.Y.; Fock, K.M.; Hongo, M.; Hou, X.H.; Kachintorn, U.; et al. Asian Consensus Report on Functional Dyspepsia. *J. Neurogastroenterol. Motil.* **2012**, *18*, 150–168. [CrossRef] [PubMed]
86. Ford, A.C. Eradicating Helicobacter pylori in functional dyspepsia. *Gastroenterology* **2012**, *142*, 1613–1614. [CrossRef] [PubMed]
87. Moayyedi, P.; Shelly, S.; Deeks, J.J.; Delaney, B.; Innes, M.; Forman, D. Pharmacological interventions for non-ulcer dyspepsia (Withdrawn Paper, 2011, Art. No. CD001960). *Cochrane Database Syst. Rev.* **2011**, *2*, CD001960.
88. Talley, N.J.; Vakil, N.B.; Moayyedi, P. American gastroenterological association technical review on the evaluation of dyspepsia. *Gastroenterology* **2005**, *129*, 1756–1780. [CrossRef]
89. Pleyer, C.; Bittner, H.; Locke, G.R.; Choung, R.S.; Zinsmeister, A.R.; Schleck, C.D.; Herrick, L.M.; Talley, N.J. Overdiagnosis of gastro-esophageal reflux disease and underdiagnosis of functional dyspepsia in a USA community. *Neurogastroenterol. Motil.* **2014**, *26*, 1163–1171. [CrossRef]
90. Fujiwara, Y.; Arakawa, T. Overlap in Patients with Dyspepsia/Functional Dyspepsia. *J. Neurogastroenterol. Motil.* **2014**, *20*, 447–457. [CrossRef]
91. Nakano, M.; Kitano, S.; Nanri, M.; Kiniwa, M. Lafutidine, a unique histamine H_2-receptor antagonist, inhibits distention-induced gastric acid secretion through an H_2 receptor-independent mechanism. *Eur. J. Pharmacol.* **2011**, *658*, 236–241. [CrossRef]
92. Kaptchuk, T.J.; Friedlander, E.; Kelley, J.M.; Sanchez, M.N.; Kokkotou, E.; Singer, J.P.; Kowalczykowski, M.; Miller, F.G.; Kirsch, I.; Lembo, A.J. Placebos without Deception: A Randomized Controlled Trial in Irritable Bowel Syndrome. *PLoS ONE* **2010**, *5*, e15591. [CrossRef] [PubMed]
93. Orive, M.; Barrio, I.; Orive, V.M.; Matellanes, B.; Padierna, J.A.; Cabriada, J.; Orive, A.; Escobar, A.; Quintana, J.M. A randomized controlled trial of a 10 week group psychotherapeutic treatment added to standard medical treatment in patients with functional dyspepsia. *J. Psychosom. Res.* **2015**, *78*, 563–568. [CrossRef] [PubMed]
94. Tack, J.; Camilleri, M. New developments in the treatment of gastroparesis and functional dyspepsia. *Curr. Opin. Pharmacol.* **2018**, *43*, 111–117. [CrossRef] [PubMed]
95. Agah, S.; Akbari, A.; Heshmati, J.; Sepidarkish, M.; Morvaridzadeh, M.; Adibi, P.; Mazidi, M.; Farsi, F.; Ofori-Asenso, R.; Talley, N.J. Systematic review with meta-analysis: Effects of probiotic supplementation on symptoms in functional dyspepsia. *J. Funct. Foods.* **2020**, *68*, 103902. [CrossRef]
96. Ford, A.C.; Quigley, E.M.M.; Lacy, B.E.; Lembo, A.J.; Saito, Y.A.; Schiller, L.R.; Soffer, E.E.; Spiegel, B.M.R.; Moayyedi, P. Efficacy of Prebiotics, Probiotics, and Synbiotics in Irritable Bowel Syndrome and Chronic Idiopathic Constipation: Systematic Review and Meta-analysis. *Am. J. Gastroenterol.* **2014**, *109*, 1547–1561. [CrossRef] [PubMed]
97. Paone, P.; Cani, P.D. Mucus barrier, mucins and gut microbiota: The expected slimy partners? *Gut* **2020**, *69*, 2232–2243. [CrossRef] [PubMed]
98. Ford, A.C.; Talley, N.J. Mucosal inflammation as a potential etiological factor in irritable bowel syndrome: A systematic review. *J. Gastroenterol.* **2011**, *46*, 421–431. [CrossRef]
99. Vanheel, H.; Vicario, M.; Martinez, C.; Vanuytsel, T.; Pardon, N.; Santos, J.; Söder-Holm, J.; Tack, J.; Farré, R. Duodenal low-grade inflammation and impaired mucosal integrity in functional dyspepsia patients. *Neurogastroenterol. Motil.* **2012**, *24*, 24–25.
100. Terciolo, C.; Dapoigny, M.; Andre, F. Beneficial effects of Saccharomyces boulardii CNCM I-745 on clinical disorders associated with intestinal barrier disruption. *Clin. Exp. Gastroenterol.* **2019**, *12*, 67–82. [CrossRef]
101. Barbara, G.; Barbaro, M.R.; Fuschi, D.; Palombo, M.; Falangone, F.; Cremon, C.; Marasco, G.; Stanghellini, V. Corrigendum: Inflammatory and Microbiota-Related Regulation of the Intestinal Epithelial Barrier. *Front. Nutr.* **2021**, *8*, 790387. [CrossRef]
102. Llewellyn, A.; Foey, A. Probiotic Modulation of Innate Cell Pathogen Sensing and Signaling Events. *Nutrients* **2017**, *9*, 1156. [CrossRef] [PubMed]
103. Groschwitz, K.R.; Ahrens, R.; Osterfeld, H.; Gurish, M.F.; Han, X.N.; Åbrink, M.; Finkelman, F.D.; Pejler, G.; Hogan, S.P. Mast cells regulate homeostatic intestinal epithelial migration and barrier function by a chymase/Mcpt4-dependent mechanism. *Proc. Natl. Acad. Sci. USA* **2009**, *106*, 22381–22386. [CrossRef] [PubMed]
104. Jin-Bo, L.; Jian, J.U. Research Progress in Pathogenesis of Functional Dyspepsia. *Med. Recapitul.* **2011**, *50*, 161–164.
105. Zhang, T.; Zhang, B.; Ma, X.; Zhang, J.; Wei, Y.; Wang, F.; Tang, X. Research trends in the field of the gut-brain interaction: Functional dyspepsia in the spotlight—An integrated bibliometric and science mapping approach. *Front. Neurosci.* **2023**, *17*, 1109510. [CrossRef]
106. Castelli, V.; D'Angelo, M.; Quintiliani, M.; Benedetti, E.; Cifone, M.G.; Cimini, A. The emerging role of probiotics in neurodegenerative diseases:new hope for Parkinson's disease? *J. Biomed. Sci.* **2021**, *16*, 7.
107. Neufeld, K.A.M.; Mao, Y.K.; Bienenstock, J.; Foster, J.A.; Kunze, W.A. The microbiome is essential for normal gut intrinsic primary afferent neuron excitability in the mouse. *Neurogastroenterol. Motil.* **2013**, *25*, 183–188. [CrossRef]

108. Simrén, M.; Törnblom, H.; Palsson, O.S.; van Tilburg, M.A.L.; Van Oudenhove, L.; Tack, J.; Whitehead, W.E. Visceral hypersensitivity is associated with GI symptom severity in functional GI disorders: Consistent findings from five different patient cohorts. *Gut* **2018**, *67*, 255–262. [CrossRef]
109. Adak, A.; Khan, M.R. An insight into gut microbiota and its functionalities. *Cell. Mol. Life Sci.* **2019**, *76*, 473–493. [CrossRef]
110. Amaral, F.A.; Sachs, D.; Costa, V.V.; Fagundes, C.T.; Cisalpino, D.; Cunha, T.M.; Ferreira, S.H.; Cunha, F.Q.; Silva, T.A.; Nicoli, J.R.; et al. Commensal microbiota is fundamental for the development of inflammatory pain. *Proc. Natl. Acad. Sci. USA* **2008**, *105*, 2193–2197. [CrossRef]
111. Li, Q.; Chen, C.; Su, J.B.; Qiu, Y.H.; Wu, H.; Song, T.Q.; Mao, X.H.; He, Y.; Cheng, Z.J.; Li, J.D.; et al. Recurrence and prognosis in intrahepatic cholangiocarcinoma patients with different etiology after radical resection: A multi-institutional study. *BMC Cancer* **2022**, *22*, 329. [CrossRef]
112. Tan, V.P.Y.; Liu, K.S.H.; Lam, F.Y.F.; Hung, I.F.N.; Yuen, M.F.; Leung, W.K. Randomised clinical trial: Rifaximin versus placebo for the treatment of functional dyspepsia. *Aliment. Pharmacol. Ther.* **2017**, *45*, 767–776. [CrossRef]
113. Martin, C.R.; Osadchiy, V.; Kalani, A.; Mayer, E.A. The Brain-Gut-Microbiome Axis. *Cell. Mol. Gastroenterol. Hepatol.* **2018**, *6*, 133–148. [CrossRef]
114. Barbara, G.; Stanghellini, V.; Brandi, G.; Cremon, C.; Di Nardo, G.; De Giorgio, R.; Corinaldesi, R. Interactions between commensal bacteria and gut sensorimotor function in health and disease. *Am. J. Gastroenterol.* **2005**, *100*, 2560–2568. [CrossRef] [PubMed]
115. Liu, H.N.; Ohya, S.; Nishizawa, Y.; Sawamura, K.; Iino, S.; Syed, M.M.; Goto, K.; Imaizumi, Y.; Nakayama, S. Serotonin Augments Gut Pacemaker Activity via 5-HT3 Receptors. *PLoS ONE* **2011**, *6*, e24928. [CrossRef] [PubMed]
116. Tack, J.; Janssen, P.; Masaoka, T.; Farré, R.; Van Oudenhove, L. Efficacy of Buspirone, a Fundus-Relaxing Drug, in Patients with Functional Dyspepsia. *Clin. Gastroenterol. Hepatol.* **2012**, *10*, 1239–1245. [CrossRef]
117. Fukui, H.; Xu, X.; Miwa, H. Role of Gut Microbiota-Gut Hormone Axis in the Pathophysiology of Functional Gastrointestinal Disorders. *J. Neurogastroenterol. Motil.* **2018**, *24*, 367–386. [CrossRef]
118. Singh, R.; Zogg, H.; Wei, L.; Bartlett, A.; Ghoshal, U.C.; Rajender, S.; Ro, S. Gut Microbial Dysbiosis in the Pathogenesis of Gastrointestinal Dysmotility and Metabolic Disorders. *J. Neurogastroenterol. Motil.* **2021**, *27*, 19–34. [CrossRef] [PubMed]
119. Tan, R.Q.; Zhang, Z.; Ju, J.; Ling, J.H. Effect of Chaihu Shugan Powder-Contained Serum on Glutamate-Induced Autophagy of Interstitial Cells of Cajal in the Rat Gastric Antrum. *Evid. Based Complement. Altern. Med.* **2019**, *2019*, 7318616. [CrossRef]
120. Wang, X.Z.; Zhang, P.; Zhang, X. Probiotics Regulate Gut Microbiota: An Effective Method to Improve Immunity. *Molecules* **2021**, *26*, 6076. [CrossRef] [PubMed]
121. Childs, C.E.; Röytiö, H.; Alhoniemi, E.; Fekete, A.A.; Forssten, S.D.; Hudjec, N.; Lim, Y.N.; Steger, C.J.; Yaqoob, P.; Tuohy, K.M.; et al. Xylo-oligosaccharides alone or in synbiotic combination with *Bifidobacterium animalis* subsp *lactis* induce bifidogenesis and modulate markers of immune function in healthy adults: A double-blind, placebo-controlled, randomised, factorial cross-over study. *Br. J. Nutr.* **2014**, *111*, 1945–1956. [CrossRef]
122. Afzaal, M.; Saeed, F.; Shah, Y.A.; Hussain, M.; Rabail, R.; Socol, C.T.; Hassoun, A.; Pateiro, M.; Lorenzo, J.M.; Rusu, A.V.; et al. Human gut microbiota in health and disease: Unveiling the relationship. *Front. Microbiol.* **2022**, *13*, 999001. [CrossRef] [PubMed]
123. Isoulari, E.; Sutas, Y.; Kankaanpaa, P.; Arvilommi, H.; Salminen, S. Probiotics: Effects on immunity 1,2,3. *Am. J. Clin. Nutr.* **2001**, *73*, 444s–450s. [CrossRef] [PubMed]
124. Kelly, D.; King, T.; Aminov, R. Importance of microbial colonization of the gut in early life to the development of immunity. *Mutat. Res.* **2007**, *622*, 58–69. [CrossRef] [PubMed]
125. Galdeano, C.M.; de Moreno de LeBlanc, A.; Vinderola, G.; Bonet, M.E.B.; Perdigon, G. Proposed model: Mechanisms of immunomodulation induced by probiotic bacteria. *Clin. Vaccine Immunol. CVI* **2007**, *14*, 485–492. [CrossRef]
126. Park, S.O.; Seo, K.H. Digital livestock systems and probiotic mixtures can improve the growth performance of swine by enhancing immune function, cecal bacteria, short-chain fatty acid, and nutrient digestibility. *Front. Vet. Sci.* **2023**, *10*, 1126064. [CrossRef]
127. Lin, S.S. Mucosal immunity mediated modulation of the gut microbiome by oral delivery of probiotics into Peyer's patches (vol 7, eabf0677, 2021). *Sci. Adv.* **2022**, *8*, f0677.
128. Vitetta, L.; Saltzman, E.T.; Thomsen, M.; Nikov, T.; Hall, S. Adjuvant Probiotics and the Intestinal Microbiome: Enhancing Vaccines and Immunotherapy Outcomes (vol 5, 50, 2017). *Vaccines* **2018**, *6*, 26. [CrossRef]
129. Kye, Y.J.; Lee, S.Y.; Kim, H.R.; Lee, B.H.; Park, M.S.; Ji, G.E.; Sung, M.K. Lactobacillus acidophilus PIN7 paraprobiotic supplementation ameliorates DSS-induced colitis through anti-inflammatory and immune regulatory effects. *J. Appl. Microbiol.* **2022**, *132*, 3189–3200. [CrossRef]
130. Jang, Y.J.; Kim, W.K.; Han, D.H.; Lee, K.; Ko, G. *Lactobacillus fermentum* species ameliorate dextran sulfate sodium-induced colitis by regulating the immune response and altering gut microbiota. *Gut Microbes* **2019**, *10*, 696–711. [CrossRef]
131. Silva, Y.P.; Bernardi, A.; Frozza, R.L. The Role of Short-Chain Fatty Acids from Gut Microbiota in Gut-Brain Communication. *Front. Endocrinol.* **2020**, *11*, 25. [CrossRef]
132. Ríos-Covián, D.; Ruas-Madiedo, P.; Margolles, A.; Gueimonde, M.; de los Reyes-Gavilán, C.G.; Salazar, N. Intestinal Short Chain Fatty Acids and their Link with Diet and Human Health. *Front. Microbiol.* **2016**, *7*, 185. [CrossRef] [PubMed]
133. Li, J.; Sung, C.Y.J.; Lee, N.; Ni, Y.Q.; Pihlajamäki, J.; Panagiotou, G.; El-Nezami, H. Probiotics modulated gut microbiota suppresses hepatocellular carcinoma growth in mice. *Proc. Natl. Acad. Sci. USA* **2016**, *113*, E1306–E1315. [CrossRef] [PubMed]
134. Sun, M.M.; Wu, W.; Liu, Z.J.; Cong, Y.Z. Microbiota metabolite short chain fatty acids, GPCR, and inflammatory bowel diseases. *J. Gastroenterol.* **2017**, *52*, 1–8. [CrossRef] [PubMed]

135. Flint, H.J.; Duncan, S.H.; Scott, K.P.; Louis, P. Links between diet, gut microbiota composition and gut metabolism. *Proc. Nutr. Soc.* **2015**, *74*, 13–22. [CrossRef] [PubMed]
136. Aoudia, N.; Rieu, A.; Briandet, R.; Deschamps, J.; Chluba, J.; Jego, G.; Garrido, C.; Guzzo, J. Biofilms of *Lactobacillus plantarum* and *Lactobacillus fermentum*: Effect on stress responses, antagonistic effects on pathogen growth and immunomodulatory properties. *Food Microbiol.* **2016**, *53*, 51–59. [CrossRef]
137. Markowiak-Kopec, P.; Slizewska, K. The Effect of Probiotics on the Production of Short-Chain Fatty Acids by Human Intestinal Microbiome. *Nutrients* **2020**, *12*, 11. [CrossRef]
138. Ragsdale, S.W.; Pierce, E. Acetogenesis and the Wood-Ljungdahl pathway of CO(2) fixation. *Biochim. Biophys. Acta* **2008**, *1784*, 1873–1898. [CrossRef]
139. Yan, F.; Liu, L.P.; Dempsey, P.J.; Tsai, Y.H.; Raines, E.W.; Wilson, C.L.; Cao, H.L.; Cao, Z.; Liu, L.S.; Polk, D.B. A *Lactobacillus rhamnosus* GG-derived Soluble Protein, p40, Stimulates Ligand Release from Intestinal Epithelial Cells to Transactivate Epidermal Growth Factor Receptor. *J. Biol. Chem.* **2013**, *288*, 30742–30751. [CrossRef]
140. Wang, Y.; Li, N.; Yang, J.J.; Zhao, D.M.; Chen, B.; Zhang, G.Q.; Chen, S.; Cao, R.F.; Yu, H.; Zhao, C.Y.; et al. Probiotics and fructo-oligosaccharide intervention modulate the microbiota-gut brain axis to improve autism spectrum reducing also the hyper-serotonergic state and the dopamine metabolism disorder. *Pharmacol. Res.* **2020**, *157*, 104784. [CrossRef]
141. Bellono, N.W.; Bayrer, J.R.; Leitch, D.B.; Castro, J.; Zhang, C.C.; O'Donnell, T.A.; Brierley, S.M.; Ingraham, H.A.; Julius, D. Enterochromaffin Cells Are Gut Chemosensors that Couple to Sensory Neural Pathways. *Cell* **2017**, *170*, 185–198. [CrossRef]
142. Engevik, M.A.; Luck, B.; Visuthranukul, C.; Ihekweazu, F.D.; Engevik, A.C.; Shi, Z.C.; Danhof, H.A.; Chang-Graham, A.L.; Hall, A.; Endres, B.T.; et al. Human-Derived *Bifidobacterium dentium* Modulates the Mammalian Serotonergic System and Gut-Brain Axis. *Cell. Mol. Gastroenterol. Hepatol.* **2021**, *11*, 221–248. [CrossRef] [PubMed]
143. Rupp, S.K.; Stengel, A. Bi-Directionality of the Microbiota-Gut-Brain Axis in Patients with Functional Dyspepsia: Relevance of Psychotherapy and Probiotics. *Front. Neurosci.* **2022**, *16*, 844564. [CrossRef] [PubMed]
144. Tominaga, K.; Fujikawa, Y.; Tsumoto, C.; Kadouchi, K.; Tanaka, F.; Kamata, N.; Yamagami, H.; Tanigawa, T.; Watanabe, T.; Fujiwara, Y.; et al. Disorder of autonomic nervous system and its vulnerability to external stimulation in functional dyspepsia. *J. Clin. Biochem. Nutr.* **2016**, *58*, 161–165. [CrossRef] [PubMed]
145. Xiang, J.W.; Zhang, Z.Y.; Xie, H.Y.; Zhang, C.C.; Bai, Y.; Cao, H.; Che, Q.S.; Guo, J.; Su, Z.Q. Effect of different bile acids on the intestine through enterohepatic circulation based on FXR. *Gut Microbes* **2021**, *13*, 1949095. [CrossRef] [PubMed]
146. Jiao, N.; Baker, S.S.; Chapa-Rodriguez, A.; Liu, W.S.; Nugent, C.A.; Tsompana, M.; Mastrandrea, L.; Buck, M.; Baker, R.D.; Genco, R.J.; et al. Suppressed Hepatic Bile Acid Signaling Despite Elevated Production of Primary and Secondary Bile Acids in NAFLD. *Gastroenterology* **2017**, *152*, S1068. [CrossRef]
147. Zhang, J.Q.; Wu, H.M.; Wang, X.; Xie, J.Y.; Li, X.; Ma, J.X.; Wang, F.Y.; Tang, X.D. Efficacy of prebiotics and probiotics for functional dyspepsia: A systematic review and meta-analysis. *Medicine* **2020**, *99*, e19107. [CrossRef]
148. Wauters, L.; Talley, N.J.; Walker, M.M.; Tack, J.; Vanuytsel, T. Novel concepts in the pathophysiology and treatment of functional dyspepsia. *Gut* **2020**, *69*, 591–600. [CrossRef]
149. Kaji, I.; Iwanaga, T.; Watanabe, M.; Guth, P.H.; Engel, E.; Kaunitz, J.D.; Akiba, Y. SCFA transport in rat duodenum. *Am. J. Physiol. Gastrointest. Liver Physiol.* **2015**, *308*, G188–G197. [CrossRef]
150. Verdu, E.F.; Collins, S.M. Irritable bowel syndrome and probiotics: From rationale to clinical use. *Curr. Opin. Gastroenterol.* **2005**, *21*, 697–701. [CrossRef]
151. Kerkhoven, L.A.S.V.; Laheij, R.J.F.; Meineche-Schmidt, V.; Zanten, V.V.; Jansen, J.B.M.J. Functional dyspepsia: Not all roads seem to lead to rome. *J. Clin. Gastroenterol.* **2009**, *43*, 118–122. [CrossRef]

foods

MDPI

Review

A Review of Plant-Based Drinks Addressing Nutrients, Flavor, and Processing Technologies

Aijun Xie [2], Yushi Dong [4], Zifei Liu [3], Zhiwei Li [5], Junhua Shao [1], Mohan Li [1,3,*] and Xiqing Yue [1,*]

1 College of Food Science, Shenyang Agricultural University, Shenyang 110866, China; shaojh024@163.com
2 Department of Chemical and Biomolecular Engineering, National University of Singapore, Singapore 119077, Singapore; axie26@wisc.edu
3 Department of Food Science and Technology, National University of Singapore, Singapore 117542, Singapore; e0506474@u.nus.edu
4 Department of Nutritional Sciences, King's College London, London SE19NH, UK; k21070993@kcl.ac.uk
5 Jiangsu Key Laboratory of Oil & Gas Storage and Transportation Technology, Changzhou University, Changzhou 213164, China; lzw@cczu.edu.cn
* Correspondence: fslimohan@126.com (M.L.); yxqsyau@126.com (X.Y.)

Abstract: Plant-based drinks have garnered significant attention as viable substitutes for traditional dairy milk, providing options for individuals who are lactose intolerant or allergic to dairy proteins, and those who adhere to vegan or vegetarian diets. In recent years, demand for plant-based drinks has expanded rapidly. Each variety has unique characteristics in terms of flavor, texture, and nutritional composition, offering consumers a diverse range of choices tailored to meet individual preferences and dietary needs. In this review, we aimed to provide a comprehensive overview of the various types of plant-based drinks and explore potential considerations including their nutritional compositions, health benefits, and processing technologies, as well as the challenges facing the plant-based drink processing industry. We delve into scientific evidence supporting the consumption of plant-based drinks, discuss their potential roles in meeting dietary requirements, and address current limitations and concerns regarding their use. We hope to illuminate the growing significance of plant-based drinks as sustainable and nutritious alternatives to dairy milk, and assist individuals in making informed choices regarding their dietary habits, expanding potential applications for plant-based drinks, and providing necessary theoretical and technical support for the development of a plant-based drink processing industry.

Keywords: plant-based drink; flavor; nutrients; processing technology

Citation: Xie, A.; Dong, Y.; Liu, Z.; Li, Z.; Shao, J.; Li, M.; Yue, X. A Review of Plant-Based Drinks Addressing Nutrients, Flavor, and Processing Technologies. *Foods* **2023**, *12*, 3952. https://doi.org/10.3390/foods12213952

Academic Editor: Francesco Visioli

Received: 30 May 2023
Revised: 24 October 2023
Accepted: 26 October 2023
Published: 29 October 2023

1. Introduction

The demand for plant-based alternatives to dairy milk has surged tremendously in recent years, driven by factors including ethical concerns, environmental sustainability, and health-conscious lifestyles [1]. Consequently, plant-based drinks have emerged as a diversity of popular choices for individuals seeking a dairy-free alternative that aligns with their values and offers a range of nutritional benefits [2].

Plant-based drinks, also known as non-dairy or alternative drinks, refer to beverages derived from plant sources, including nuts, grains, legumes, and seeds [3]. These milk alternatives have gained significant attention as viable substitutes for traditional dairy milk, providing options for those who are lactose intolerant or allergic to dairy proteins, and those who adhere to a vegan or vegetarian diet [4–6]. In recent years, as the market for plant-based drinks has rapidly expanded, a wide variety of options, including almond, soy, oat, rice, and coconut drinks, have become available to meet the demand. Each variety possesses unique flavor, texture, and nutritional composition characteristics, offering consumers a diverse array of choices to suit their individual preferences and dietary needs [7].

One of the key factors driving plant-based drink popularity is its perceived sustainability [8]. The environmental impact of dairy milk production, including land and water use, greenhouse gas emissions, and animal welfare concerns, has prompted consumers to seek alternative options [9]. Plant-based drink production generally requires fewer resources and generates fewer emissions than dairy milk production, making it a more environmentally friendly option [10]. Plant-derived drinks offer numerous nutritional benefits. While dairy milk is commonly associated with calcium and protein content [11–13], plant-based alternatives also provide essential nutrients including vitamins, minerals, healthy fats, and fiber [2]. The nutritional profile varies across different plant-based drink varieties, and fortified options are available to enhance the nutrient content and mimic the benefits of dairy milk [14]. The choice between plant-based drinks and dairy milk for infants and adults depends on individual dietary preferences and nutritional needs, with plant-based drinks offering lactose-free and vegan-friendly options, while dairy milk provides a nutrient-rich source of essential nutrients like calcium, vitamin D, and high-quality protein. Caregivers should prioritize breast milk or infant formula for infants, and adults should select the option that aligns with their dietary and health goals, considering factors such as lactose tolerance and environmental impact.

In this review, we aimed to provide a comprehensive overview of plant-based drinks, exploring the available types and their nutritional compositions, health benefits, potential considerations when choosing a product, processing technologies, and challenges facing the plant-based drink processing industry. We delve into the scientific evidence supporting the consumption of plant-based drinks, discuss their potential roles in meeting dietary requirements, and address limitations and concerns associated with their use. In doing so, we hope to expand awareness of the growing significance of plant-based drinks as sustainable and nutritious alternatives to dairy milk, thereby assisting individuals in making informed choices regarding their dietary habits, expanding potential applications for plant-based drinks, and providing necessary theoretical and technical support for the development of a plant-based drink processing industry.

2. Classification of Plant-Based Drinks

Plant-based drinks can be classified according to the raw materials used, such as almonds, oats, soy, walnuts, peanuts, coconuts, cashews, rice, hemp, and flax (Table 1). Plant-based drinks can be broadly classified into six types based on raw materials: cereals (oats and rice), legumes (soybeans and peas), pseudo-cereals (quinoa), seeds (peanuts, sesame, and sunflower), nuts (walnuts and almonds), and high-protein or fatty fruits (coconut). Beverages made from plant-derived ingredients and plant proteins are also referred to as plant-based drinks [7]. Plant-based drinks provide supplemental proteins, calcium, and various other nutrients. They can be consumed in pure form or used as a companion for coffee and tea, and can serve as an ingredient in processed foods (baked products and plant-based ice cream).

Table 1. Plant-based drink classification by raw material source.

Classification	Raw Material Sources	References
Nut-based plant drinks	Almonds, pistachios, walnuts, cashews, etc.	[7,15]
Soy-based plant drinks	Peas, chickpeas, soybeans, peanuts, cowpeas, etc.	[7]
Seed-based plant drinks	Sesame, flax, hemp, pumpkin seeds, etc.	[7,15]
Cereal-based plant drinks	Millet, corn, barley, sorghum, wheat, etc.	[7,15]
Pseudo-cereal-based plant drinks	Quinoa, moss bran, etc.	[7]

3. Plant-Based Drink Nutrients

3.1. Protein

In the absence of added exogenous protein, soy-based plant drink has the highest protein content, similar to that of cow milk. Soy-based plant drinks are generally considered to be a complete protein source for adults, containing all essential amino acids. Other plant-based drinks have lower protein contents, with rice-based plant drinks having the lowest protein content [16,17]. Compared to milk protein, plant-based drinks may exhibit slight quantitative deficiencies in some essential amino acids, such as methionine and lysine [18]. Methionine and cysteine are the limiting amino acids in pea and soy tissue proteins, whereas lysine is the limiting amino acid in flax proteins and other cereal proteins, including rice [19]. Due to the limited presence of these amino acids, plant proteins are generally considered to provide a lower nutritional value than animal proteins. Additionally, the presence of plant-derived anti-nutritional factors, including phytic acid and saponins, results in the lower digestibility of plant proteins relative to milk protein. Overall, the biological value (BV) and digestible indispensable amino acid scores (DIAASs) of plant proteins are slightly lower than those of milk proteins. For example, BV scores for milk protein and casein are 104 and 80, respectively, whereas BV scores for soy, pea, and flaxseed proteins are 74, 65, and 77.4, respectively. DLAASs for milk protein and casein are 115 and 111, respectively, whereas DLAASs for soy and pea proteins are 89 and 80, respectively [16,19]. Besides protein, other compositions of plant-based drinks can be seen in Table 2.

Table 2. General composition of different plant-based drinks.

Drink Type	Protein (g)	Total Fat (g)	Ash (g)	Total Carbohydrates (g)	Fiber (g)	References
Soy	2.78	1.96	0.75	3	<0.75	[20,21]
Coconut	2.02	21.3	0.97	2.81	-	[20,22]
Oat	0.8	2.75	0.79	5.1	<0.75	[20,23]
Flaxseed	-	1.04	-	0.42	-	[20,24]
Rice	0.42	1.04	0.34	9.58	-	[20,25]
Cashew	2.2	5.29	-	5.73	0.4	[20,26]
Almond	0.66	1.56	0.6	0.67	<0.75	[20,26]

3.2. Dietary Fiber

Dietary fiber refers to the non-digestible carbohydrates present in plant-based foods [27]. It primarily exists in plant cell walls and includes cellulose, hemicellulose, and soluble fibers. Dietary fibers remain structurally intact as they pass through the digestive tract because they are not broken down by digestive enzymes in the human body [28]. Dietary fibers offer various health benefits, including promoting gastrointestinal health, preventing constipation, controlling blood sugar and cholesterol levels, providing a feeling of fullness, and regulating body weight. Common sources of dietary fiber include whole grains, vegetables, fruits, legumes, nuts, and seeds. It is recommended that adults consume 25–30 g of dietary fiber per day to maintain good health [29–32]. In plant-based drinks, various soluble or insoluble dietary fibers present in the cell walls of seeds, cereals, or fruit-based ingredients, such as flaxseed gum, almond polysaccharides, and soy polysaccharides, exhibit potential prebiotic characteristics that are beneficial to human health. This prebiotic function is not inherent to dairy milk. Prebiotics are compounds resistant to digestion in the small intestine, and select dietary fibers, such as inulin, oligofructose, and fructooligosaccharides (FOSs), fall into this category. They play a prominent role in regulating the gut microbiota and promoting gut ecosystem health [33,34]. This microbial fermentation process generates short-chain fatty acids (SCFAs), notably butyrate, acetate, and propionate, which confer various health advantages, including trophic effects on colonic epithelial cells and the mitigation of inflammatory responses. Additionally, prebiotic fibers foster gut microbiome diversity by providing diverse nutritional substrates for a spectrum of bacterial strains. A balanced and diverse gut microbiome is associated with improved gastrointestinal function, enhanced immune responses, and potential implications for neurological health. More specifically, beta-glucan in oat drinks increases satiety and lowers blood glucose and cholesterol levels [7]. Flaxseed drink contains flaxseed gum, which improves the gut microbiota, controls weight, enhances satiety, and protects gut and cardiovascular health. Soy drinks containing fiber can lower plasma cholesterol in animals or humans without affecting the absorption of essential mineral elements, including zinc and copper. It also helps maintain gut health and controls blood sugar and lipid levels [35].

3.3. *Fats and Fat-Soluble Ingredients*

Plant-based drinks primarily contain unsaturated fatty acids with generally low levels of saturated fatty acids and no cholesterol. This composition is beneficial for lowering low-density lipoprotein and cholesterol levels, thus providing positive effects to individuals addressing high blood cholesterol and cardiovascular diseases [36]. Polyunsaturated fatty acids (PUFAs), a subgroup of dietary fats encompassing omega-3 and omega-6 fatty acids, confer a spectrum of health benefits when judiciously incorporated into the diet. These benefits are underpinned by intricate biochemical mechanisms. Omega-3 PUFAs, notably found in fatty fish and flaxseeds, substantiate reduced risks of cardiovascular disease through the mitigation of triglyceride levels, blood pressure regulation, and anti-inflammatory actions. Furthermore, omega-3 PUFAs exhibit anti-arrhythmic properties, modulating heart rhythm and averting arrhythmias. In terms of cognitive well-being, these PUFAs, particularly docosahexaenoic acid (DHA), assume a pivotal role in brain development and function, enhancing cognitive faculties and conferring protection against neurodegenerative conditions, such as Alzheimer's disease. Additionally, mood regulation is influenced by omega-3 PUFAs, which may ameliorate symptoms of depression and anxiety. Their robust anti-inflammatory attributes extend to the management of chronic inflammation, impacting the pathogenesis of prevalent ailments including cardiovascular disorders, arthritis, and certain cancers. Beyond these, omega-3 PUFAs underpin various other health benefits, including skin barrier fortification, immune system augmentation, eye health preservation, potential cancer risk reduction, appetite regulation for weight management, and vital contributions to pregnancy and child development. To optimize these benefits, maintaining an appropriate balance between omega-3 and omega-6 intake is pivotal, as excessive omega-6 and insufficient omega-3 consumption can contribute to inflammatory states and health complications. Consequently, a balanced diet integrating sources of both omega-3 and omega-6 PUFAs is advocated for comprehensive health promotion. Overall, higher proportions of PUFAs, especially n-3 PUFA, other bioactive essential fatty acids, and fat-soluble bioactive components, contribute to the prominent health benefits of plant-based drinks (Table 3).

Table 3. Main fatty acid components of plant-based drinks.

Drink Type	**Main Fatty Acid Components %**							**Reference**
	Saturated Fatty Acid				**Unsaturated Fatty Acid**			
	C12:0	**C14:0**	**C16:0**	**C18:0**	**C18:1n − 9**	**C18:2n − 6**	**C18:3n − 3**	
Soy	-	-	9.8	3.7	21.9	53.7	9.9	[37]
Coconut	50.0	17.3	7.5	2.7	0.01	0.77	-	[38]
Oat	-	-	5.09	1.81	45.1	23.7	0.234	[39]
Walnut	-	-	8.0	3.0	18.0	59.0	6.0	[40]
Peanut	-	-	16.63	4.88	42.63	15.56	-	[41]
Flaxseed	-	4.59	20.4	6.65	27.59	24.24	7.01	[42]
Almond	-	-	14.6	10.8	54.0	15.4	-	[39]

3.4. *Vitamins and Minerals*

Most plant-based drinks are rich in minerals including calcium, magnesium, selenium, potassium, zinc, phosphorus, copper, and manganese (Table 4). For example, the respective almond and soybean calcium contents are approximately 269 and 277 mg/100 g, whereas the respective magnesium contents are approximately 270 and 280 mg/100 g. The respective potassium contents are approximately 733 and 1797 mg/100 g. The endogenous calcium content of plant-based drinks varies widely among products, but when fortified with external sources of calcium, the calcium content in plant-based drinks is typically higher than that in cow's milk [43]. In general, the absorption rate of calcium carbonate used for calcium fortification of plant-based drinks is high, although calcium carbonate is prone to precipitation, which reduces calcium bioavailability in plant-based drinks [44,45]. Cow's milk is considered an excellent source of vitamins; however, it has a low vitamin D content and usually requires external supplementation for fortification (Table 5). Plant-based drinks also tend to have low vitamin D content. Therefore, commercially available plant-based drink products are often fortified with vitamin D. Soy drinks and almond drinks also have an advantage over cow's milk in vitamin E content, with levels reaching approximately 4.0 and 3.84 mg/100 mL, respectively. Coconut, almond, and cashew drinks have vitamin A contents exceeding 60 μg/100 mL, approximately twice the amount found in cow's milk [46]. Overall, plant-based drinks have an advantage over cow's milk in terms of fat-soluble vitamin content, whereas water-soluble vitamins may require fortification.

Table 4. Main mineral components of plant-based drinks.

Drink Type	Calcium (mg)	Iron (mg)	Phosphorus (mg)	Magnesium (mg)	Potassium (mg)	Sodium (mg)	Zinc (mg)	Reference
Soy	155	0.37	46	17.5	118	39	0.26	[20,21]
Coconut	18	3.3	96	46	220	13	0.56	[20,22]
Oat	148	0.26	89	5.9	148	42	0.09	[20,23]
Flaxseed	125	0.15	62	-	-	33	-	[20,24]
Rice	125	0.3	62	-	-	42	-	[20,25]
Cashew	9	1.51	66	35	84	51	1.26	[20,26]
Almond	158	0.12	19	8.2	49	59	0.08	[20,26]

Table 5. Main vitamin components of plant-based drinks.

Drink Type	Thiamin (mg)	Riboflavin (mg)	Vitamin B-6 (mg)	Folate (μg)	Vitamin B-12 (μg)	Retinol (μg)	Vitamin D (μg)	Reference
Soy	0.044	0.331	0.036	16	1.33	89	4.63	[20,21]
Coconut	0.022	-	0.028	14	-	-	-	[20,22]
Oat	0.04	0.281	0.006	<6	0.51	85	1.7	[20,23]
Flaxseed	-	-	-	-	0.62	-	1.05	[20,24]
Rice	-	-	-	-	0.62	-	1.05	[20,25]
Cashew	-	0.015	0.053	-	-	-	-	[20,26]
Almond	0.005	0.083	<0.01	<6	0.45	61′	1.59	[20,26]

3.5. Bioactive Molecules

Plant-based drinks generally contain beneficial bioactive molecules, including flavonoids, phenolic acids, lignans, and phytosterols (Table 6) [47]. Plant polyphenols exhibit excellent antioxidant properties and provide significant health benefits, including anticancer effects, protection against radiation damage, antimicrobial activity against pathogenic bacteria, lipid-lowering effects, and the prevention of cardiovascular diseases. Resveratrol in peanut drinks reportedly possesses antioxidant, antibacterial, hepatoprotective, cardiovascular, radioprotective, and anti-HIV activities [48]. Sesamol in sesame drinks has been reported to inhibit obesity and insulin resistance in mice fed a high-fat, high-fructose diet. It reduces hepatic fat synthesis, inhibits lipid accumulation and inflammatory responses in white adipose tissue, and decreases adipocyte size while promoting the conversion of white adipose tissue to brown adipose tissue by improving mitochondrial lipid metabolism [49]. Lignans, such as enterolignans, are natural plant estrogens found in high amounts in flaxseed drinks. They can control the production of three types of estrogen, inhibit ovarian estrogen production, reduce the risk of breast cancer, and exhibit significant anti-colorectal cancer effects. Their antioxidant function primarily centers on the neutralization of free radicals, highly reactive molecules capable of inducing cellular damage, including DNA, protein, and lipid harm, potentially culminating in cancer and other chronic maladies. The mechanistic underpinnings of lignan-mediated anti-cancer effects encompass several facets: free radical scavenging via electron donation, the modulation of inflammatory processes, participation in estrogen metabolism pathways (particularly enterolignans), interference with key cellular signaling pathways implicated in tumorigenesis, the induction of apoptosis, and the regulation of angiogenesis. Moreover, lignans may enhance detoxification processes via the upregulation of detoxifying enzymes. Consequently, a growing body of research suggests that lignans may be associated with a diminished risk of specific cancer types, particularly hormone-related malignancies like breast, prostate, and ovarian cancers [50]. The role of flavonoids within plant-based drinks hinges upon their natural occurrence in constituent ingredients or their introduction during processing. Certain plant-based drink varieties, such as almond and soy drinks, may contain inherent flavonoids derived from their foundational ingredients, endowing these beverages with antioxidant capabilities. These flavonoids possess the capacity to mitigate oxidative stress and confer potential health benefits. However, it is important to note that flavonoid enrichment is not a prevailing practice in commercially available fortified plant-based drinks, which predominantly focus on the fortification of vitamins and minerals to replicate the nutritional profile of dairy milk. Therefore, individuals with specific health risks, such as those at a high risk of cardiovascular diseases, may experience enhanced benefits from consuming plant-based drinks.

Table 6. Total phenolic compounds and phytosterol compositions of plant-based drinks.

Drink Type	Total Phenolic Compounds	β-Sitosterol (mg/100 mL)	β-Sitosterol-β-D-Glucoside (mg/100 mL)	Campesterol (μg/100 mL)	Stigmasterol (μg/100 mL)	Reference
Almond	1.24 mg GAE/L	2.5 ± 0.1	13 ± 2	62 ± 4	1915 ± 109	[51–53]
Hazelnut	130.42 mg GAE/100 mL	-	-	-	-	[54,55]
Sesame	4 mg GAE/g	-	-	-	-	[56,57]
Soy	8.79 mg GAE/100 g	2.5 ± 0.5	4.9 ± 2.1	1290 ± 291	998 ± 111	[51,58,59]
Rice	122.05 mg GAE/100 mL	0.51 ± 0.07	2.4 ± 0.6	260 ± 28	234 ± 23	[51,60]
Cashew	-	2.7 ± 0.4	>60	279 ± 44	15 ± 1	[51]
Oat	15 mg GAE/100 mL	2.1 ± 0.2	26 ± 4	475 ± 30	182 ± 16	[51,61]

4. Key Substances That Contribute to Aroma in Plant-Based Drink

Flavor is an important quality of plant-based drink that determines its market acceptance [62]. Flavor is determined through small molecules formed via the decomposition of proteins, fats, or carbohydrates in a food. These substances can stimulate the sensory neurons responsible for smell and taste, creating a comprehensive physiological sensation. Molecules contributing to flavor can be divided into odorants and tastants. Tastants are nonvolatile food compounds that are perceived by taste buds in the oral cavity, whereas odorants are volatile food compounds that are sensed by olfactory receptors in the nasal mucosa [63–65]. Enhancing aroma and reducing or mitigating the generation of off-flavors are key aspects of flavor modulation in plant-based drinks. Selecting high-quality ingredients and employing environmentally friendly processing techniques to prepare plant-based drinks with excellent flavor significantly impacts market potential. With advancements in flavor extraction and separation technologies, the most widely used techniques for volatile compound extraction include headspace solid-phase microextraction (HS-SPME), dynamic headspace sampling (DHS), solvent-assisted flavor evaporation (SAFE), stir bar sorptive extraction (SBSE), liquid–liquid extraction (LLE), and simultaneous distillation extraction (SDE). These techniques are often combined with gas chromatography–mass spectrometry (GC-MS) and olfactory techniques, such as gas chromatography–olfactometry (GC-O), aroma extract dilution analysis (AEDA), and dynamic headspace dilution analysis (DHDA), for aroma and off-flavor identification in plant-based drinks [66–68].

Different volatile compounds exhibit distinct odor characteristics and contribute to the overall flavor profile of a plant-based drink when combined in certain proportions. Based on the existing literature, the main aromatic compounds contributing to the fragrance of plant-based drinks include aldehydes, ketones, alcohols, esters, and acids [67,69].

4.1. Aldehyde Compounds

Aldehyde compounds are primarily formed through the degradation and auto-oxidation of unsaturated fatty acids such as oleic and linoleic acids. Aldehydes generally have low odor thresholds and contribute significantly to flavor profiles, making them important aromatic compounds [70]. Medium-chain aldehydes exhibit fatty, fresh, and greasy aromas, whereas aldehydes with higher carbon numbers produce a citrus peel aroma [71]. During soybean growth and processing, over 20 volatile compounds are associated with a beany off-flavor, with hexanal being the most important [72,73]. The hexanal content in walnut kernels correlates negatively with nutty and sweet aromas, and correlates positively with bitterness and acidity. Therefore, storing walnuts at 5 °C in a light-protected environment is considered optimal [74]. Benzaldehyde, a metabolite of phenylalanine, imparts a fruity aroma to plant-based drinks. In soy drinks, 3-methylbutanal contributes to a dark-chocolate flavor, whereas decanal provides sweet, citrus, and floral tastes. In oat acidophilus drinks, nonanal exhibits a honeywax floral aroma, and octanal produces a strong fruity fragrance [75].

4.2. Ketone and Alcohol Compounds

Ketones are another product of fat oxidation. Ketones produce less intense aromas in walnut drinks, contributing less to their flavor. Ketones are generally associated with creamy and fruity aromas. For example, 2-methyl-3-hydroxy-4-pyranone is present in both soy drinks and vanilla soy drinks and evokes a caramel aroma [76]. 2,3-butanedione and its degradation product, 3-hydroxy-2-butanone, generate a sweet aroma in prepared soy drinks and a creamy aroma in roasted walnut drinks, possibly due to prolonged exposure to high temperatures [77]. During the lyophilization of peas, significant amounts of 3,5-octadien-2-one and-ionone are produced, possibly through the oxidative and condensation reactions of carotenoids [78]. Alcohol compounds are primarily produced through the oxidative degradation of fatty acids, such as hexanol, which exhibits a floral aroma and is a product of linoleic acid auto-oxidation. 2,3-butanediol contributes to the fruity aroma of walnut drinks, whereas the nonanol present exclusively in walnut drinks roasted for 25 min imparts a fresh fatty note [77]. Nonanol in soy drinks has a strong rose and orange throat aroma, along with a fresh fatty note [79]. Vanillin and linalool are present in vanilla drinks, and have distinct and unique aromatic profiles.

4.3. Ester and Acid Compounds

Esters are formed via the esterification of carboxylic acid derivatives with alcohols and are primarily derived from the oxidation of lipid precursors [80]. Esters are characterized by their typical fruity aromas, which contribute to the delicate fragrances of plant-based drinks. The key aromatic ester compounds in roasted walnut drinks are ethyl acetate, triethyl phosphate, and ethyl benzoate, all of which impart fruity notes. Cinnamyl acetate contributes sweet orange and grape aromas, whereas ethyl palmitate imparts a creamy fragrance. The quantity of esters in roasted walnut drinks increases with prolonged roasting time [81]. Acidic compounds contribute significantly to the taste profiles of plant-based drinks. Butyric acid, valeric acid, and l-lactic acid are characteristic acidic flavor compounds in fermented coconut drinks that impart a creamy aroma [82]. In vanilla soy drinks, key acidic compounds include 17-octadecynoic acid, palmitic acid, and oleic acid, whereas in soy drinks, 3-butyne-1-acid, acetic acid, and 15-hydroxydecanoic acid are predominant.

4.4. Pyrazines and Other Compounds

Pyrazines are formed primarily via Maillard reactions between reducing sugars and amino acids and via Strecker degradation of carbonyl compounds [83]. Glucose degradation leads to the formation of carbonyl compounds that react with free amino acids to produce α-amino ketones under alkaline conditions. Subsequently, these compounds undergo condensation to form various pyrazine compounds. During plant-based drink preprocessing steps, including microwaving, roasting, and radiofrequency treatments, pyrazines and alkylated pyrazines are generated. These volatile substances can partially mask off-flavors in plant-based drinks [84]. Because of their high concentrations and low odor thresholds, pyrazine compounds exhibit strong roasted, nutty, and caramel flavors. For example, 2,5-dimethylpyrazine imparts a roasted and nutty aroma, whereas 2, 3, 5-trimethylpyrazine imparts a nutty fragrance [77]. Additionally, 2,4-di-tert-butylphenol is a major phenolic aroma compound in soy drinks and is present in significant proportions among the phenolic aroma components. In addition to 2-ethylfuran, soy drinks and vanilla soy drinks also contain other aromatic compounds, including 2-methyl-1-propene, octadecynol, and 8-octadecyne. Phenolic compounds such as 4-ethenylguaiacol have been identified as important aroma contributors in walnut drinks, exhibiting a clove-like aroma. Additionally, 4-allylanisole, which is present in roasted walnut drinks, imparts a fennel-like aroma and is one of the major aromatic components in almond oil [77,85].

5. Flavor Formation Pathways

5.1. Maillard Reaction

The Maillard reaction is a complex non-enzymatic reaction that involves the condensation of reducing sugars with amino groups in amino acids, peptides, or proteins. It can generate a variety of compounds with different colors and flavors via multiple pathways (Figure 1). This process, also known as non-enzymatic browning, is generally divided into three stages [86–88]. In the initial stage, the carbonyl group of the reducing sugar undergoes nucleophilic addition to the amino group of the amino acid to form a Schiff base. Owing to their instability, Schiff bases undergo cyclization to form N-substituted aldimines, which then undergo Amadori rearrangements to form reactive intermediates. To reduce sugars to produce these intermediates, the sugars must be converted into an open-chain structure, which occurs slowly during this stage. Therefore, this stage does not cause significant changes in the color and flavor of food but generates flavor precursors [89]. During the intermediate stage, the Maillard reaction becomes more complex. Intermediates formed in the initial stage, such as Amadori rearrangement products and Heyns compounds, are further degraded to produce reducing ketones, furfural, and unsaturated carbonyl compounds. In the final stage, the numerous reactive intermediates formed in the intermediate stage, including reducing ketones, unsaturated aldimines, and glucosone aldehydes, undergo further condensations, polymerization, or reactions with amino acids, ultimately leading to the formation of melanoidins [90].

As Maillard reaction substrates, amino acids serve as the major determinants of flavor in plant-based drinks. The most significant mechanism of amino acid degradation is Strecker degradation. Different amino acids undergo Strecker degradation to produce specific aldehydes, which are important compounds responsible for distinct flavors in foods, and which serve as intermediates in further reactions. Dimethyl sulfide is formed through the thermal degradation of its precursor, S-methylmethionine, and exhibits a strong boiled cabbage flavor [91]. In addition, during thermal processing of soybeans, their proteins can undergo Maillard reactions with reducing sugars, resulting in food browning and the formation of various volatile compounds. Concurrently, the activity of endogenous oxidative enzymes decreases, inhibiting fatty acid oxidation reactions and reducing the generation of beany-flavor compounds [92,93]. In the absence of lipoxidase, the production of volatile compounds in soy drinks is associated with the Maillard reaction. When lipoxidase activity is lost, major unsaturated fatty acids, including linoleic acid, oleic acid, and linolenic acid, are the primary sources of volatile compounds. This indicates a direct relationship between the Maillard reaction and major unsaturated fatty acids in soy drinks [94].

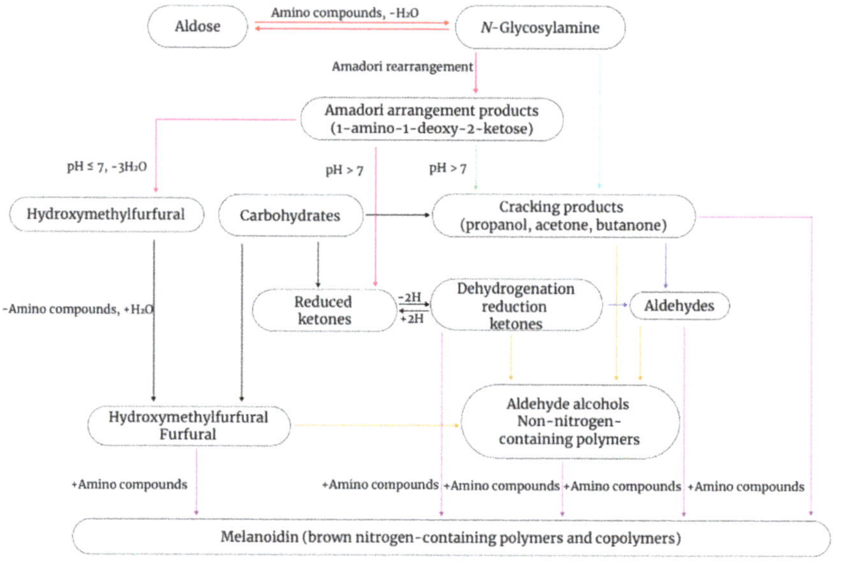

Figure 1. Schematic diagram of the Maillard reaction.

5.2. *Oxidative Degradation of Fats*

Oxidative fat degradation is the main cause of off-flavors in plant-based drinks. When plant-based drink ingredients are stored under normal conditions, drink quality is relatively stable, but endogenous lipoxidases can be activated during processing, leading to the development of rancidity and other undesirable flavors [95]. Lipid oxidation can be divided into two main pathways: enzymatic and non-enzymatic.

Linear-chain saturated fatty acids undergo oxidation, forming flavor compounds including short-chain or medium-chain fatty acids, aldehydes, alcohols, esters, lactones, and methyl ketones. The oxidation reactions of unsaturated fatty acids, such as linoleic acid and linolenic acid, yield various aldehydes, alcohols, and esters [96,97]. The main off-flavor compounds in soy drinks are C6 and C9 aldehydes and their corresponding alcohols. These compounds are significant constituents of the flavors found in vegetables, fruits, leaves, and leguminous plants, and are primarily derived from the lipoxygenase (LOX) enzyme-catalyzed oxidation pathway [98,99]. LOX oxidizes linoleic acid to form 13- or 9-hydroperoxy-octadecadienoic acid (13-/9-HPOD), which is further cleaved via hydroperoxide lyase to generate hexanal; (E,E)-2,4-decadienal; (E)-2-octenal; and other compounds. When linolenic acid serves as the lipid precursor, (E)-2-hexenal, (E,Z)-3,6-nonadienal, and other compounds are formed [100]. The generated C6 and C9 aldehydes can be further metabolized via alcohol dehydrogenase to produce the corresponding alcohols, such as hexanol and (E)-2-hexenol. Non-enzymatic oxidations, including photooxidation and autoxidation, are primarily caused via heat, light, photosensitizers, oxygen, and transition-metal ions. They can generate volatile compounds including aldehydes and furans, thereby influencing the flavors of plant-based drinks. Light exposure during food processing is unavoidable. For example, after storing soy flour under light for a certain period, its 2-pentylfuran content increases significantly [101]. Singlet oxygen can be generated in the presence of riboflavin in soy drinks, and through specific oxidative mechanisms, riboflavin catalyzes the formation of 2-pentylfuran from linoleic acid. Chlorophyll can induce the formation of singlet oxygen through a similar process, and singlet oxygen content is positively correlated with the duration of exposure to light and air [102].

6. Processing Technology

A plant-based drink is a colloidal suspension or emulsion consisting of dissolved and decomposed plant material [7]. Although the exact processes vary, the same general outline applies to all plant-based drink processing methods (Figure 2). Water extraction of plant-based drinks can be divided into two types: dry processes (the dry milling of raw materials and the extraction of flour from water) and wet processes (the soaking and wet milling of plant sources) [1].

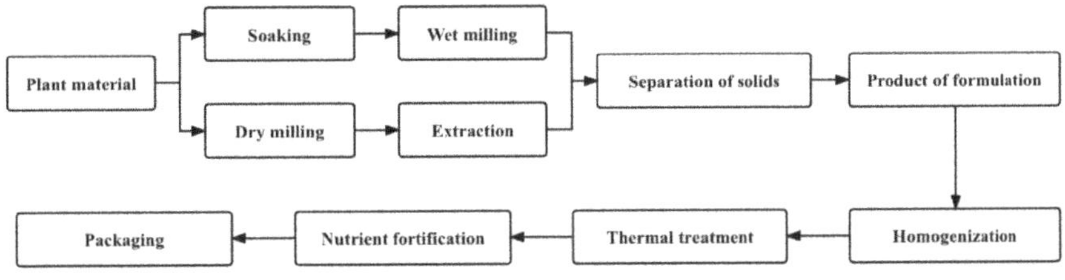

Figure 2. The general workflow of plant-based drink manufacturing processes (adapted from [1]).

6.1. Pre-Treatment

Before making plant-based drinks, some pretreatment of raw materials may be required, including soaking, blanching, steaming, and baking. The purpose of pretreatment is to enhance extraction, increase nutritional quality, enhance organoleptic characteristics, and eliminate off-odors [6,16].

Soaking is an important pretreatment process in the production of plant-based drinks. The main purpose of soaking is to promote softening and expansion of raw materials including grains and nuts, and to facilitate the breaking of raw materials during the grinding process [103]. This process reduces resistance to mechanical grinding and fully hydrates proteins, rendering them easier to leach and thereby increasing the extraction rate. Soaking also helps reduce the initial microbial load, eliminate off-odors, improve organoleptic properties, and enhance nutritional value [77]. Some raw materials, including cowpeas, cashews, soybeans, oats, and sesame seeds, usually need to be soaked in water for 3–18 h. Some raw materials, including soybeans, peanuts, and tiger nuts, can be soaked in alkaline solutions with 0.2–2% NaHCO3 at ratios ranging from 1:2 to 1:12 to eliminate nutty or beany flavors [7,14,104,105].

Starch in raw materials is gelatinized during steaming or baking, and the peculiar smell of the raw plant materials can be effectively controlled. This process can remove heat-sensitive anti-nutritional factors and toxic and harmful substances (such as cyanogenic glycosides) from raw materials, as well as partially inactivate enzymes. In addition, Maillard and other reactions can occur during baking, which can change plant material flavors and impact aroma development [106,107]. Heating can also reduce protein solubility and extraction rates [108].

6.2. Extraction

To improve the yield, extraction efficiency can be increased by increasing the temperature, adding enzymes, fermenting, germinating, or increasing the pH (using NaOH or bicarbonate). During the extraction process, an alkaline medium is used to increase the protein yield. Higher extraction temperatures also increase fat extraction, although protein denaturation reduces fat solubility and yield [1].

Microwaves, such as ultra-high-frequency electromagnetic waves, promote high-frequency reciprocating dipole moments, generating "internal frictional heat", which is absorbed by food and water to generate heat [109]. It can realize simultaneous heating and temperature increase without fast and uniform heat conduction, and energy consumption is a small fraction of that required in traditional heating [110]. Compared to the steam injection extraction method, microwave-assisted extraction can significantly improve the extraction yield of plant-based drinks, including protein content, total soluble solids, and protein solubility. It completely destroys plant cell and subcellular structure integrity, promoting the release of proteins, lipids, and other soluble solids [111]. Microwave-assisted extraction can also help remove safety risk factors, including cyanogenic glycosides, facilitate aroma enhancement through Maillard reactions, and passivate endogenous oxidases in raw plant materials, to improve the oxidative stability of plant-based drinks [112].

6.3. Enzymatic Processing

Enzymolysis technology functions under mild reaction conditions and exhibits low energy consumption, high efficiency, and low solvent consumption, promoting the release of intracellular compounds, increasing soluble sugar and protein contents, and improving product biological activity, thereby improving the stability and sensory qualities of plant-based drinks [113]. Enzymatic hydrolysis is a key step in promoting nutrient dissolution during plant-based drink production. Carbohydrases and proteases are widely used in plant-based food processing [114]. Carbohydrases hydrolyze glycosidic bonds in plant cell wall layers, promoting the breakdown of insoluble fibers, generating low-molecular-weight sugars, and releasing proteins and other intracellular compounds [115]. For example, cellulase hydrolyzes primary cell walls, whereas pectinase hydrolyzes secondary cell walls [116]. The disruption of the network structure of plant cell wall components via pectinase increases the release rate of proteins and fats into cereal beverages. Second, enzymatic hydrolysis can improve the stability and flavor of plant-based drinks. Carbohydrase treatment (1.2% Celluclast 1.5 L, 3 h) improves the physical stability of soybean drinks during storage and ameliorates bean flavor [117]. It is generally believed that proteolysis will release low-molecular-weight peptides, resulting in bitter taste, but when Sahoo et al. immobilized two non-commercial proteases derived from sunflower seeds, added them to soybean drink, and incubated them at 30 °C for 1 h, they found that the beany smell decreased and the pleasant smell increased [118]. Low-molecular-weight peptides produced after proteolysis are primarily composed of hydrophobic amino acids. Hydrophobicity, primary sequence, spatial structure, peptide chain length, and molecular size are all determinants of peptide bitterness. Therefore, specific proteolysis is required to prevent bitterness in plant-based drinks [114]. Enzymatic hydrolysis can also enhance biological activities in plant-based products because this technology can degrade macromolecular substances into smaller molecular compounds with higher biological activity and increase the release of bioactive components (polyphenols, flavonoids, etc.). For example, after protease treatment of soybean drinks and flaxseed protein, hydrolysate showed antioxidant, anti-inflammatory, anti-obesity, and immunomodulatory activities [119]. Additionally, because phenolic compounds in plant materials may form protein–phenolic complexes with globular proteins through hydrophobic interactions and hydrogen bonds, the release of total phenols increases after proteolysis of sesame seed extracts [120].

6.4. Fermentation

The production of plant-based drinks using microbial fermentation generally involves mixed-culture fermentation using two or more microbial strains to enhance the fermentation effect and improve the quality of the final product [82]. Lactic acid bacteria, bacilli, and yeasts are the most widely used microorganisms for this purpose [121]. Fermentation can improve plant-based drink flavors by reducing undesirable odors or by creating pleasant aromas. After peanut drink fermentation, the beany flavor was mitigated by the reduction of n-hexanal and n-hexanol, and diacetyl and 2,3-butanedione, with a butterscotch aroma, were produced during fermentation [114]. Fermentation can also improve the nutritional value of plant-based drinks by increasing nutrient content, improving nutrient bioavailability, and removing antinutritional factors [122]. Fermentation can increase protein and vitamin contents [82] as well as improve calcium and vitamin bioavailability. Significant increases in crude protein content and B vitamins, such as riboflavin and thiamine, have been observed in soybean drinks fermented with bifidobacteria [123]. Additionally, vitamins (including vitamins B and K) are produced during the fermentation process. Yeast fermentation promotes vitamin B2 production [82]. In addition, since lactic acid bacteria can produce phytase to catalyze the hydrolysis of phytate into inositol and phosphate, fermentation technology has the potential to reduce the content of anti-nutritional factors in raw plant materials and increase their utilization of minerals [124]. Fermentation also enhances biological activities in plant-based drinks. Fermentation of almond drinks using lactic acid bacteria increases the content of phenolic compounds, conferring higher antioxidant capacity [125]. Soybeans fermented using lactic acid bacteria can produce bioactive peptides that inhibit angiotensin-converting enzymes and display antihypertensive effects [126]; β-glucosidase produced via lactic acid bacteria during fermentation can convert conjugated isoflavones in soy drinks into more bioactive glucoside ketones [127,128].

6.5. Germination

Germination is a common method for releasing the nutrients and phytonutrients of plant-based drinks, making them more readily available to digestive enzymes [129]. Sprouted grains are more nutritious than raw grains, and rich in digestible energy, bioavailable vitamins, minerals, amino acids, proteins, and phytochemicals [130]. Germination promotes the enzymatic decomposition of carbohydrates into simple sugars by activating endogenous enzymes such as α-amylase, thereby increasing the digestibility of starch degradation and providing energy for seed development [131]. The effect of sprouting on carbohydrates is largely dependent on the activation of hydrolytic and amylolytic enzymes, resulting in a reduction in starch and an increase in simple sugars [131]. The duration of the process is an important factor. When the amylase activity is at its maximum, the maximum hydrolysis time of starch is 48 to 72 h [132]. Studies have shown that after germination, the phenolic activity and antioxidant activity of millet drink increased by 92 ± 1.99 and $33.42 \pm 0.55\%$, respectively. The amalgamation of sonication with germination reduced the average antinutrient concentration to $23.31 \pm 0.36\%$ [133]. The germination process enhances the functional properties of plant-based drinks. The application of germination processes combined with other technologies would be a potential processing technique to acquire nutritious plant-based drinks.

6.6. Separation

Separation is a necessary step in plant-based drink production. Its purpose is to separate solid particles and suspended impurities in plant-based drinks. It mainly removes insoluble fibers to ensure that the particles reach a certain fineness, which is beneficial for the stability of the product. According to final product particle size requirements, separation can be achieved via filtration, decantation, or centrifugation to remove coarse particles from the slurry [1]. When using ingredients with a high-fat content, such as peanuts, excess fat can be removed via skimmers during dairy processing. The separated creamy product can be heat-treated to obtain oil or used as an ingredient in ice cream or baking recipes [108].

6.7. Product Formulation

The protein, vitamin, and mineral contents in plant-based drinks are very important. Therefore, some plant-based drinks are selectively fortified with exogenous nutrients during processing to increase the total content of specific nutrients. To improve the taste and flavor of plant-based drinks, sugar, salt, acidity adjustment, flavors, and fragrances are also added from exogenous sources during the blending process [3,134]. To maintain a stable plant-based drink system after high-temperature sterilization, emulsion stabilizers, such as monoglyceride fatty acid ester, diglyceride fatty acid ester, sucrose fatty acid ester, carboxymethyl cellulose sodium, and xanthan gum, are used. Although plant-based drinks are similar to animal milk in terms of appearance and consistency, significant differences in nutritional quality and bioavailability exist. Therefore, these products must be fortified to improve their nutritional quality [1]. The nutrients used must be bioavailable and sufficiently stable [108].

6.8. Homogenization

Plant-based drinks contain insoluble particles including proteins, starches, fibers, and other cellular materials. These particles are denser than water and, therefore, settle, making the product unstable. Suspension stability can be improved by reducing particle size to increase solubility, or by using hydrocolloids and emulsifiers. Homogenization improves plant-based drink stability by breaking up aggregates and lipid droplets, thereby reducing particle size and increasing their distribution [16]. Mechanical devices used for homogenizing raw materials include high-shear mixers, colloid mills, high-pressure valve homogenizers, and microfluidizers [1].

Ultra-high-pressure homogenization (UHPH) is a common homogenization method used in food processing [135] that is widely used with plant-based drinks. UHPH can significantly improve the physical stability of almond and soy drinks and extend their shelf lives [136]. It offers significant advantages, including the effective inactivation of microorganisms, the retention of micronutrients, green energy saving, and uniform processing [137,138]. In UHPH, the liquid sample is subjected to shear force and cavitation explosion force in the cavity of the homogenizer to produce high-speed fluid impact and vortex action. The structure within the liquid sample is destroyed, and particles within the emulsion become smaller, allowing a more stable suspension, thereby effectively improving the stability of plant-based drinks [139]. Because high pressures can damage microorganisms, UHPH treatment can inhibit the growth of harmful microorganisms and achieve sterilization [126]. Compared with UHT treatment, soybean drinks treated with UHPH (300 MPa, 80 °C) have higher colloidal stability, a stable primary oxidation level, and a significantly lower hexanal value, and can be stored at room temperature for up to 6 months [140]. Both 200 and 300 MPa UHPH reduce the number of spores and Enterobacteriaceae and reduce particle size in plant-based drinks. At 200 MPa, soy drink proteins are partially denatured, whereas at 300 MPa, the degree of denaturation is the same as that observed after UHT treatment [126]. After UHPH treatment, the product has a longer shelf life and better quality characteristics. It is, thus, expected to replace heat treatment in the production of plant-based drinks. UHPH equipment must be combined with sterile filling, requiring strict production equipment monitoring and process control procedures.

6.9. Heat Treatment

Heat treatment is applied to kill bacteria and thereby extend shelf life by reducing microorganism numbers and inactivating enzymes [141]. An appropriate combination of time and temperature can ensure the destruction of microorganisms and enzymes in plant-based drinks while avoiding the aggregation of oil-coated droplets at high temperatures [142]. Plant-based drinks usually undergo pasteurization, ultra-high-temperature instantaneous sterilization, high-temperature high-pressure sterilization, or other methods to extend shelf life, but high temperatures can cause changes in the structure and physical and chemical properties of plant-based drinks. In particular, the denaturation of proteins and polysaccharides in a stable solution will destabilize the system, and heating of starch in a system will significantly increase viscosity, which may also negatively impact the quality of plant-based drinks [143,144]. Therefore, the use of new food processing technologies, including pulsed electric fields, to extend shelf life is recommended [108]. The use of emerging food processing techniques, including pulsed electric fields and ohmic heating, has been proposed to prolong the shelf life of plant-based drinks [1].

Pulsed electric field (PEF) is a non-thermal food sterilization technique that exposes samples to pulses of high pressure at temperatures in the range of 30–40 °C [145]. The application of high-intensity PEF (10–80 kV/cm) causes electroporation to increase the permeability of microbial cell membranes, eventually leading to cell damage or death [146]. PEF can inactivate endogenous food enzymes and kill microorganisms, but exerts a less negative impact on food nutrition, texture, taste, and color [147,148]. It is mainly suitable for liquid foods with low conductivity and low viscosity, and has been gradually applied to an increasing number of liquid products, including plant-based drinks, to extend their shelf life [45]. The application of PEF to soy drinks can effectively inactivate Escherichia coli and Staphylococcus aureus without affecting soy drink quality characteristics. This technology may, thus, be a favorable alternative to heat treatment for soy drink pasteurization [149]. PEF treatment and parameter optimization (treatment time, pulse intensity, pulse frequency, and pulse width) also affect enzyme inactivation. Studies have shown that soybean lipoxygenase activity decreases with increasing treatment time, pulse intensity, pulse frequency, and pulse width, and stronger treatment parameters result in a higher degree of soybean lipoxygenase inactivation. The maximum inactivation of soybean lipoxygenase via PEF was 88% at 42 kV/cm (duration, 1036 μs; pulse frequency, 400 Hz; pulse width, 2 μs; temperature, 25 °C) [150]. Additionally, studies have revealed that pulse type and treatment (soaking or cooking) have significant effects ($p < 0.0001$) on soybean trypsin inhibitory activity, and the interactive effect of pulse type-by-treatment was also significant ($p < 0.0001$) [151]. While PEF does not inactivate spores, it can be inactivated by adding organic acids or nisin, adjusting pH, and other methods. Currently, the industrialization of this technology faces certain limitations, including the high cost and limited development of industrial PEF equipment [152].

Ohmic heating is an advanced heat treatment technology that uses a low-frequency electric current to heat food, which can kill spoilage microorganisms and prolong shelf life [153]. When a 50–60 Hz electric current is applied to the food matrix, electrical energy is transported through the resistive medium to promote ion recombination and increase the molecular motion speed to facilitate the release of thermal energy [110]. Ohmic heating can extend the retention of heat-sensitive components, increase production and energy efficiency, and heat quickly and uniformly. Electric field strength, temperature, and time all influence the post-treatment effect [154]. Saxena et al. studied the effect of ohmic heating on polyphenol oxidase activity in sugarcane beverages under three electric field intensities (24, 32, and 48 V/cm) and four temperatures (60–90 °C) with a treatment time of 5–20 min. At 60 °C, polyphenol oxidase activity decreased with increasing electric field strength, while at 70–90 °C, an increase in enzyme activity was observed at 32 and 48 V/cm [155]. This technology has also been used in the production of soybean drinks, where it reduces the beany smell and reduces the activity of endogenous trypsin and chymotrypsin inhibitors due to combined electrochemical and thermal effects [156,157]. Studies have shown that ohmic heating (220 V, 50 Hz) for periods over 3 min efficiently inactivates TI when compared to induction cooker or electric stove methods over 3 min. The residual trypsin inhibitory activity was 13% (ohmic heating), which is significantly lower compared to 19% using an induction cooker and an electric stove [157]. Additionally, lipoxygenase inactivation followed first-order kinetics during ohmic heating and conventional heating. However, a significant variation in rate constants was observed. Studies have shown that the electric field has an additional effect on lipoxygenase inactivation, with approximately 5 times lower D values. This means that for the same inactivation degree, the time required for thermal treatment is much lower when an ohmic heating process is applied, thus reducing negative thermal effects on the other food components [158]. To date, risks including "cold spots" caused by uneven resistivity remain, complicating the industrial application of this process [159]. The selection of electrode materials is a very important factor when considering the industrial application of ohmic heating. An electrode can be designed for specific ohmic heating conditions only, as the electro-chemical interactions of food and electrode may depend on the electric field strength as well as the frequency and type of wave. The composition of food material is also a challenge for food processors while considering ohmic heating because different types of plant-based drinks may contain different components with different properties, e.g., electrical conductivity. The ohmic heating behavior of the food material in such cases becomes complex, and may lead to the underheating of components with low electrical conductivity or vice versa [160].

6.10. Packaging and Shelf Life

Plant-based drinks are packaged as needed for storage and distribution after processing, usually in plastic bottles or carton systems [1]. Plant-based drinks can also be drum-dried or spray-dried to produce stable powders that can be reconstituted into the desired product [161,162]. However, liquid plant-based drinks must be stabilized before drying to obtain a stable product [163]. For example, a calcium-fortified soy drink (200 mg/100 g) was formulated by adding water (85–90 °C), full-fat soy flour (10%), sucrose (2.75%), and soy protein isolate (2.25%). Following homogenization, the blend was twice clarified and pasteurized at 65 °C/30 min before refrigeration. The samples of the soy drink (45 °C) were adjusted to a pH of 8 before adding calcium lactogluconate (1.55%) and potassium citrate (1.25%). For successful calcium fortification, it is recommended to maintain a calcium-to-protein ratio < 38 mg/g and to use an appropriate sequestering agent at a molar ratio of 0.8/mole calcium [164].

Plant-based drinks are rich in nutrients and are an ideal medium for microbial growth. Therefore, their quality will be adversely affected by the rapid growth of microorganisms. Heat treatment has been used to extend the shelf life of plant-based drinks while increasing the total solid output and improving the flavor. Overheating will adversely affect the development of nutrients (vitamins and amino acids), browning, and the development of cooked flavors [165]. In order to eliminate or reduce the destructive impact on plant-based drinks, various time and temperature combinations have been practiced to obtain the best quality products. Different heat treatments, such as pasteurization (heating to below 100 °C to destroy pathogenic microorganisms), container sterilization (121 °C for 15–20 min to achieve commercial sterility), and ultra-high-temperature treatment (at high temperatures of 135–150 °C for several seconds) have been widely studied [166]. Ultra-high-temperature treatment involves direct heating methods, including steam injection, or indirect heating in plate or tube heat exchangers. After any of the above treatments, packaging needs to be carried out under sterile conditions to maintain sterility. After pasteurization, plant-based drinks need to be stored under refrigerated conditions, while after sterilization in a container or ultra-high-temperature sterilization treatment, the plant-based drinks can be stored at room temperature for several weeks [7]. The effect of pasteurization on soybean drinks has been studied, and the results showed that pasteurized soybean drinks can be stored for 3 days after being heated at 60 °C for 30 min without obvious deterioration, and sterilized soybean drinks can be stored for 1 year after being heat treated at 120 °C for 5 min [167]. Heat treatment is widely applied to extend the shelf life of plant-based drinks, such as soybean drinks and peanut drinks. However, the existence of medium and high starch concentrations in oat drinks and rice drinks restricts their application [7]. Therefore, these types of plant-based drinks need to apply non-heat treatment technology to extend their shelf life. Some non-heat treatment technologies, including high-pressure throttling, ultra-high-pressure homogenization (UHPH), and high-pressure processing, have been studied to extend the shelf life of plant-based drinks [126,168–170]. The effect of ultra-high-pressure homogenization combined with heat treatment on the microbial stability of almond drinks has been studied. The microbiological analysis, physical stability, and chemical analysis results showed that when comparing UHPH treatment with pasteurization and UHT treatment, the product quality was higher compared to samples treated with pasteurization or UHT. After incubation at 30 °C for 20 days at a pressure of 300 MPa, a temperature of 65 and/or 75 °C, and a holding time of less than 0.7 s, the product showed no bacterial growth [171]. These results show the potential of non-heat treatment for extending the shelf life of plant-based drinks. However, the antiseptic effects and mechanisms of pulsed electric fields and other non-thermal technologies, such as pulsed light and ultrasound, on different types of plant-based drinks need to be further explored.

7. Challenges Facing Plant-Based Drink Processing

The plant-based drink processing industry must address challenges, including the beany smell and urease in soy drinks, which reduce their culinary appeal, and starch granules in oat drinks, which affect their smooth taste. The challenges faced during plant-based drink processing include product safety, nutrition, and stability.

7.1. Plant Cell Wall Tissue Limits the Dissolution of Endogenous Nutrients

The health benefits of foods depend on their individual components and their structure (or matrix). Food structure plays a regulatory role in digestion and subsequent physiological metabolic reactions and affects nutrient dissolution during processing. Plant-based food structure is determined by the presence of a cell wall, which is composed of a polymeric structure with a skeleton composed of cellulose as the core and a hydrated gel matrix composed of polysaccharides [172]. The intercellular layer is rich in pectin. As the outermost layer of a plant cell, it also connects juxtaposed plant cells together [172]. Plant cell wall structure and composition vary depending on plant species, tissue distribution, and growth stage [173]. Legumes and other dicot seed cell walls are rich in pectin and xylan, whereas cereal cell walls and those of other monocots are low in pectin but contain arabinoxylan and/or mixed junctions containing b-D-glucan [174]. The dissolution of proteins, oils, dietary fibers, and small-molecule phytochemicals in plant cells is greatly limited by cell walls. Therefore, to increase the solid content of plant-based drinks (by increasing nutrient dissolution), it is crucial to optimize and innovate technologies applied during processing stages, including heat treatment, extrusion, fermentation, grinding, and homogenization [175]. By changing the structure of plant-derived foods and using treatments such as heat and pressure, cell walls can be broken into porous structures, releasing endogenous nutrients from inside cells, thereby improving the dissolution rate of proteins, lipids, polyphenols, and flavonoids in plant-based drinks.

7.2. The Challenge of Efficiently Removing Consumer Health Risk Factors from Plant Sources

Heat-sensitive consumer health risk factors (including cyanogenic glycosides and antivitamin factors) can be eliminated via heat treatment [176]. Major allergens in plant-based drinks include soybean proteins 7S and 11S; peanut proteins 7S, 11S, and 2S; and gluten in grain seeds. Currently, the most widely used methods of elimination include enzymatic hydrolysis, fermentation, high static pressure, and irradiation [177]. In grain seeds, phytic acid is mainly concentrated in the outer shell and germ part and is relatively low in the endosperm. The phytic acid content in soybean seed coats is low, with only 1% located in the germ and 99% existing in the cotyledons. Therefore, depending on the raw material, phytic acid can be selectively reduced using different methods, including shelling, water blanching, and removal by soaking. Phytic acid can also be removed by cooking, germination, microbial fermentation, and the addition of exogenous phytase [178].

7.3. The Control of System Stability

As multiphase dispersion systems, plant-based drinks contain various colloidal substances, including proteins, lipids, oil bodies, polysaccharides, polyphenols, phytic acid, derivatives of these compounds, plant tissue fragments, and other particles. These systems are thermodynamically unstable and prone to phase separation [179]. The main factors underlying the physical instability of plant-based drinks include the following: (A) Force-induced separation: the dispersed particles in plant-based drinks have different densities than those in the water phase, causing them to move in response to gravitational forces. (B) Particulate matter less dense than water, such as grease bodies or fat droplets, tends to float, whereas denser particulate matter, such as plant cell debris, starch granules, protein aggregates, and calcium carbonate granules, tends to sink. (C) Aggregation in plant-based drinks: electrostatic and hydrophobic interactions occur between lipids, protein-embedded fat droplets, protein particles, and/or plant cell debris. When the interaction force between colloids changes, aggregation occurs easily, mainly through flocculation and coalescence. Plant-based drinks are, thus, prone to stratification and precipitation during storage, which can be detrimental to taste. Plant-based drinks can also experience system instability owing to various chemical or biochemical processes, including oxidation, hydrolysis, and microbial action, which reduce the stability and safety of plant-based drinks and produce an unpleasant volatile odor [3,180].

7.4. Adjustment of System Flavor

Plant-based drinks, which are commercially popular, are generally roasted, giving them nutty, burnt, and sweet aromas. Owing to the unique properties of their raw materials, different plant-based drinks, including coconut, bean, flax, and grain, also have unique flavors and aromas. Plant-based drinks may also contain unappealing flavor attributes. In addition to rancid flavors caused via the oxidation of polyunsaturated fatty acids, they may also have unique grass, raw, and earthy flavors derived from plant raw materials and mainly attributable to low-molecular-weight alcohols, aldehydes, ketones, and furans. Typical off-flavor substances include 1-octen-3-ol (mushroom flavor), hexanal (grass flavor), (E)-2-octenal (cucumber flavor), and (E,E)-2,4 -decadienal (oily), and 2-pentylfuran (grass flavor) [181]. Some substances do not individually produce obvious odors, but their odorant capacities are enhanced in mixtures. Therefore, the material basis of odors is also difficult to define in studies of plant-based drinks. Regarding the formation mechanism of off-flavor components, using soy drink as an example, it is generally believed that C6 aldehyde is an important component of the flavor of legume foods and is derived from the enzymatic oxidation pathway catalyzed via lipoxygenase, with linoleic acid and linolenic acid being the main precursors [182]. Linoleic acid is oxidized via lipoxygenase to produce 9- or 13-hydroxy-linoleic acid hydroperoxide, which further reacts with hydroperoxide lyase to form hexanal, (E)-octenal, and (E,E)-2,4-Decadienal, etc. When linolenic acid is used as a substrate, (E)-2-hexenal and (E,Z)-3,6-nonadienal are formed [183].

Hexanal and (E)-2-hexenal are the main grassy-flavor-contributing components produced via the lipoxygenase pathway [184]. Non-enzymatic reactions also occur in soybean pulping. Lipids, proteins, carbohydrates, and other precursor substances can form different oxidation intermediates, including hydroperoxides and free radicals, ultimately leading to the production of odors. These all require the regulation of systemic flavor in plant-based mill production processes [185].

7.5. Control of Spoilage

Because plant-based drinks are rich in carbon and nitrogen sources, microorganisms can easily reproduce and cause spoilage. The main reasons include the following: (A) Mildewed or spoiled plant-based drink raw materials make it difficult to meet the microbial indicator specifications of the final product. (B) Improperly selected plant-based drink sterilization methods. Pasteurization can kill pathogenic bacteria and most non-pathogenic bacteria in plant-based drinks, but these products retain small proportions of heat-resistant bacteria and spores, resulting in shorter shelf lives. Ultra-high-temperature instantaneous sterilization methods coupled with aseptic canning, canning first, and then high-temperature and high-pressure sterilization are commonly used and are safer sterilization methods for plant-based drink processing [186]. (C) Problems arise during sterilization, and improper process control, such as excessive product accumulation during high-temperature and high-pressure sterilization, incomplete sterilization, ultra-high-temperature instantaneous sterilization coupled with aseptic canning process, and improper cleaning of sterilizers and pipelines, may cause corrosion and deterioration of final products during storage [187]. Therefore, it is necessary to develop green and efficient methods and technologies to control the production of harmful microorganisms in plant-based drinks.

8. Conclusions

Plant-based drinks represent an emerging healthy food with a high nutritional value, which exert a positive impact on human health through long-term consumption. The flavor quality of plant-based drinks is an important evaluation standard for consumers, but most plant-based drinks are difficult for consumers to accept because of their bitter and astringent flavors. Improving the flavor of plant-based drinks is an important bottleneck that the industry needs to overcome. Therefore, the scientific issues in using food flavoromics research methods to conduct in-depth research into plant-based drink flavors should focus on new extraction technologies for plant-based drink flavors, the analysis of substances contributing to aromas, pleasant aroma and off-odor formation mechanisms, aroma optimization, and odor control technologies. Additionally, raw materials containing low levels of odor components should be screened, and novel processing technologies should be explored in depth, combined with the improvement and optimization of odor removal or burial processes, to create new plant-based drink options rich in nutrition and good flavor, expand plant-based drink marketing opportunities, and provide necessary theoretical and technical support for development within the plant-based drink processing industry.

In the burgeoning landscape of plant-based drinks, several impending challenges merit careful consideration for the sustainable evolution of this dairy alternative. Environmental sustainability remains a paramount concern, demanding the judicious management of resource-intensive crop production. Supply chain resilience and nutritional adequacy require meticulous attention to mitigate vulnerabilities associated with disruptions and nutritional disparities. Furthermore, allergen labeling precision, product quality enhancement, and adherence to evolving regulatory frameworks necessitate ongoing vigilance. Competition-driven market dynamics, coupled with consumer education imperatives, underscore the need for strategic differentiation and consumer enlightenment. Infrastructure development, waste management solutions, and the recognition of regional variations in preferences further compound the multifaceted nature of these challenges. Addressing these complexities while upholding ethical and labor standards constitutes an imperative dimension of sustainable plant-based drink development, necessitating a collaborative effort among stakeholders to navigate these intricacies effectively.

Author Contributions: Conceptualization, A.X., Y.D. and M.L.; writing—original draft preparation, A.X., M.L., Z.L. (Zifei liu), Z.L. (Zhiwei Li), J.S. and X.Y.; writing—review and editing, M.L. and X.Y.; supervision, M.L.; project administration, X.Y. All authors have read and agreed to the published version of the manuscript.

Funding: This research received no external funding.

Data Availability Statement: Data is contained within the article.

Conflicts of Interest: The authors declare no conflict of interest.

References

1. Reyes-Jurado, F.; Soto-Reyes, N.; Dávila-Rodríguez, M.; Lorenzo-Leal, A.; Jiménez-Munguía, M.; Mani-López, E.; López-Malo, A. Plant-based milk alternatives: Types, processes, benefits, and characteristics. *Food Rev. Int.* **2021**, *6*, 1–32. [CrossRef]
2. Chalupa-Krebzdak, S.; Long, C.J.; Bohrer, B.M. Nutrient density and nutritional value of milk and plant-based milk alternatives. *Int. Dairy J.* **2018**, *87*, 84–92. [CrossRef]
3. McClements, D.J.; Newman, E.; McClements, I.F. Plant-based milks: A review of the science underpinning their design, fabrication, and performance. *Compr. Rev. Food Sci. Food Saf.* **2019**, *18*, 2047–2067. [CrossRef] [PubMed]
4. Li, M.; Kang, S.; Zheng, Y.; Shao, J.; Zhao, H.; An, Y.; Cao, G.; Li, Q.; Yue, X.; Yang, M. Comparative metabolomics analysis of donkey colostrum and mature milk using ultra-high-performance liquid tandem chromatography quadrupole time-of-flight mass spectrometry. *J. Dairy Sci.* **2020**, *103*, 992–1001. [CrossRef] [PubMed]
5. Li, M.; Li, W.; Wu, J.; Zheng, Y.; Shao, J.; Li, Q.; Kang, S.; Zhang, Z.; Yue, X.; Yang, M. Quantitative lipidomics reveals alterations in donkey milk lipids according to lactation. *Food Chem.* **2020**, *310*, 125866. [CrossRef]
6. Haas, R.; Schnepps, A.; Pichler, A.; Meixner, O. Cow milk versus plant-based milk substitutes: A comparison of product image and motivational structure of consumption. *Sustainability* **2019**, *11*, 5046. [CrossRef]
7. Sethi, S.; Tyagi, S.K.; Anurag, R.K. Plant-based milk alternatives an emerging segment of functional beverages: A review. *J. Food Sci. Technol.* **2016**, *53*, 3408–3423. [CrossRef] [PubMed]
8. Pimentel, D.; Pimentel, M. Sustainability of meat-based and plant-based diets and the environment. *Am. J. Clin. Nutr.* **2003**, *78*, 660S–663S. [CrossRef]
9. McClements, D.J.; Grossmann, L. A brief review of the science behind the design of healthy and sustainable plant-based foods. *NPJ Sci. Food* **2021**, *5*, 17. [CrossRef]

10. Grossmann, L.; McClements, D.J. The science of plant-based foods: Approaches to create nutritious and sustainable plant-based cheese analogs. *Trends Food Sci. Technol.* **2021**, *118*, 207–229. [CrossRef]
11. Li, M.; Chen, J.; Shen, X.; Abdlla, R.; Liu, L.; Yue, X.; Li, Q. Metabolomics-based comparative study of breast colostrum and mature breast milk. *Food Chem.* **2022**, *384*, 132491. [CrossRef] [PubMed]
12. Li, M.; Li, Q.; Kang, S.; Cao, X.; Zheng, Y.; Wu, J.; Wu, R.; Shao, J.; Yang, M.; Yue, X. Characterization and comparison of lipids in bovine colostrum and mature milk based on UHPLC-QTOF-MS lipidomics. *Food Res. Int.* **2020**, *136*, 109490. [CrossRef] [PubMed]
13. Li, M.; Yu, H.; Chen, J.; Abdlla, R.; Liu, A.; Song, W.; Zhang, J.; Zhang, X.; Yue, X.; Li, Q. Novel insights into whey protein differences between donkey and bovine milk. *Food Chem.* **2021**, *365*, 130397. [CrossRef] [PubMed]
14. Paul, A.A.; Kumar, S.; Kumar, V.; Sharma, R. Milk Analog: Plant based alternatives to conventional milk, production, potential and health concerns. *Crit. Rev. Food Sci. Nutr.* **2020**, *60*, 3005–3023. [CrossRef] [PubMed]
15. Qamar, S.; Manrique, Y.J.; Parekh, H.; Falconer, J.R. Nuts, cereals, seeds and legumes proteins derived emulsifiers as a source of plant protein beverages: A review. *Crit. Rev. Food Sci. Nutr.* **2020**, *60*, 2742–2762. [CrossRef] [PubMed]
16. Mäkinen, O.E.; Wanhalinna, V.; Zannini, E.; Arendt, E.K. Foods for special dietary needs: Non-dairy plant-based milk substitutes and fermented dairy-type products. *Crit. Rev. Food Sci. Nutr.* **2016**, *56*, 339–349. [CrossRef] [PubMed]
17. Jeske, S.; Zannini, E.; Arendt, E.K. Evaluation of physicochemical and glycaemic properties of commercial plant-based milk substitutes. *Plant Foods Hum. Nutr.* **2017**, *72*, 26–33. [CrossRef] [PubMed]
18. Schaafsma, G. The protein digestibility–corrected amino acid score. *J. Nutr.* **2000**, *130*, 1865S–1867S. [CrossRef]
19. Boye, J.; Wijesinha-Bettoni, R.; Burlingame, B. Protein quality evaluation twenty years after the introduction of the protein digestibility corrected amino acid score method. *Br. J. Nutr.* **2012**, *108*, S183–S211. [CrossRef]
20. U.S. Department of Agriculture, Agricultural Research Service. FoodData Central. Available online: https://fdc.nal.usda.gov/ (accessed on 24 September 2023).
21. Giri, S.; Mangaraj, S. Processing influences on composition and quality attributes of soymilk and its powder. *Food Eng. Rev.* **2012**, *4*, 149–164. [CrossRef]
22. Belewu, M.; Belewu, K. Comparative physico-chemical evaluation of tiger-nut, soybean and coconut milk sources. *Int. J. Agric. Biol.* **2007**, *5*, e787.
23. Zhou, S.; Jia, Q.; Cui, L.; Dai, Y.; Li, R.; Tang, J.; Lu, J. Physical–Chemical and Sensory Quality of Oat Milk Produced Using Different Cultivars. *Foods* **2023**, *12*, 1165. [CrossRef] [PubMed]
24. Meng, C.; Chen, Y.; Wang, X.; Chen, H.; Deng, Q. Effect of Different Temperatures on the Storage Stability of Flaxseed Milk. *Foods* **2023**, *12*, 3571. [CrossRef] [PubMed]
25. Seong, G.-U.; Moon, K.-D.; Park, D.-S.; Shin, D.; Lee, S.-B.; Lee, J.-Y.; Cho, J.-H.; Park, S.; Kim, J.; Kim, J. Quality characteristics of plant-based rice milk prepared with different rice varieties. *Korean J. Food Preserv.* **2022**, *29*, 395–406. [CrossRef]
26. Zheng, B.; Zhou, H.; McClements, D.J. Nutraceutical-fortified plant-based milk analogs: Bioaccessibility of curcumin-loaded almond, cashew, coconut, and oat milks. *LWT* **2021**, *147*, 111517. [CrossRef]
27. Xie, A.-J.; Lee, D.-J.; Lim, S.-T. Characterization of resistant waxy maize dextrins prepared by simultaneous debranching and crystallization followed by acidic or enzymatic hydrolysis. *Food Hydrocoll.* **2021**, *121*, 106942. [CrossRef]
28. Wang, L.; Liu, H.-M.; Xie, A.-J.; Zhu, C.-Y.; Qin, G.-Y. Dietary fiber extraction from defatted corn hull by hot-compressed water. *Pol. J. Food Nutr. Sci.* **2018**, *68*. [CrossRef]
29. Xie, A.-J.; Yin, H.-S.; Liu, H.-M.; Zhu, C.-Y.; Yang, Y.-J. Chinese quince seed gum and poly (N, N-diethylacryl amide-co-methacrylic acid) based pH-sensitive hydrogel for use in drug delivery. *Carbohydr. Polym.* **2018**, *185*, 96–104. [CrossRef]
30. Xie, A.; Zhu, C.; Hua, L. Food gum based hydrogel polymers. *MOJ Food Process. Technol.* **2017**, *4*, 192–196.
31. Wang, L.; Liu, H.-M.; Zhu, C.-Y.; Xie, A.-J.; Ma, B.-J.; Zhang, P.-Z. Chinese quince seed gum: Flow behaviour, thixotropy and viscoelasticity. *Carbohydr. Polym.* **2019**, *209*, 230–238. [CrossRef]
32. Wang, L.; Liu, H.-M.; Xie, A.-J.; Wang, X.-D.; Zhu, C.-Y.; Qin, G.-Y. Chinese quince (*Chaenomeles sinensis*) seed gum: Structural characterization. *Food Hydrocoll.* **2018**, *75*, 237–245. [CrossRef]
33. Xie, A.; Zhao, S.; Liu, Z.; Yue, X.; Shao, J.; Li, M.; Li, Z. Polysaccharides, proteins, and their complex as microencapsulation carriers for delivery of probiotics: A review on carrier types and encapsulation techniques. *Int. J. Biol. Macromol.* **2023**, 124784. [CrossRef] [PubMed]
34. Li, M.; Li, Q.; Dong, H.; Zhao, S.; Ning, J.; Bai, X.; Yue, X.; Xie, A. Pilose antler polypeptides enhance chemotherapy effects in triple-negative breast cancer by activating the adaptive immune system. *Int. J. Biol. Macromol.* **2022**, *222*, 2628–2638. [CrossRef] [PubMed]
35. Swallah, M.S.; Fan, H.; Wang, S.; Yu, H.; Piao, C. Prebiotic impacts of soybean residue (Okara) on eubiosis/dysbiosis condition of the gut and the possible effects on liver and kidney functions. *Molecules* **2021**, *26*, 326. [CrossRef]
36. Yanai, H.; Katsuyama, H.; Hamasaki, H.; Abe, S.; Tada, N.; Sako, A. Effects of dietary fat intake on HDL metabolism. *J. Clin. Med. Res.* **2015**, *7*, 145. [CrossRef]
37. Li, H.; Liu, Y.; Bao, Y.; Liu, X.; Zhang, H. Conjugated linoleic acid conversion by six Lactobacillus plantarum strains cultured in MRS broth supplemented with sunflower oil and soymilk. *J. Food Sci.* **2012**, *77*, M330–M336. [CrossRef] [PubMed]
38. Patil, U.; Benjakul, S. Coconut milk and coconut oil: Their manufacture associated with protein functionality. *J. Food Sci.* **2018**, *83*, 2019–2027. [CrossRef] [PubMed]
39. Institute, N.F. Frida Food Data. Available online: http://frida.fooddata.dk (accessed on 25 September 2023).

40. Hussain, M.; Sun, Y.; Pan, Y.; Liu, L.; Zhang, X.; Wang, Q.; Lin, S.; Qayum, A.; Hussain, K.; Li, X. Formulation, invitro digestive study, and comparative fatty acid analysis of walnut oil-based infant formula, with human milk, animal milk, and commercial infant formula. *Innov. Food Sci. Emerg. Technol.* **2023**, *84*, 103279. [CrossRef]
41. Isanga, J.; Zhang, G. Production and evaluation of some physicochemical parameters of peanut milk yoghurt. *LWT-Food Sci. Technol.* **2009**, *42*, 1132–1138. [CrossRef]
42. Caroprese, M.; Marzano, A.; Marino, R.; Gliatta, G.; Muscio, A.; Sevi, A. Flaxseed supplementation improves fatty acid profile of cow milk. *J. Dairy Sci.* **2010**, *93*, 2580–2588. [CrossRef]
43. Vanga, S.K.; Raghavan, V. How well do plant based alternatives fare nutritionally compared to cow's milk? *J. Food Sci. Technol.* **2018**, *55*, 10–20. [CrossRef] [PubMed]
44. Heaney, R.P.; Rafferty, K. The settling problem in calcium-fortified soybean drinks. *J. Am. Diet. Assoc.* **2006**, *106*, 1753. [CrossRef] [PubMed]
45. Heaney, R.P.; Dowell, M.S.; Rafferty, K.; Bierman, J. Bioavailability of the calcium in fortified soy imitation milk, with some observations on method. *Am. J. Clin. Nutr.* **2000**, *71*, 1166–1169. [CrossRef] [PubMed]
46. Romulo, A. Nutritional Contents and Processing of Plant-Based Milk: A Review. In Proceedings of the IOP Conference Series: Earth and Environmental Science, Online, 17–18 November 2022; p. 012054.
47. Ferragut, V.; Valencia-Flores, D.C.; Pérez-González, M.; Gallardo, J.; Hernández-Herrero, M. Quality characteristics and shelf-life of ultra-high pressure homogenized (UHPH) almond beverage. *Foods* **2015**, *4*, 159–172. [CrossRef] [PubMed]
48. Patel, B.Y.; Volcheck, G.W. Food allergy: Common causes, diagnosis, and treatment. *Mayo Clin. Proc.* **2015**, *90*, 1411–1419. [CrossRef] [PubMed]
49. Yin, F.; Sancheti, H.; Liu, Z.; Cadenas, E. Mitochondrial function in ageing: Coordination with signalling and transcriptional pathways. *J. Physiol.* **2016**, *594*, 2025–2042. [CrossRef] [PubMed]
50. Qiu, S.-X.; Lu, Z.-Z.; Luyengi, L.; Lee, S.K.; Pezzuto, J.M.; Farnsworth, N.R.; Thompson, L.U.; Fong, H.H. Isolation and characterization of flaxseed (*Linum usitatissimum*) constituents. *Pharm. Biol.* **1999**, *37*, 1–7. [CrossRef]
51. Decloedt, A.I.; Van Landschoot, A.; Watson, H.; Vanderputten, D.; Vanhaecke, L. Plant-based beverages as good sources of free and glycosidic plant sterols. *Nutrients* **2017**, *10*, 21. [CrossRef]
52. Bolling, B.W.; Dolnikowski, G.; Blumberg, J.B.; Chen, C.-Y.O. Polyphenol content and antioxidant activity of California almonds depend on cultivar and harvest year. *Food Chem.* **2010**, *122*, 819–825. [CrossRef]
53. Bernat, N.; Cháfer, M.; Chiralt, A.; Laparra, J.M.; González-Martínez, C. Almond milk fermented with different potentially probiotic bacteria improves iron uptake by intestinal epithelial (Caco-2) cells. *Int. J. Food Stud.* **2015**, *4*, 49–60. [CrossRef]
54. Pelvan, E.; Olgun, E.Ö.; Karadağ, A.; Alasalvar, C. Phenolic profiles and antioxidant activity of Turkish Tombul hazelnut samples (natural, roasted, and roasted hazelnut skin). *Food Chem.* **2018**, *244*, 102–108. [CrossRef] [PubMed]
55. Kamiloglu, S.; Pasli, A.A.; Ozcelik, B.; Capanoglu, E. Evaluating the in vitro bioaccessibility of phenolics and antioxidant activity during consumption of dried fruits with nuts. *LWT-Food Sci. Technol.* **2014**, *56*, 284–289. [CrossRef]
56. Fitrotin, U.; Utami, T.; Hastuti, P.; Santoso, U. Antioxidant properties of fermented sesame milk using Lactobacillus plantarum Dad 13. *Int. Res. J. Biol. Sci.* **2015**, *4*, 56–61.
57. Bopitiya, D.; Madhujith, T. Antioxidant activity and total phenolic content of sesame (*Sesamum indicum* L.) seed oil. *Trop. Agric. Res.* **2013**, *24*, 296.
58. Hamza, A.H.; Mahmoud, R.H. Soy milk and Sesame seeds (Phytoestrogens) Ameliorate cardiotoxcity induced by adriamycin in experimental animals. *Am. J. Res. Commun.* **2013**, *1*, 1–15.
59. Alu'datt, M.H.; Rababah, T.; Ereifej, K.; Alli, I. Distribution, antioxidant and characterisation of phenolic compounds in soybeans, flaxseed and olives. *Food Chem.* **2013**, *139*, 93–99. [CrossRef] [PubMed]
60. Ismail, M.; Abou-Dobara, M.; Nawal, M. Functional rice rayeb milk: Chemical, microbiological and sensory properties. *J. Nutr. Health Sci* **2018**, *5*, 1–12.
61. Moretto, L.; Tonolo, F.; Folda, A.; Scalcon, V.; Bindoli, A.; Bellamio, M.; Feller, E.; Rigobello, M.P. Comparative analysis of the antioxidant capacity and lipid and protein oxidation of soy and oats beverages. *Food Prod. Process. Nutr.* **2021**, *3*, 1–10. [CrossRef]
62. Grossmann, L.; Kinchla, A.J.; Nolden, A.; McClements, D.J. Standardized methods for testing the quality attributes of plant-based foods: Milk and cream alternatives. *Compr. Rev. Food Sci. Food Saf.* **2021**, *20*, 2206–2233. [CrossRef]
63. Schifferstein, H.N.; Verlegh, P.W. The role of congruency and pleasantness in odor-induced taste enhancement. *Acta Psychol.* **1996**, *94*, 87–105. [CrossRef]
64. Noble, A.C. Taste-aroma interactions. *Trends Food Sci. Technol.* **1996**, *7*, 439–444. [CrossRef]
65. Frank, R.A.; Byram, J. Taste–smell interactions are tastant and odorant dependent. *Chem. Senses* **1988**, *13*, 445–455. [CrossRef]
66. Wang, S.; Chen, H.; Sun, B. Recent progress in food flavor analysis using gas chromatography–ion mobility spectrometry (GC–IMS). *Food Chem.* **2020**, *315*, 126158. [CrossRef] [PubMed]
67. Song, H.; Liu, J. GC-O-MS technique and its applications in food flavor analysis. *Food Res. Int.* **2018**, *114*, 187–198. [CrossRef] [PubMed]
68. Snow, N.H.; Slack, G.C. Head-space analysis in modern gas chromatography. *TrAC Trends Anal. Chem.* **2002**, *21*, 608–617. [CrossRef]
69. Lv, Y.C.; Song, H.L.; Li, X.; Wu, L.; Guo, S.T. Influence of blanching and grinding process with hot water on beany and non-beany flavor in soymilk. *J. Food Sci.* **2011**, *76*, S20–S25. [CrossRef] [PubMed]

70. Saison, D.; De Schutter, D.P.; Uyttenhove, B.; Delvaux, F.; Delvaux, F.R. Contribution of staling compounds to the aged flavour of lager beer by studying their flavour thresholds. *Food Chem.* **2009**, *114*, 1206–1215. [CrossRef]
71. Curioni, P.; Bosset, J. Key odorants in various cheese types as determined by gas chromatography-olfactometry. *Int. Dairy J.* **2002**, *12*, 959–984. [CrossRef]
72. Yu, H.; Liu, R.; Hu, Y.; Xu, B. Flavor profiles of soymilk processed with four different processing technologies and 26 soybean cultivars grown in China. *Int. J. Food Prop.* **2017**, *20*, S2887–S2898. [CrossRef]
73. Yoo, S.-H.; Chang, Y.H. Volatile compound, physicochemical, and antioxidant properties of beany flavor-removed soy protein isolate hydrolyzates obtained from combined high temperature pre-treatment and enzymatic hydrolysis. *Prev. Nutr. Food Sci.* **2016**, *21*, 338. [CrossRef]
74. Jensen, P.N.; Sørensen, G.; Engelsen, S.B.; Bertelsen, G. Evaluation of quality changes in walnut kernels (*Juglans regia* L.) by Vis/NIR spectroscopy. *J. Agric. Food Chem.* **2001**, *49*, 5790–5796. [CrossRef]
75. Raftaniamiri, Z.; Khandelwal, P.; Aruna, B. Development of acidophilus milk via selected probiotics & prebiotics using artificial neural network. *Adv. Biosci. Biotechnol.* **2010**, *1*, 224–231.
76. Chetschik, I.; Granvogl, M.; Schieberle, P. Comparison of the key aroma compounds in organically grown, raw West-African peanuts (*Arachis hypogaea*) and in ground, pan-roasted meal produced thereof. *J. Agric. Food Chem.* **2008**, *56*, 10237–10243. [CrossRef] [PubMed]
77. Aydar, E.F.; Tutuncu, S.; Ozcelik, B. Plant-based milk substitutes: Bioactive compounds, conventional and novel processes, bioavailability studies, and health effects. *J. Funct. Foods* **2020**, *70*, 103975. [CrossRef]
78. De Girolamo, A.; Lattanzio, V.; Schena, R.; Visconti, A.; Pascale, M. Effect of alkaline cooking of maize on the content of fumonisins B1 and B2 and their hydrolysed forms. *Food Chem.* **2016**, *192*, 1083–1089. [CrossRef]
79. Kaneko, S.; Kumazawa, K.; Nishimura, O. Studies on the key aroma compounds in soy milk made from three different soybean cultivars. *J. Agric. Food Chem.* **2011**, *59*, 12204–12209. [CrossRef]
80. Sun, W.; Zhao, Q.; Zhao, H.; Zhao, M.; Yang, B. Volatile compounds of Cantonese sausage released at different stages of processing and storage. *Food Chem.* **2010**, *121*, 319–325. [CrossRef]
81. Combet, E.; Eastwood, D.C.; Burton, K.S.; Henderson, J. Eight-carbon volatiles in mushrooms and fungi: Properties, analysis, and biosynthesis. *Mycoscience* **2006**, *47*, 317–326. [CrossRef]
82. Tangyu, M.; Muller, J.; Bolten, C.J.; Wittmann, C. Fermentation of plant-based milk alternatives for improved flavour and nutritional value. *Appl. Microbiol. Biotechnol.* **2019**, *103*, 9263–9275. [CrossRef] [PubMed]
83. Pripis-Nicolau, L.; De Revel, G.; Bertrand, A.; Maujean, A. Formation of flavor components by the reaction of amino acid and carbonyl compounds in mild conditions. *J. Agric. Food Chem.* **2000**, *48*, 3761–3766. [CrossRef] [PubMed]
84. Ma, Z.; Boye, J.I.; Azarnia, S.; Simpson, B.K. Volatile flavor profile of Saskatchewan grown pulses as affected by different thermal processing treatments. *Int. J. Food Prop.* **2016**, *19*, 2251–2271. [CrossRef]
85. Degon, J.G.; Zheng, C.; Elkhedir, A.; Yang, B.; Zhou, Q.; Li, W. Effect of microwave pre-treatment on physical quality, bioactive compounds, safety risk factor, and storage stability of peanut butter. *Oil Crop Sci.* **2021**, *6*, 137–144. [CrossRef]
86. Van Boekel, M. Kinetic aspects of the Maillard reaction: A critical review. *Food/Nahr.* **2001**, *45*, 150–159. [CrossRef]
87. Martins, S.I.; Jongen, W.M.; Van Boekel, M.A. A review of Maillard reaction in food and implications to kinetic modelling. *Trends Food Sci. Technol.* **2000**, *11*, 364–373. [CrossRef]
88. Ellis, G. The maillard reaction. In *Advances in Carbohydrate Chemistry*; Elsevier: Amsterdam, The Netherlands, 1959; Volume 14, pp. 63–134.
89. Fu, M.-X.; Wells-Knecht, K.J.; Blackledge, J.A.; Lyons, T.J.; Thorpe, S.R.; Baynes, J.W. Glycation, glycoxidation, and cross-linking of collagen by glucose: Kinetics, mechanisms, and inhibition of late stages of the Maillard reaction. *Diabetes* **1994**, *43*, 676–683. [CrossRef] [PubMed]
90. Van Boekel, M. Effect of heating on Maillard reactions in milk. *Food Chem.* **1998**, *62*, 403–414. [CrossRef]
91. Dai, X.; Wang, P.; Wu, W.; Wang, H.; Xu, Q.; Li, Z. Separation of Dimethyl Sulfide from Wort by Multi-Layer Centrifugal Film Method. *Foods* **2022**, *11*, 2901. [CrossRef] [PubMed]
92. Mottram, D.S.; Elmore, J.S. Control of the Maillard reaction during the cooking of food. *Control. Mail. Pathw. Gener. Flavors* **2010**, *14*, 143–155.
93. Matsui, K.; Takemoto, H.; Koeduka, T.; Ohnishi, T. 1-Octen-3-ol is formed from its glycoside during processing of soybean [*Glycine max* (L.) Merr.] seeds. *J. Agric. Food Chem.* **2018**, *66*, 7409–7416. [CrossRef]
94. Lokuruka, M. Effects of processing on soybean nutrients and potential impact on consumer health: An overview. *Afr. J. Food Agric. Nutr. Dev.* **2011**, *11*, 4. [CrossRef]
95. Klensporf, D.; Jeleń, H.H. Analysis of volatile aldehydes in oat flakes by SPME-GC/MS. *Pol. J. Food Nutr. Sci.* **2005**, *14*, 389.
96. Matsumura, S.; Takahashi, S.; Nishikitani, M.; Kubota, K.; Kobayashi, A. The role of diglycosides as tea aroma precursors: Synthesis of tea diglycosides and specificity of glycosidases in tea leaves. *J. Agric. Food Chem.* **1997**, *45*, 2674–2678. [CrossRef]
97. Hüsnü, K.; Başer, C.; Demirci, F. Chemistry of essential oils. In *Flavours and Fragrances: Chemistry, Bioprocessing and Sustainability*; John Wiley & Sons Ltd.: Hoboken, NJ, USA, 2007; pp. 43–86. [CrossRef]
98. Taurino, M.; De Domenico, S.; Bonsegna, S.; Santino, A. The hydroperoxide lyase branch of the oxylipin pathway and green leaf volatiles in plant/insect interaction. *J. Plant Biochem. Physiol.* **2013**, *1*, 2. [CrossRef]

99. Matsui, K. Green leaf volatiles: Hydroperoxide lyase pathway of oxylipin metabolism. *Curr. Opin. Plant Biol.* **2006**, *9*, 274–280. [CrossRef] [PubMed]
100. Moreira, M.A.; Tavares, S.R.; Ramos, V.; De Barros, E.G. Hexanal production and TBA number are reduced in soybean [*Glycine max* (L.) Merr.] seeds lacking lipoxygenase isozymes 2 and 3. *J. Agric. Food Chem.* **1993**, *41*, 103–106. [CrossRef]
101. Lee, J.; Min, S.; Choe, E.; Min, D. Formation of volatile compounds in soy flour by singlet oxygen oxidation during storage under light. *J. Food Sci.* **2003**, *68*, 1933–1937. [CrossRef]
102. Bradley, D.; Lee, H.; Min, D. Singlet oxygen detection in skim milk by electron spin resonance spectroscopy. *J. Food Sci.* **2003**, *68*, 491–494. [CrossRef]
103. Padma, M.; Jagannadarao, P.; Edukondalu, L.; Ravibabu, G.; Aparna, K. Physico-chemical analysis of milk prepared from broken rice. *Int. J. Curr. Microbiol. Appl. Sci.* **2018**, *7*, 426–428. [CrossRef]
104. Gajdoš Kljusurić, J.; Benković, M.; Bauman, I. Classification and processing optimization of barley milk production using NIR spectroscopy, particle size, and total dissolved solids analysis. *J. Chem.* **2015**, *2015*, 896051. [CrossRef]
105. Bridges, M. Moo-ove over, cow's milk: The rise of plant-based dairy alternatives. *Pract. Gastroenterol.* **2018**, *21*, 20–27.
106. Xiao, X.; Zou, P.-R.; Hu, F.; Zhu, W.; Wei, Z.-J. Updates on Plant-Based Protein Products as an Alternative to Animal Protein: Technology, Properties, and Their Health Benefits. *Molecules* **2023**, *28*, 4016. [CrossRef]
107. Boukid, F.; Hassoun, A.; Zouari, A.; Tülbek, M.Ç.; Mefleh, M.; Aït-Kaddour, A.; Castellari, M. Fermentation for Designing Innovative Plant-Based Meat and Dairy Alternatives. *Foods* **2023**, *12*, 1005. [CrossRef] [PubMed]
108. Yadav, D.N.; Bansal, S.; Jaiswal, A.K.; Singh, R. Plant based dairy analogues: An emerging food. *Agri. Res. Tech.* **2017**, *10*, 2471–6774. [CrossRef]
109. Martins, C.P.; Cavalcanti, R.N.; Rocha, R.S.; Esmerino, E.A.; Freitas, M.Q.; Pimentel, T.C.; Silva, M.C.; Cruz, A.G. Microwave heating impacts positively on the physical properties of orange juice-milk beverage. *Int. J. Dairy Technol.* **2022**, *75*, 129–138. [CrossRef]
110. Torki-Harchegani, M.; Ghanbarian, D.; Pirbalouti, A.G.; Sadeghi, M. Dehydration behaviour, mathematical modelling, energy efficiency and essential oil yield of peppermint leaves undergoing microwave and hot air treatments. *Renew. Sustain. Energy Rev.* **2016**, *58*, 407–418. [CrossRef]
111. Varghese, T.; Pare, A. Effect of microwave assisted extraction on yield and protein characteristics of soymilk. *J. Food Eng.* **2019**, *262*, 92–99. [CrossRef]
112. Wong, W.L.E.; Gupta, M. Using microwave energy to synthesize light weight/energy saving magnesium based materials: A review. *Technologies* **2015**, *3*, 1–18. [CrossRef]
113. Jacob, M.; Jaros, D.; Rohm, H. Recent advances in milk clotting enzymes. *Int. J. Dairy Technol.* **2011**, *64*, 14–33. [CrossRef]
114. Penha, C.B.; Santos, V.D.P.; Speranza, P.; Kurozawa, L.E. Plant-based beverages: Ecofriendly technologies in the production process. *Innov. Food Sci. Emerg. Technol.* **2021**, *72*, 102760. [CrossRef]
115. Vong, W.C.; Liu, S.-Q. Biovalorisation of okara (soybean residue) for food and nutrition. *Trends Food Sci. Technol.* **2016**, *52*, 139–147. [CrossRef]
116. Pandey, S.; Poonia, A. Plant-based milk substitutes: A novel non-dairy source. *Innov. Food Technol. Curr. Perspect. Future Goals* **2020**, 63–71. [CrossRef]
117. Rosenthal, A.; Deliza, R.; Cabral, L.M.; Cabral, L.C.; Farias, C.A.; Domingues, A.M. Effect of enzymatic treatment and filtration on sensory characteristics and physical stability of soymilk. *Food Control* **2003**, *14*, 187–192. [CrossRef]
118. Sahoo, A.; Gaikwad, V.; Ranveer, R.; Dandge, P.; Waghmare, S. Improvement of shelf life of soymilk using immobilized protease of Oerskovia xanthineolytica NCIM 2839. *3 Biotech* **2016**, *6*, 1–7. [CrossRef] [PubMed]
119. Karamać, M.; Kosińska-Cagnazzo, A.; Kulczyk, A. Use of different proteases to obtain flaxseed protein hydrolysates with antioxidant activity. *Int. J. Mol. Sci.* **2016**, *17*, 1027. [CrossRef] [PubMed]
120. Görgüç, A.; Bircan, C.; Yılmaz, F.M. Sesame bran as an unexploited by-product: Effect of enzyme and ultrasound-assisted extraction on the recovery of protein and antioxidant compounds. *Food Chem.* **2019**, *283*, 637–645. [CrossRef] [PubMed]
121. Jeske, S.; Zannini, E.; Arendt, E.K. Past, present and future: The strength of plant-based dairy substitutes based on gluten-free raw materials. *Food Res. Int.* **2018**, *110*, 42–51. [CrossRef] [PubMed]
122. Rasika, D.; Vidanarachchi, J.K.; Luiz, S.F.; Azeredo, D.R.P.; Cruz, A.G.; Ranadheera, C.S. Probiotic delivery through non-dairy plant-based food matrices. *Agriculture* **2021**, *11*, 599. [CrossRef]
123. Hou, J.-W.; Yu, R.-C.; Chou, C.-C. Changes in some components of soymilk during fermentation with bifidobacteria. *Food Res. Int.* **2000**, *33*, 393–397. [CrossRef]
124. Rekha, C.; Vijayalakshmi, G. Bioconversion of isoflavone glycosides to aglycones, mineral bioavailability and vitamin B complex in fermented soymilk by probiotic bacteria and yeast. *J. Appl. Microbiol.* **2010**, *109*, 1198–1208. [CrossRef]
125. Wansutha, S.; Yuenyaow, L.; Jantama, K.; Jantama, S.S. Antioxidant activities of almond milk fermented with lactic acid bacteria. *Thaiphesatchasan* **2018**, *42*, 115–119.
126. Cruz, N.; Capellas, M.; Hernández, M.; Trujillo, A.; Guamis, B.; Ferragut, V. Ultra high pressure homogenization of soymilk: Microbiological, physicochemical and microstructural characteristics. *Food Res. Int.* **2007**, *40*, 725–732. [CrossRef]
127. Penha, C.B.; Falcão, H.G.; Ida, E.I.; Speranza, P.; Kurozawa, L.E. Enzymatic pretreatment in the extraction process of soybean to improve protein and isoflavone recovery and to favor aglycone formation. *Food Res. Int.* **2020**, *137*, 109624. [CrossRef] [PubMed]

128. Zhao, D.; Shah, N.P. Changes in antioxidant capacity, isoflavone profile, phenolic and vitamin contents in soymilk during extended fermentation. *LWT-Food Sci. Technol.* **2014**, *58*, 454–462. [CrossRef]
129. Criste, A.D.; Urcan, A.C.; Coroian, C.O.; Copolovici, L.; Copolovici, D.M.; Burtescu, R.F.; Oláh, N.K. Plant-Based Beverages from Germinated and Ungerminated Seeds, as a Source of Probiotics, and Bioactive Compounds with Health Benefits—Part 1: Legumes. *Agriculture* **2023**, *13*, 1185. [CrossRef]
130. Aboulfazli, F.; Shori, A.B.; Baba, A.S. Effects of the replacement of cow milk with vegetable milk on probiotics and nutritional profile of fermented ice cream. *LWT* **2016**, *70*, 261–270. [CrossRef]
131. Oghbaei, M.; Prakash, J. Effect of primary processing of cereals and legumes on its nutritional quality: A comprehensive review. *Cogent Food Agric.* **2016**, *2*, 1136015. [CrossRef]
132. Nirmala, M.; Rao, M.S.; Muralikrishna, G. Carbohydrates and their degrading enzymes from native and malted finger millet (Ragi, Eleusine coracana, Indaf-15). *Food Chem.* **2000**, *69*, 175–180. [CrossRef]
133. Saxena, S.; Vasudevan, H.; Saini, S.; Sasmal, S. Comparative Nutritional Assessment of Millet-Based Milk Produced by Ultrasound, Germination, and a Combined Approach. *ACS Food Sci. Technol.* **2023**, *3*, 600–607. [CrossRef]
134. Özer, B.H.; Kirmaci, H.A. Functional milks and dairy beverages. *Int. J. Dairy Technol.* **2010**, *63*, 1–15. [CrossRef]
135. BALCI, A.T.; Wilbey, R.A. High pressure processing of milk-the first 100 years in the development of a new technology. *Int. J. Dairy Technol.* **1999**, *52*, 149–155. [CrossRef]
136. Poliseli-Scopel, F.H.; Hernández-Herrero, M.; Guamis, B.; Ferragut, V. Comparison of ultra high pressure homogenization and conventional thermal treatments on the microbiological, physical and chemical quality of soymilk. *LWT-Food Sci. Technol.* **2012**, *46*, 42–48. [CrossRef]
137. Zamora, A.; Guamis, B. Opportunities for ultra-high-pressure homogenisation (UHPH) for the food industry. *Food Eng. Rev.* **2015**, *7*, 130–142. [CrossRef]
138. Mollakhalili-Meybodi, N.; Nejati, R.; Sayadi, M.; Nematollahi, A. Novel nonthermal food processing practices: Their influences on nutritional and technological characteristics of cereal proteins. *Food Sci. Nutr.* **2022**, *10*, 1725–1744. [CrossRef] [PubMed]
139. Araujo, A.J.D. Method for Preparing a Beanless-Flavor Soymilk and/or Okara Using Carbon Dioxide in a State of Sublimation. Google Patents EP1830658A2 12 September 2007.
140. Poliseli-Scopel, F.H.; Hernández-Herrero, M.; Guamis, B.; Ferragut, V. Sterilization and aseptic packaging of soymilk treated by ultra high pressure homogenization. *Innov. Food Sci. Emerg. Technol.* **2014**, *22*, 81–88. [CrossRef]
141. Rysstad, G.; Kolstad, J. Extended shelf life milk—Advances in technology. *Int. J. Dairy Technol.* **2006**, *59*, 85–96. [CrossRef]
142. Panchal, H.; Patel, J.; Chaudhary, S. A comprehensive review of solar milk pasteurization system. *J. Sol. Energy Eng.* **2018**, *140*, 010801. [CrossRef]
143. McCarthy, K.; Parker, M.; Ameerally, A.; Drake, S.; Drake, M. Drivers of choice for fluid milk versus plant-based alternatives: What are consumer perceptions of fluid milk? *J. Dairy Sci.* **2017**, *100*, 6125–6138. [CrossRef] [PubMed]
144. Watts, S. A mini review on technique of milk pasteurization. *J. Pharmacogn. Phytochem.* **2016**, *5*, 99–101.
145. Gentès, M.C.; Caron, A.; Champagne, C.P. Potential applications of pulsed electric field in cheesemaking. *Int. J. Dairy Technol.* **2022**, *75*, 270–288. [CrossRef]
146. Ross, A.I.; Griffiths, M.W.; Mittal, G.S.; Deeth, H.C. Combining nonthermal technologies to control foodborne microorganisms. *Int. J. Food Microbiol.* **2003**, *89*, 125–138. [CrossRef]
147. Silva, A.R.; Silva, M.M.; Ribeiro, B.D. Health issues and technological aspects of plant-based alternative milk. *Food Res. Int.* **2020**, *131*, 108972. [CrossRef]
148. Zhao, W.; Tang, Y.; Lu, L.; Chen, X.; Li, C. Pulsed electric fields processing of protein-based foods. *Food Bioprocess Technol.* **2014**, *7*, 114–125. [CrossRef]
149. Li, Y.-Q.; Tian, W.-L.; Mo, H.-Z.; Zhang, Y.-L.; Zhao, X.-Z. Effects of pulsed electric field processing on quality characteristics and microbial inactivation of soymilk. *Food Bioprocess Technol.* **2013**, *6*, 1907–1916. [CrossRef]
150. Li, Y.-Q.; Chen, Q.; Liu, X.-H.; Chen, Z.-X. Inactivation of soybean lipoxygenase in soymilk by pulsed electric fields. *Food Chem.* **2008**, *109*, 408–414. [CrossRef] [PubMed]
151. Shi, L.; Mu, K.; Arntfield, S.D.; Nickerson, M.T. Changes in levels of enzyme inhibitors during soaking and cooking for pulses available in Canada. *J. Food Sci. Technol.* **2017**, *54*, 1014–1022. [CrossRef] [PubMed]
152. Pataro, G.; Ferrari, G. Limitations of pulsed electric field utilization in food industry. In *Pulsed Electric Fields to Obtain Healthier and Sustainable Food for Tomorrow*; Elsevier: Amsterdam, The Netherlands, 2020; pp. 283–310.
153. Korhonen, H. Technology options for new nutritional concepts. *Int. J. Dairy Technol.* **2002**, *55*, 79–88. [CrossRef]
154. Wattanayon, W.; Udompijitkul, P.; Kamonpatana, P. Ohmic heating of a solid-liquid food mixture in an electrically conductive package. *J. Food Eng.* **2021**, *289*, 110180. [CrossRef]
155. Saxena, J.; Ahmad Makroo, H.; Srivastava, B. E ffect of ohmic heating on Polyphenol Oxidase (PPO) inactivation and color change in sugarcane juice. *J. Food Process Eng.* **2017**, *40*, e12485. [CrossRef]
156. Saini, A.; Morya, S. A Review based study on Soymilk: Focuses on production technology, Prospects and Progress Scenario in last Decade. *Pharma Innov.* **2021**, *5*, 10.
157. Lu, L.; Zhao, L.; Zhang, C.; Kong, X.; Hua, Y.; Chen, Y. Comparative effects of ohmic, induction cooker, and electric stove heating on soymilk trypsin inhibitor inactivation. *J. Food Sci.* **2015**, *80*, C495–C503. [CrossRef]

158. Castro, I.; Macedo, B.; Teixeira, J.; Vicente, A. The effect of electric field on important food-processing enzymes: Comparison of inactivation kinetics under conventional and ohmic heating. *J. Food Sci.* **2004**, *69*, C696–C701. [CrossRef]
159. Shim, J.; Lee, S.H.; Jun, S. Modeling of ohmic heating patterns of multiphase food products using computational fluid dynamics codes. *J. Food Eng.* **2010**, *99*, 136–141. [CrossRef]
160. Makroo, H.; Rastogi, N.; Srivastava, B. Ohmic heating assisted inactivation of enzymes and microorganisms in foods: A review. *Trends Food Sci. Technol.* **2020**, *97*, 451–465. [CrossRef]
161. Aidoo, H.; Sakyi-Dawson, E.; Tano-Debrah, K.; Saalia, F.K. Development and characterization of dehydrated peanut–cowpea milk powder for use as a dairy milk substitute in chocolate manufacture. *Food Res. Int.* **2010**, *43*, 79–85. [CrossRef]
162. Zungur Bastıoğlu, A.; Tomruk, D.; Koç, M.; Ertekin, F.K. Spray dried melon seed milk powder: Physical, rheological and sensory properties. *J. Food Sci. Technol.* **2016**, *53*, 2396–2404. [CrossRef] [PubMed]
163. Abdullah, Z.; Taip, F.S.; Mustapa Kamal, S.M.; Abdul Rahman, R.Z. Effect of sodium caseinate concentration and sonication amplitude on the stability and physical characteristics of homogenized coconut milk. *J. Food Process. Preserv.* **2018**, *42*, e13773. [CrossRef]
164. Yazici, F.; Alvarez, V.; Mangino, M.; Hansen, P. Formulation and processing of a heat stable calcium-fortified soy milk. *J. Food Sci.* **1997**, *62*, 535–538. [CrossRef]
165. Kwok, K.; Qin, W.; Tsang, J. Heat inactivation of trypsin inhibitors in soymilk at ultra-high temperatures. *J. Food Sci.* **1993**, *58*, 859–862. [CrossRef]
166. Sizer, C. Aseptic packaging of soymilk. In *Food Uses of Whole Oil and Protein Seeds*; Lucas, E.W., Erickson, D.R., Nip, W.K., Eds.; American Oil Chemists' Society: Champaign, IL, USA, 1989; pp. 98–101, Ch. 6.
167. Han, T.B. Technology of Soymilk and Some Derivatives. 1958. Available online: https://library.wur.nl/WebQuery/wurpubs/fulltext/180242 (accessed on 25 September 2023).
168. Arise, A.K.; Malomo, S.A.; Jacob, O.A.; Olagunju, O.F.; Esan, O.T. Comparative studies of yoghurt produced from animal and selected imitation vegetable milk. *Nutrire* **2022**, *47*, 33. [CrossRef]
169. SV, R.; Shameena Beegum, P.; Pandiselvam, R.; Manikantan, M.; Hebbar, K. Plant-Based Milk Alternatives: Nutritional Potential and Challenges. In *Conceptualizing Plant-Based Nutrition: Bioresources, Nutrients Repertoire and Bioavailability*; Springer: Berlin/Heidelberg, Germany, 2022; pp. 91–106.
170. Anand, S.P.; Awasti, N. Novel Dairy-Based Drinks: Changing Scenario. *Dairy Process. Adv. Res. Appl.* **2020**, 301–325. [CrossRef]
171. Valencia-Flores, D.C.; Hernández-Herrero, M.; Guamis, B.; Ferragut, V. Comparing the effects of ultra-high-pressure homogenization and conventional thermal treatments on the microbiological, physical, and chemical quality of almond beverages. *J. Food Sci.* **2013**, *78*, E199–E205. [CrossRef]
172. Scheller, H.V.; Ulvskov, P. Hemicelluloses. *Annu. Rev. Plant Biol.* **2010**, *61*, 263–289. [CrossRef]
173. Colosimo, R.; Warren, F.J.; Edwards, C.H.; Ryden, P.; Dyer, P.S.; Finnigan, T.J.; Wilde, P.J. Comparison of the behavior of fungal and plant cell wall during gastrointestinal digestion and resulting health effects: A review. *Trends Food Sci. Technol.* **2021**, *110*, 132–141. [CrossRef]
174. Burton, R.A.; Fincher, G.B. Evolution and development of cell walls in cereal grains. *Front. Plant Sci.* **2014**, *5*, 456. [CrossRef] [PubMed]
175. Bocker, R.; Silva, E.K. Innovative technologies for manufacturing plant-based non-dairy alternative milk and their impact on nutritional, sensory and safety aspects. *Future Foods* **2022**, *5*, 100098. [CrossRef]
176. Chain, E.P.o.C.i.t.F.; Schrenk, D.; Bignami, M.; Bodin, L.; Chipman, J.K.; del Mazo, J.; Grasl-Kraupp, B.; Hogstrand, C.; Hoogenboom, L.; Leblanc, J.C. Evaluation of the health risks related to the presence of cyanogenic glycosides in foods other than raw apricot kernels. *EFSA J.* **2019**, *17*, e05662.
177. Abd Rahim, M.H.; Hazrin-Chong, N.H.; Harith, H.H.; Wan, W.A.A.Q.I.; Sukor, R. Roles of fermented plant-, dairy-and meat-based foods in the modulation of allergic responses. *Food Sci. Hum. Wellness* **2023**, *12*, 691–701. [CrossRef]
178. Liang, J.; Han, B.-Z.; Nout, M.R.; Hamer, R.J. Effects of soaking, germination and fermentation on phytic acid, total and in vitro soluble zinc in brown rice. *Food Chem.* **2008**, *110*, 821–828. [CrossRef] [PubMed]
179. Ahrens, S.; Venkatachalam, M.; Mistry, A.M.; Lapsley, K.; Sathe, S.K. Almond (*Prunus dulcis* L.) protein quality. *Plant Foods Hum. Nutr.* **2005**, *60*, 123–128. [CrossRef] [PubMed]
180. Sarangapany, A.K.; Murugesan, A.; Annamalai, A.S.; Balasubramanian, A.; Shanmugam, A. An overview on ultrasonically treated plant-based milk and its properties–A Review. *Appl. Food Res.* **2022**, *2*, 100130. [CrossRef]
181. Feng, X.; Li, X.; Zhang, C.; Kong, X.; Chen, Y.; Hua, Y. Formation mechanism of hexanal and (E)-2-hexenal during soybean [*Glycine max* (L.) Merr] processing based on the subcellular and molecular levels. *J. Agric. Food Chem.* **2021**, *70*, 289–300. [CrossRef]
182. Trawatha, S.; TeKrony, D.; Hildebrand, D. Relationship of Soybean Seed Quality to Fatty Acid and C6-Aldehyde Levels during Storage. *Crop Sci.* **1995**, *35*, 1415–1422. [CrossRef]
183. Noordermeer, M.A.; Van Der Goot, W.; Van Kooij, A.J.; Veldsink, J.W.; Veldink, G.A.; Vliegenthart, J.F. Development of a biocatalytic process for the production of C6-aldehydes from vegetable oils by soybean lipoxygenase and recombinant hydroperoxide lyase. *J. Agric. Food Chem.* **2002**, *50*, 4270–4274. [CrossRef]
184. Zhang, C.; Hua, Y.; Li, X.; Kong, X.; Chen, Y. Key volatile off-flavor compounds in peas (*Pisum sativum* L.) and their relations with the endogenous precursors and enzymes using soybean (*Glycine max*) as a reference. *Food Chem.* **2020**, 333, 127469. [CrossRef]

185. Wang, B.; Zhang, Q.; Zhang, N.; Bak, K.H.; Soladoye, O.P.; Aluko, R.E.; Fu, Y.; Zhang, Y. Insights into formation, detection and removal of the beany flavor in soybean protein. *Trends Food Sci. Technol.* **2021**, *112*, 336–347. [CrossRef]
186. Rico-Munoz, E.; dos Santos, J.L.P. The fungal problem in thermal processed beverages. *Curr. Opin. Food Sci.* **2019**, *29*, 80–87. [CrossRef]
187. Agalloco, J.; Akers, J. The future of aseptic processing. In *Advanced Aseptic Processing Technology*; CRC Press: Thames, UK, 2016; pp. 465–469.

foods

MDPI

Review

A Comprehensive Review of Bioactive Compounds from Lactic Acid Bacteria: Potential Functions as Functional Food in Dietetics and the Food Industry

Bibi Nabihah Abdul Hakim [1], Ng Jia Xuan [1] and Siti Nur Hazwani Oslan [1,2,*]

1 Faculty of Food Science and Nutrition, Universiti Malaysia Sabah, Jalan UMS, Kota Kinabalu 88400, Sabah, Malaysia; bibinabihah@ums.edu.my (B.N.A.H.); ng_jia_bn20@iluv.ums.edu.my (N.J.X.)

2 Innovative Food Processing and Ingredients Research Group, Faculty of Food Science and Nutrition, Universiti Malaysia Sabah, Jalan UMS, Kota Kinabalu 88400, Sabah, Malaysia

* Correspondence: snhazwanioslan@ums.edu.my

Abstract: Lactic acid bacteria (LAB) are beneficial microbes known for their health-promoting properties. LAB are well known for their ability to produce substantial amounts of bioactive compounds during fermentation. Peptides, exopolysaccharides (EPS), bacteriocins, some amylase, protease, lipase enzymes, and lactic acid are the most important bioactive compounds generated by LAB activity during fermentation. Additionally, the product produced by LAB is dependent on the type of fermentation used. LAB derived from the genera *Lactobacillus* and *Enterococcus* are the most popular probiotics at present. Consuming fermented foods has been previously connected to a number of health-promoting benefits such as antibacterial activity and immune system modulation. Furthermore, functional food implementations lead to the application of LAB in therapeutic nutrition such as prebiotic, immunomodulatory, antioxidant, anti-tumor, blood glucose lowering actions. Understanding the characteristics of LAB in diverse sources and its potential as a functional food is crucial for therapeutic applications. This review presents an overview of functional food knowledge regarding interactions between LAB isolated from dairy products (dairy LAB) and fermented foods, as well as the prospect of functioning LAB in human health. Finally, the health advantages of LAB bioactive compounds are emphasized.

Keywords: lactic acid bacteria; fermentation; bioactive compound; functional properties; application of LAB

Citation: Abdul Hakim, B.N.; Xuan, N.J.; Oslan, S.N.H. A Comprehensive Review of Bioactive Compounds from Lactic Acid Bacteria: Potential Functions as Functional Food in Dietetics and the Food Industry. *Foods* **2023**, *12*, 2850. https://doi.org/10.3390/foods12152850

Academic Editor: Xiqing Yue

Received: 31 May 2023
Revised: 22 June 2023
Accepted: 26 June 2023
Published: 27 July 2023

1. Introduction

Over time, the purpose of food has evolved beyond mere taste and nutrition, now serving as a powerful means to enhance human health through added functionality. Diet plays a crucial role in overall human health, serving as a frontline defense against numerous diseases [1]. As the interest in the relationship between food and health continues to rise, the demand for functional foods is also increasing. While there is no universally accepted definition, functional foods are generally described as food products that offer various health benefits when incorporated into one's diet. Functional foods can be classified into four main categories according to their definition: conventional foods, modified foods, foods designed for special dietary needs, and medicinal foods [2]. There is a growing interest in characterizing and incorporating bioactive constituents into foods in order to satisfy medically defined criteria and nourish populations.

A recent strategy that has gained traction is the use of fermentates, which are powdered formulations formed from fermentation reactions. Fermentates can be made up of either the bacteria that are fermenting or the metabolites and bioactive compounds that are excreted in the fermentation broth. Milk and dairy products are consumed by more than six billion people worldwide, as they are a food group with a wide variety in terms of taste, texture, and nutritional value [3]. Milk is enriched with useful components, such as minerals and vitamins [4]. In particular, fermented dairy products are a good source of different species of live lactic acid bacteria (LAB) [5].

Peptides, exopolysaccharides (EPS), bacteriocins, some amylase, protease, lipase enzymes, and lactic acid are the major bioactive molecules produced by LAB activity during fermentation [6]. However, not all LAB strains can generate these compounds. The health benefits of LAB have made them a popular ingredient in therapeutic nutrition. One of the most common ways that LAB is used in therapeutic nutrition is in the form of probiotic supplements [7]. Fermented dairy products may show their health-promoting effects due to the influences of microbial metabolites (biogenic or bioactive effect) formed during the fermentation process, as well as the probiotic effects of certain LAB strains isolated from their composition [8]. Many studies have stated that the consumption of probiotic-containing dairy products such as yogurt, cultured fermented milk, and kefir has been linked to a variety of health benefits, such as cholesterol metabolism and angiotensin-converting enzyme (ACE) inhibition, antimicrobial activity, tumor suppression, faster wound healing, and immune system modulation [9,10]. Moreover, the consumption of probiotics balances the gut and urinary tract microbiome by promoting the growth of beneficial bacteria that outcompete pathogens for food and binding sites and locally generate antimicrobial metabolites. As a side benefit, probiotics help the mucosal barrier function by influencing the host immune system [11].

This article summarizes the complex relationships between dairy LAB and human health, and suggests an innovative approach to describing and incorporating bioactive compounds into foods in order to serve as a crucial functional food. These bioactive compounds have been studied for their potential in treating food intolerance, gastrointestinal complications, diabetes mellitus, inflammatory bowel disease, liver disease, and cancer, and the article emphasizes the fermentation and/or probiotic potentials of dairy LAB.

1.1. Characteristics of LAB

Lactic acid bacteria (LAB) play a pivotal role in the food processing industry, serving as a vital group of bacteria with substantial significance. Most of these microorganisms are "generally recognized as safe" (GRAS) because they are nonpathogenic, useful in technological and industrial processes, acid and bile tolerant, and are able to produce antimicrobial substances; they have also been consumed by people all over the world for a long time in dairy products [12]. LAB is a well-known class of microorganisms used in the food industry due to their wide range of phylogenetic and functional diversity. LAB are defined as a taxonomic order of bacteria that is both phylogenetically and functionally diversified [13]. Lactic acid bacteria from the genera *Lactobacillus* and *Enterococcus* are currently the most popular probiotics. These LAB strains include *L. acidophilus*, *L. fermentum*, *L. casei*, *L. reuteri*, *L. rhamnosus*, *L. helveticus*, *L. lactis*, *L. crispatus*, *L. gasseri*, *L. plantarum*, and *E. faecalis* [14].

In the last 20 years, *Lactobacillus* has emerged as the preeminent nomenclature for probiotics, gaining widespread popularity in scientific discourse and research. Therefore, it is important to approach the probiotic potential of LAB genera with caution and subject them to individual assessment and scrutiny to determine their suitability as probiotics. Some LAB genera are probiotics, although scientists disagree. The most researched strains of probiotic LAB include *Lactobacillus acidophilus* NCFM, *Lactobacillus acidophilus* LA-5, *Lactobacillus* casei DN-114 001, *Lacticaseibacillus casei* strain Shirota, *Lacticaseibacillus casei* Zhang, and *Lactobacillus rhamnosus* GG (ATCC53103) [11]. Recently, novel probiotic LAB such as *Limosilactobacillus reuteri* and *Lactobacillus johnsonii* have been employed in the production of functional dairy products [15]. More research is required to screen and unravel the probiotic potentials of novel LAB strains with unique favorable health effects on both humans and animals, and that are of scientific and industrial value, because probiotic traits and features are strain-specific [16]. Moreover, the attributes of probiotics can include various characteristics, such as hemolytic activity and antibiotic resistance [17,18]. However, it is important to carefully evaluate these attributes, as excessive hemolytic activity can be harmful and antibiotic resistance may have implications for antibiotic effectiveness. Other features include the ability to adhere, the capacity to inhibit or eliminate harmful microbes, to auto- and co-aggregate, and to be harmless to animals. Li et al. [19] demonstrated that all five isolates had significant adhesion potential, extraordinary aggregation capacity, and antibacterial properties.

1.2. Source of LAB in Dairy Products

Milk and other dairy products are generally thought to be the principal dietary sources for LAB. Throughout the world, people drink either fresh or fermented cow and goat milk. Table 1 has shown source of LAB in dairy and non-dairy products. High numbers of LAB as beneficial bacteria in milk suggest a source for biological materials with great public health value and extensive applicability in the dairy sector [20]. According to Agagunduz et al. [11], the milk-based sources employed (kind of animal, diet, age, length of the lactation period, etc.) and food processing techniques (temperature, storage conditions, etc.) are the two primary elements that determine the nutritional value of dairy products. The beneficial health effects of fermented milk and dairy products are mostly attributed to the presence of LAB, which can be naturally found in some dairy products. The most common dairy products that contain LAB are fermented milk, yogurt, cheese, and other milk products [21]. They may be included as a starter culture or occasionally as novel ingredients or additives for the purpose of boosting the functionality of the product, and their ability to increase the nutritional value of fermented milk products [22]. Due to their long history of usage in food and milk fermentation, LAB starter cultures are now classified as GRAS [23].

While there is no definitive cell count number that can ensure the health effects of the probiotic strain in a food product, it has been shown that at least 10^6–10^8 cfu/g is adequate to benefit from the advantageous effects of probiotics [24]. This very clearly demonstrates that just because a culture that has the potential to show probiotic potential is present in a product does not necessarily mean that the product itself will have probiotic properties. Probiotics are only effective against certain strains of bacteria; thus, even various strains of the same species might have wildly diverse effects on the host. As a result, it is indicated that more research is needed to understand the probiotic potential of new LAB strains as well as well-known dairy product starting cultures [25].

Table 1. Source of LAB in dairy and non-dairy products.

LAB Source	Family	Genus	Gram	Shape	Acid-Resistant	Respiration	References
Dairy Product	*Lactobacillaceae*	*Lactobacillus*	+	Rod shaped	Changeable	Facultative anaerobic	[26]
		Pediococcus	+	Spherical shaped	High acid resistant	Facultative anaerobic	[27]
	Steptococcaceae	*Streptococcus*	+	Coccoid shaped	Low acid resistant	Facultative anaerobic	[28]
		Lactococcus	+	Coccoid	Changeable	Facultative anaerobic	[29]
	Leuconostocaecae	*Leuconostoc*	+	Spherical, oval shaped	Changeable	Facultative anaerobic	[30]
	Bifidobacteriaceae	*Bifidobacterium*	+	Rod-branch-shaped	High acid resistant	Anaerobic	[31]
	Enterococcaceae	*Enterococcus*	+	Coccoid shape	Moderate acid resistant	Facultative anaerobic	[31]
	Propionibacteriaceae	*Propionibacterium*	+	Rod shaped	Low acid resistant	Anaerobic	[30]

Table 1. *Cont.*

LAB Source	Family	Genus	Gram	Shape	Acid-Resistant	Respiration	References
Non-diary	*Aerococcaceae*	*Aerococcus*	+	Coccoid shaped	Low acid resistant	Facultative anaerobic	[32]
	Carbobacteriaceae	*Carnobacterium*	+	Rod shaped	Not available	Facultative anaerobic	[33]
	Leuconostocaecae	*Oenococcus*	+	Spherical shaped	Changeable	Facultative anaerobic	[34]
		Weissella	+	Coccoid or rod shaped	Changeable	Facultative anaerobic	[35]
		Fructobacillus	+	Elongated and slightly cylindrical shaped	Not available	Facultative anaerobic	[36]
	Enterococcaceae	*Tetragenococcus*	+	Coccoid shaped	Changeable	Facultative anaerobic	[37]
		Vagococcus	+	Coccoid shaped	Changeable	Facultative anaerobic	[38]

1.3. Source of LAB in Fermented Food

Fermented Food

Fermented foods have become an important branch of the food industry as these foods are abundant sources of potential beneficial microbes that extend the shelf life and increase the nutritional variety and organoleptic properties of the food [39]. Historically, fermented food has been consumed as a staple food since the development of human civilizations. The functional microorganisms naturally present in fermented food offer unique properties to the consumer, including antimicrobial and antioxidant properties and bioactive compound production [40]. Certain strains of probiotics, such as *Lactobacillus, Leuconostoc,* and *Enterococcus,* have the ability to thrive and remain viable throughout the fermentation process. These beneficial bacteria can be commonly found in various fermented foods, including yogurt, sauerkraut, kimchi, and kefir [41]. By regulating the immune function of the host mucosa or by regulating the balance of intestinal flora, it can promote nutrient absorption and maintain intestinal health. The LAB that are widely encountered in fermented food include *Lactobacillus, Leuconostoc* and *Enterococcus, Weissella, Pediocossu,* etc. [40]. LAB are indeed involved in producing a wide range of fermented food products, including alcoholic drinks, fermented bread and noodles, fermented fish and meat, fermented dairy products, and fermented vegetables [42]. According to Sudhanshu et al. [43], *Lactobacillus plantarum* is commonly found in fermented vegetables due to the acid and salt resistance in the specific fermentation conditions. Kimchi contains *Leuconostoc, Lactobacillus,* and *Lactococcus,* which are responsible for the creation of unique sensory properties and nutritional properties [43]. Fish that has been fermented frequently contains *L. plantarum,* which has qualities that make it safe for consumption.

The presence of a live culture is dependent on the processing method and the specific food. An unsuitable process may affect the viability of the LAB. Commercial yogurt contains live cultures such as *Lactobacillus delbrueckii* subsp. bulgaricus that are intentionally added during production to create unique texture, flavor, and nutritional value. The survivability of probiotics in the gastrointestinal tract (GIT) is often regarded as critical for their potential health effects [44]. In the context of postbiotics, however, the vitality of probiotic bacteria may not be as important as the existence and activity of the bioactive compounds they create, such as organic acids, enzymes, peptides, polysaccharides, and other metabolites [45]. In the case of conventional probiotic treatments that ingest live cells, viability is critical for delivering the desired effects in the gastrointestinal system. According to Sahadeva et al. [46], the ability of the LAB to resist acid and bile is vital to indicate the survival rate of the bacteria in the intestinal transit and exert their potential benefits. As a result, some factors such as the types of strains and fermentation conditions need to be considered during fermentation [44].

Postbiotics can be characterized as metabolite byproducts produced by beneficial microorganisms throughout the growth and fermentation process that have a positive impact on a consumer's health [47]. Numerous bioactive metabolites including organic acid (lactic acid, acetic acid), carbohydrates, enzymes, bacteriocins, vitamins (vitamin B12, riboflavin, and folate), and short-chain fatty acids are present in the postbiotics prepared from LAB [47,48]. The procedure strains of postbiotics can be naturally found in fermented food, which plays an important role in the production of bioactive metabolites, including those from bacterial (*Lactobacillus*, *Streptococcus*, and *Bifidobacterium*) and fungal species (*Saccharomyces*) [47]. The consumption of postbiotics may help in the enhancement of gut health, anti-inflammatory effects, and prevent respiratory infections.

1.4. Metabolism Characteristics of LAB

LAB have a number of vital metabolic characteristics that support their function in fermentation, including metabolizing sugar (glucose, lactose, and fructose) into lactic acid, bile tolerance, hydrolyzing protein, and antimicrobial properties [49]. Numerous beneficial compounds including organic acid, antibacterial, and exopolysaccharides are produced by metabolism. Lactic acid bacteria can indeed differ across distinct strains in terms of their specific metabolic characteristics and abilities. The genetic composition, growth conditions, external environment, and their adaptation to different environments determine the metabolism characteristics of LAB. For example, *Lactobacillus delbrueckii* subp. bulgaricus commonly used in yogurt production is associated with lactose metabolism, whereas the *Lactobacillus plantarum* found in fermented vegetables is able to metabolize a wide range of sugars [43]. Furthermore, the utilization of specific strains with known metabolic characteristics and improved control over the fermentation parameters are important for producing the desired quality of the product.

1.5. Product Synthesized by LAB

LAB are well known for their ability to decompose macromolecules in various food substances and synthesize lactic acid as the main product. Lactic acid is a significant bio-based compound that contributes to texture, flavor, and nutritional enhancement, and also reduces the pH value of the environment, which inhibits harmful substances. The product synthesized by lactic acid bacteria depends on the types of fermentation carried out [50]. On the other hand, these bacteria are also associated with the potential health attributed to the bioactive peptides, bacteriocins, vitamins, and exopolysaccharides [49].

LAB can yield byproducts that possess bioactivity and contribute to various health-promoting effects, including anti-allergic, modulate respiratory immunity, anti-gastric activity, anti-inflammatory, antimicrobial activity, and anti-oxidant effects [51]. EPS can be produced by several strains of LAB that have been demonstrated in numerous studies to lead to health modulation, such as anti-diabetic, cholesterol-lowering, anti-oxidant, anti-ulcer, and immunomodulatory properties [6]. Aside from these benefits, several strains have the ability to produce enzymes (proteases, lipases, and amylases) with various functionalities that increase nutrient absorption. LAB have been found to produce metabolites that exhibit antimicrobial properties. Finally, the organic acid (acetic acid and lactic acid) and bacteriocins produced by LAB exhibit anti-microbial activities.

1.5.1. Organic Acids

Certain metabolisms, including sugar metabolism, can synthesize various types of organic acid, including lactic acid, acetic acid, butyric acid, and propionic acid, depending on the metabolic pathway. Lactic acid is the main product produced along the metabolic pathway, which is divided into L-lactic acid and D-lactic acid based on the different configurations around the chiral atom. The anaerobic condition throughout the glycolysis pathway results in the production of lactic acid which contributes to the sour flavor of fermented food, such as yogurt and pickles [52]. The fermentation can be divided into homo-lactic fermentation and hetero-lactic fermentation depending on the final product produced [26]. According to Thomas Bintsis [53], homo-lactic fermentation is the process in which lactic acid is the only type of acid, whereas hetero-lactic acid is involved in the production of lactic acid with other byproducts, such as carbon dioxide, ethanol, and acetic acid.

In the process of homo-lactic acid fermentation, glucose acts as the carbon source to create pyruvate through the glycolysis process, which is then subsequently converted to lactic acid by lactate dehydrogenase. The energy was previously generated in the form of NADH. As a result, only lactic acid is produced (one mole of glucose produces two moles of lactic acid and two ATP molecules) [54]. Lactobacillus and Lactococcus are examples of LAB during homo-lactic acid fermentation. Some homo-fermentative microbes can create formic acid under stressful conditions through mixed acid fermentation, including different carbon sources, pH values, or temperatures [55].

In contrast, hetero-lactic acid bacteria decompose the glucose into lactic acid alongside byproducts including acetic acid, ethanol, and carbon dioxide through the phosphoketolase pathway. Leuconostoc and Oenococcus are examples of hetero-lactic acid bacteria. Theoretically in hetero-lactic fermentation, one mole of lactic acid is created when one mole of glucose is decomposed [54]. Glucose 6-phosphate has been transformed into carbon dioxide, ribulose 5-phosphate, and NADPH via the pentose phosphate (PP) pathway [54]. Lactate dehydrogenase plays an important role in the production of lactic acid from pyruvate, and the configuration of the lactic acid is determined by its stereospecificity. L-lactase dehydrogenase is responsible for the synthesis of D-lactic acid, whereas D-lactase dehydrogenase is responsible for the synthesis of D-lactic acid [49]. Other than glucose, lactic acid bacteria can also metabolize fructose, mannose, or galactose. These hexoses serve as alternative carbon sources for the fermentation process [53].

The industrial production of organic acid can be performed by chemical synthesis and fermentation methods for commercial applications. Numerous studies have been carried out by the food industry to improve the purity of lactic acid, as it is important in terms of safety, product stability, flavor, and aroma. Saccharification and fermentation (SF) and separate hydrolysis and fermentation (SHF) are commonly applied in the food industry in order to produce lactic acid with high optical purity and to reduce sugar residue [56]. *Lactobacillus*, *Leuconostoc*, and *Streptococcus* are known to produce various organic acids as end products to prevent the spoilage of food and to improve the taste [57]. Apart from as a flavor enhancer, organic acid in food can be utilized as a food preservation, cleaning, and sanitizing agent due to its antimicrobial and antioxidant properties [57]. Although the LAB mainly produces lactic acid, it can also produce 3-hydroxy propionate, acetate, and succinate. For instance, *Limosilactobacillus reuteri* are capable of producing 3-hydroxypropionic acid as a metabolic byproduct of glycerol metabolism and *Lactiplantibacillus pentosus* can produce acetic acid [49]. The metabolic capacity of LAB to generate organic acid plays a significant role in their probiotic functionality. Figure 1 has shown homolactic fermentation and heterolactic fermentation in LAB.

1.5.2. Bacteriocins

Bacteriocins are antimicrobial peptides or proteins produced by both Gram-positive and Gram-negative bacteria against different closely related bacteria [58]. Lactic acid bacteria have been extensively documented by several studies for their probiotic properties, mycotoxin degradation, and inhibition of pathogenic bacteria [59]. According to Kumariya et al. [60], the bacteriocin function comprises the target bacteria's cell integrity, impedes cellular processes, and interferes with the synthesis of DNA or protein. Various environmental factors, including pH, incubation temperature, nutritional availability, and composition in the growth medium, have a significant impact on bacteriocin synthesis.

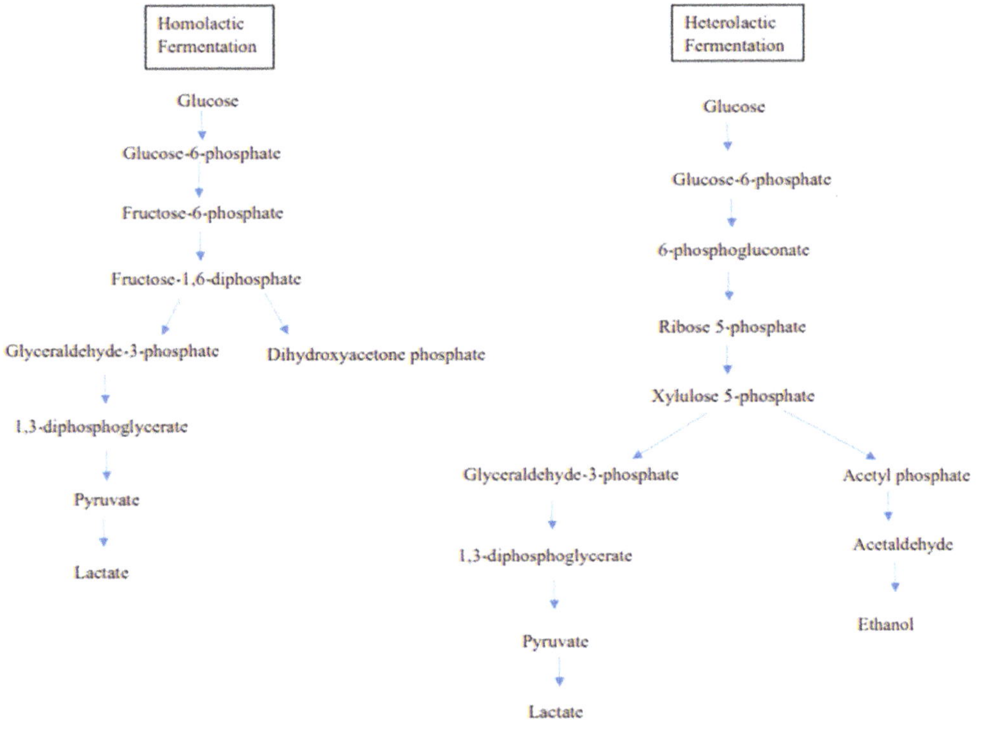

Figure 1. Homolactic fermentation and heterolactic fermentation in LAB [49].

Bacteriocins can be divided into four different classes based on their biochemical and genetic characteristics. Class I bacteriocins, also known as lantibiotics, are small post-translationally modified peptides (<5 kDa) that are characterized based on the presence of lanthionine and methyllanthionine [26]. Nisin produced by *Lactococcus lactis* is indeed one of the well-known examples of Class I bacteriocin that have been extensively studied [61]. According to Svetoslav D. Todorov [59], Class II is the non-lantibiotic, which can be divided into four subclasses depending on their characteristics: Class II a (listeria-active bacteriocins), Class II b (two-peptide complexes), Class II c (the sec-dependent bacteriocins), and Class II d (unclassified small heat-stable non-lathionine bacteriocins). This bacteriocin is small with an amphiphilic helical structure (<10 kDa) that causes cell death by disrupting the integrity of the cell [62]. Class III bacteriocins are the large bacteriocins (>30 kDa) that are synthesized by the *Lactobacillus helveticus*. Bacteriocins generated by bacteria typically need to be secreted from the cell in order to interact with target cells and exhibit their antimicrobial effect [49].

The inherent characteristics of LAB bacteriocins confirm their potential for application in the food industry. Bacteriocins have been extensively used in food preservation, and their potential for use in cancer therapy and oral care [63] is well known as a natural food preservative that is secreted by *Lactococcus lactis* and works against the Listeria monocytogenes [49]. This natural preservative is commonly used in the dairy industry and canned food industry for its antimicrobial properties, improvement of sensory properties, and food quality. For instance, nisin has been reported to inhibit the growth of Gram-positive bacteria, including *Lactilactobacillus sakei* in ham production [64]. In addition, *Lactiplantibacillus plantarum* can prevent and extend the shelf life of raw minced beef by inhibiting the growth of spoilage microorganisms [65].

1.5.3. Vitamins

The metabolites and enzymes produced during the fermentation process can contribute to the bioavailability and production of several vitamins, including vitamin B12, vitamin C, riboflavin (B2), and folate [49]. The capability of the LAB in the synthesis of various vitamins is dependent on the strains and species. According to Zhen Wu et al. [66], *L. plantarum* showed the highest folic acid production compared to other LAB. Moreover, *Lactococcus lactis* and *Streptococcus thermophilus* are common LAB that are used as the starter culture in yogurt production due to their folate synthesis capabilities [67].

Foods that contribute to the bioavailability and synthesis of vitamins during fermentation might be regarded as fortified foods, which are significant to a particular demographic. Folate is a water-soluble vitamin that is essential in the biosynthesis of nucleotides and proteins, including DNA replication [67]. The folate is synthesized from para-aminobenzoic acid (PABA) through a series of reactions. The Lactobacillus strains require the presence of the PABA in the culture medium synthesis of the folate. Several studies have shown that the capability to synthesize folate is dependent on the species, strain, and culture conditions [68]. The development of non-folate-producing LAB is determined by the amount of folate present in the medium, as this strain may need an exogenous source for growth, whereas folate-producing LAB can regulate folate biosynthesis. It can synthesize folate when the medium is deficient in it [69]. Most LAB, especially *Streptococcus* and *Lactobacillus*, are examples that have the ability to synthesize folate [70].

Riboflavin, also known as vitamin B2, is a water-soluble vitamin that serves as the precursor of the flavin adenine dinucleotide (FAD) and flavin mononucleotide (FMN), which are essential for the coenzymes in the redox reactions within the cell [71]. The genes encoding riboflavin synthase in LAB are clustered on the rib operon and contain the genes responsible for the synthesis of riboflavin. The guanosine triphosphate and 5-phosphate ribose are converted into riboflavin, catalyzed by the products of riboflavin synthase genes, namely RibC, RibB, RibA, and RibH. Vitamin synthesis during the fermentation process can be considered as the nutritional fortification of the food. LAB can enhance the nutritional content of fermented food by producing vitamin K and vitamin B12, which contribute to functional food. For example, *S. thermophilus*, used as a starter culture in the dairy product industry, can synthesize folate. Moreover, *Lactococcus laudensis* and *Lactococcus hircilactis* are added to fermented milk production to produce folate that enhances the nutritional value of the product [72].

1.5.4. Exopolysaccharides (EPS)

Exopolysaccharides are biodegradable polymers formed from sugar monosaccharides, which are synthesized and secreted by LAB into their surrounding environment [73] (Pinar Sanibaba, 2016). EPS is important in order to provide the specific texture, viscosity, and probiotic properties of fermented food. These polymers are widely used as stabilizers and emulsifying agents in the food industry due to their water-holding capacity [74]. On the other hand, EPS have been associated with the potential health benefits of existing anti-inflammatory activities, and antitumor and anticancer properties [75]. Several studies have shown that EPS contribute to gut health and promote bacterial colonization by forming a protective matrix [76]; *Lactiplantibacillus plantarum*, *Fructilactobacillus*, *Lactococcus*, *Weissella*, and *Leuconostoc* are especially capable of producing different kinds of EPS based on the strain [77,78].

These polymers can be classified into homopolysaccharides (HoPS) and heteropolysaccharides (HePS) based on the composition of the sugar unit. HoPS are polysaccharides composed of a single type of monosaccharides, whereas HePS consist of different types of monosaccharides [79]. The sugar composition and chain length of the EPS depend on the species of LAB that contribute to the wide range of applications in the food industry [80]. The biosynthesis of HoPS is considered to be a simple process compared to the other polysaccharides syntheses, as it does not involve the active transportation stage in the synthetic pathway. These polymers are synthesized by glycansucrases and fructansucrase, respectively, by allowing the glucose and fructose to act as the glycosyl donors in this synthesis [73]. In contrast, HePS biosynthesis is more complex due to the sugar composition, molecular weight, and linking pattern. It is involved in the sugar activation of the sugar nucleotide precursor formation, polymer chain elongation, branching, and the export of the EPS [49,81,82]. Environmental factors, including pH, temperature, time, and also the strain of the LAB, influence EPS production. For example, Xue Han et al. [83] showed that the combination of *Streptococcus thermophilus* and *Lactobacillus delbrueckii* subsp.bulgaricus produce higher EPS content and better sensory texture of yogurt.

EPS can be considered to be valuable additives, including thickeners, fat substitutes, and texturizers that improve the rheological properties, sensory attributes, and texture of fermented food. EPS include glucans, used as the stabilizer, thickener, and emulsifier in food production to improve the texture and consistency of products. EPS-producing starter culture can be utilized in the production of fermented food to improve the rheological properties of the product. Adding the EPS-producing strain of *Lactobacillus plantarum* improves the texture properties, sensory value, and moisture content of low-fat cheddar cheese [49]. Moreover, *Lactococcus lactic* F-mou synthesis of the EPS shows excellent water-holding capacities, antioxidant properties, and inhibitory effects against pathogenic bacteria [84].

1.5.5. Gamma-Aminobutyric Acid

Gamma-aminobutyric acid (GABA) is a neurotransmitter catalyzed by glutamate decarboxylase (GAD) and pyridoxal-5′-phosphate [49]. This substance is regarded as one of the bioactive compounds created by LAB that may be beneficial to the consumer's health. GABA can enhance the metabolism of the brain cells that regulate the growth of hormone secretion, protein synthesis, fat burning, and blood pressure by improving oxygen delivery and blood flow [85]. The potential health effects of GABA include antidepression by promoting relaxation and reducing anxiety, lowering cholesterol, blood pressure regulation, and anticarcinogenic properties. *Lactobacillus namurensis*, *Lactobacillus paracasei*, and *Lactobacillus brevis* are examples of LAB that have demonstrated the capacity to produce GABA due to the presence of GAD [85]. Moreover, some *Streptococcus thermophilus*, *Lactococcus lactis*, and *Leuconostoc* strains have recently been found to be able to produce GABA [86].

The primary mechanism of intracellular GABA production is the L-Glu decarboxylation process. The decarboxylation reaction of L-glutamate to produce aminobutyric acid is carried out by GAD and pyridoxal-5′-phosphate (cofactor). There have been a number of significant genes identified that control the production of gamma-aminobutyric acid (GABA). According to Chang Jiang Lyu [87], the mutation of the GadA gene in *Levilactobacillus brevis* makes it easier for L-monosodium glutamate (MSG) to be converted to GABA. The inhibition of the GABA aminotransferase showed an increase in GABA production [86]. Additionally, the generation of GABA is influenced by several factors, including temperature, pH, culture composition, and time [88]. The addition of glutamate in the medium shows the increasing concentration of GABA by *L. paracasei* and *L. brevius.* Some microorganisms with a high level of safety that are able to produce GABA can be added to the food to act as fortification products. Additionally, LAB can ferment cheese, yogurt, and milk to act as GABA-enriched goods [89]. Currently, *Levi Lactobacillus brevis* is typically utilized in fermentation to produce GABA, as it is able to convert monosodium glutamate and L-glutamic acid into GABA.

1.5.6. Flavor Substances

In addition to the possible health benefits, fermented food is known for its distinctive flavor. The presence of desirable flavor compounds is the key factor in determining the sensory characteristics of fermented food. According to Coolbear et al. [90], organic acids, alcohols, ketones, and esters are some of the flavoring compounds made by lactic acid bacteria. LAB can function either as the dominant bacteria or combine with other bacteria to produce flavor substances. Generally, the flavor substances are generated by biosynthesis, the enzymatic reaction by the enzyme inside the food, oxidative decomposition by the exposure of heat with oxygen, and the pyrolysis process where the organic compound decomposes because of high temperature [91]. During yogurt production, the flavor substances can be generated by amino acids, fatty acids, and carbohydrates. According to Chen Chen et al. [52], *Lactococcus lactis*, *Lactobacillus species*, and *Streptococcus thermophilus* are responsible for the production of flavor substances, including alcohol and esters. In addition, the addition of LAB in sourdough fermentation contributes to the sour aroma [49]. There are multiple metabolic pathways involved in the synthesis of flavor substances. The citric acid pathway, also known as the Krebs cycle, is one of the metabolic pathways that synthesize intermediate compounds such as citric acid and succinic acid, which then contribute to flavor formation [49]. In addition, sugar metabolism leads to the production of sugar alcohol, which contributes to the sweet taste of the food.

1.6. Application of LAB in Clinical Nutrition

LAB have been used in clinical nutrition for a range of purposes and are well known for their health-promoting qualities. LAB also possess therapeutic properties that are important to enhance human health. Because they have been demonstrated to enhance immune function, promote gut health, and lower the risk of infections, LAB are frequently used as probiotics. To increase nutrient absorption and enhance gut health, LAB are also utilized in enteral and parenteral nutrition [7]. Additionally, it has been established that LAB have anti-inflammatory and antioxidant properties, making them a possible therapeutic choice for a number of conditions, such as inflammatory bowel disease, irritable bowel syndrome, and specific types of cancer.

1.6.1. LAB in the Management of Lactose Intolerance

The symptoms of lactose intolerance, an inherited autosomal recessive trait with incomplete penetrance, are caused by the non-absorbed lactose in the small intestine moving to the colon, where it is metabolized by the intestinal flora and produces short-chain fatty acids and gas, primarily hydrogen (H_2), carbon dioxide (CO_2), and methane (CH_4). Lactose intolerance symptoms vary depending on the residual lactase activity and can cause severe digestive disorders. Colonic adaptation of probiotics is one of the treatments for lactose intolerance [92]. In probiotic preparations, the most common organisms include *Lactobacillus*, *Escherichia*, *Bifidobacterium*, *Bacillus*, *Enterococcus*, *Streptococcus*, and some fungal *Saccharomyces* strains. Cano-Contreras et al. [93] highlighted the efficacy of probiotics in reducing lactose intolerance symptoms. It was suggested that the probiotics help in modifying the pH of the intestine. Some strains of LAB also help in the secretion of bacteria lactase into digestive systems [94].

The administration of probiotic supplementation increases the concentration of β-galactosidase, which helps to alleviate the symptoms of lactose malabsorption. A previous study found the effect of *L. bulgaricus* strains increases the amount of β-galactosidase [95]. Pakdaman et al. [96] demonstrated the effectiveness of LAB in reducing the symptoms of lactose intolerance, whilst Roškar et al. [97] reported a non-significant difference between the placebo group and the probiotic group in reducing lactose intolerance symptoms, particularly diarrhea and flatulence, as compared to the baseline. Nevertheless, this study found an improvement in alleviating the symptoms after LAB consumption. A recent meta-analysis reported the effectiveness of probiotic administration in alleviating lactose intolerance symptoms among adults [98] (Ahn et al. 2023).

1.6.2. LAB in the Treatment of Diarrhea

For decades, malnutrition, particularly undernutrition in hospitalized patients, has received significant attention [99]. Critically ill patients frequently experience non-contagious diarrhea, which has been linked to hospital stay. Antibiotic-associated diarrhea is very common among critically ill patients and it has been shown that microbes are not the major source or risk factor of non-infectious diarrhea [7]. A significant number of microorganisms that are part of the gut microbiota, a complex ecosystem, play important roles in the growth, metabolism, and aging of the host. The composition and phenotype of intestinal microorganisms significantly change during critical illness and the subsequent medical interventions, making the patient more vulnerable to opportunistic infections, even developing System Inflammatory Reaction Syndrome (SIRS) or Multiple Organ Dysfunction Syndrome (MODS) [100]. A wide range of antibiotics are used, which result in the loss of beneficial bacteria from the gut. Diarrhea is a major clinical adverse effect that leads to poor prognoses, such as poor wound healing, electrolyte imbalance, the loss of fluid, hemodynamic instability, and a deficiency of nutrients.

Many beneficial bacteria have been isolated and used to treat gastrointestinal symptoms. LAB have the potential to improve gut health by producing lactic acid, bacteriocins, and short-chain fatty acids, all of which serve to keep the balance of gut microbiota and prevent the overgrowth of harmful bacteria. Probiotics work by inhibiting the action of pathogenic bacteria, aiding immunomodulation, enhancing gut barrier function, and assisting in the release of neurotransmitters. Thus, probiotics aid in the maintenance of a sound gut-brain axis [101,102]

Bacteroidetes and Firmicutes phyla probiotics, including *Lactobacillus*, *Bifidobacterium*, and *Streptococcus salivarius* subsp., have been used to treat a range of intestinal symptoms, including diarrhea-dominant irritable bowel syndrome (IBS), diarrhea, inflammatory bowel disease, and antibiotic-induced diarrhea. A recent systematic review and meta-analysis included studies (all conducted in China) that showed that probiotic significantly reduces gastrointestinal complications in severe stroke patients, according to a new systematic review and meta-analysis of studies (only conducted in China) $p < 0.0001$ [103]. Skrzydło-Radomańska et al. [104] reported that the use of multi-strain synbiotic preparations was associated with a significant improvement in symptoms of IBS. A pilot randomized study also revealed the effectiveness of sporulated bacillus in alleviating the symptoms of diarrhea among patients on enteral nutrition compared to fiber-enriched formula alone [105].

Two recent meta-analyses reported different findings. A meta-analysis study by Alsuwaylihi and McCullough [7] found a potential effect of probiotics (*Lactobacillus rhamnosus* GG and *Bacillus cereus* on clinical or diarrheal outcomes in critically ill patients. Lee et al. [106], in another meta-analysis study, did not support the beneficial effect of probiotics on the treatment of diarrhea in critically ill patients. As a result, the optimum dosage and effectiveness of probiotics on the reduction of diarrhea remains inconclusive.

1.6.3. Immunomodulatory Effects of LAB

Utilizing LAB in enteral nutrition has drawn more attention in recent years, especially in critically ill patients who receive nutrition through a feeding tube. It is also worth noting that the use of LAB in enteral nutrition should be closely monitored and tailored to each patient's medical history, health state, and other variables. In vivo evidence of probiotics' ability to suppress the generation of proinflammatory cytokines and stimulate IgA secretion has been documented in several investigations in recent years [107]. The gastrointestinal tract is an essential microbiologically active ecosystem that plays a crucial role in the working of the mucosal immune system.

LAB, including *Lactobacillus* and *Streptococcus lactis*, have shown a positive effect in terms of improving the immunity of individuals. Wei et al. [108] explored the clinical effect of compound LAB capsules with Escitalopram (a medicine used to treat depression) on small intestinal bacterial overgrowth (SIBO) in patients with depression and diabetes. CD^{3+} and CD^{+4} showed a greater increment among individuals supplemented with LAB compared to the control. *Lactobacillus* and *Streptococcus lactis* act on the body to multiply in the intestinal tract, increase lactic acid production, and inhibit the reproduction of spoilage bacteria. It was found that the modulation of the immune system by gut microbiota is via the production of molecules with immunomodulatory and anti-inflammatory functions that can stimulate immune cells. Immunomodulatory effects were produced by the probiotic interaction with epithelial cells and dendritic cells, as well as with monocytes/macrophages and lymphocytes [109].

LABs were found to synthesize low molecular weight compounds such as organic acid and large molecular weight antimicrobial compounds (known as Bacteriocins). Bacteriocins produced by LAB probiotics exhibit strong inhibitory effects against pathogenic Gram-negative bacteria, such as H. pylori. Oral administration of LAB increases Paneth cells based in the small intestine [110]. Aggravations or alterations of the normal intestinal microflora in the gastrointestinal gut are the common cause of inflammatory bowel diseases such as Crohn's disease. Additional probiotics in individuals' diets have been shown to replenish or modify gut microflora [111].

Ventilator-associated pneumonia (VAP) is the most prevalent fatal complication of nosocomial infection (NI) in intensive care unit (ICU) patients. Beneficial bacteria play an important role in maintaining the intestinal barrier and host immunity. In a meta-analysis, Batra et al. [112] found that probiotics supplementation decreased the incidence of VAP, the length of mechanical ventilation, the length of ICU stays, and in-hospital mortality among ventilated critically sick ICU patients. A previous study highlighted the supplementation of two capsules of probiotic containing LAB had a lower incidence of statistically microbiologically confirmed VAP [113]. Despite high-quality random trials on the efficacy of LAB in preventing VAP, this dietary therapy remains highly controversial in the reduction of VAP among patients with trauma or other critical illness [114].

The disease known as sepsis, on the other hand, can arise as a side effect of an infection and is potentially fatal. It happens when the body's reaction to an infection is thrown off balance, resulting in a systemic inflammatory response across the body. Numerous studies have shown that nutritional therapy for malnourished individuals reduces the chances of infection complications, wound inflammation, and mortality [115]. In critically ill patients, the commensal microbiota deteriorates, in which most ICU patients are associated with infections and mortality [116]. Sepsis among critically ill patients has been associated with lowering microbiota in the gut. The integrity of the intestinal epithelial barrier and the absorptive function of the intestinal mucosa may be compromised as a result of changes in the intestinal microbial composition during severe illness. LAB has anti-inflammatory properties and may aid in reducing the risk of infection and sepsis in critically ill patients [117]. The isolated bacteria are also termed "probiotics". The administration of synbiotics (probiotic and prebiotics containing lactic acid bacteria) containing the *B. breve* strain and the *L. casei* strain, in an amount of 3 g per day, was initiated within 3 days after admission through enteral feeding. The synbiotic is used to inhibit pathogenic bacteria and toxins through signal interaction and prevent septic complications. In this study, the numbers of *Bifidobacterium* sp. and total *Lactobacillus* sp. in the synbiotic group also showed an increment [118]. Shimizu et al. [118] reported that the use of the synbiotics of LAB had fewer complications of diarrhea and ventilator-associated diarrhea. The usage of probiotics resulted in a better outcome in terms of lowering overall ICU infection rates, particularly VAP [7].

1.6.4. LAB and Hepatoprotective Effects

Ethanol exposure is strongly linked to alcoholic liver disease (ALD), a chronic illness with the highest incidence and mortality rate in the world [119]. ALD includes alcoholic fatty liver, alcoholic steatohepatitis, alcoholic hepatitis, alcoholic fibrosis, alcoholic cirrhosis, and alcoholic hepatocellular carcinoma. In recent years, studies have found a close relationship between alcohol and gut microbiota [120]. Alcohol can increase intestinal permeability, which leads to liver damage with the release of reactive oxygen species (ROS), adhesion molecules, chemokines, and proinflammatory cytokines; therefore, the use of probiotics may limit the progression of ALD by changing the intestinal bacteria [19].

Bakhshimoghaddam et al. [121] reported a reduction in the liver function test profile including the serum concentration of alanine aminotransferase, aspartate aminotransferase, alkaline phosphatase, and γ-glutamyltransferase among non-alcoholic fatty liver disease

(NAFLD) patients receiving supplementations of *Bifidobacterium animalis* compared to the control group. Nevertheless, Mohamad Nor et al. [122] found no significant effect of combining LAB containing *Lactobacillus* and *Bifidobacterium* on the liver function test profile. Variceal bleeding has a high incidence among patients with liver cirrhosis and leads to a high risk of mortality and morbidity.

1.6.5. LAB for Prevention and as a Potential Natural Anti-tumour Drug

Exopolysaccharides produced by lactic acid bacteria, as one of the most important functional components of LAB metabolic products, have attracted considerable attention in recent years due to their unique physicochemical properties [123] and their ability to modulate cancer cell proliferation and apoptosis both in vitro and in vivo [124]. The effectiveness of LAB in clinical trials has been limited and inconclusive. More clinical trials are necessary to establish the potential benefits of LAB in the prevention and treatment of cancer. In a study by Zhao et al. [125], probiotic-enriched nutrition formula among gastric cancer patients who received enteral nutrition had a lower number of surgery side effects, such as diarrhea and intestinal disorder, compared to those with fiber-free or fiber-enriched nutrition formula.

Recent studies have indicated the beneficial role of probiotics in the prevention of carcinogenesis and have presented new promising therapeutic options. However, the safety used for cancer patients remains inconclusive [126]. More research is needed to conclude the potential benefits of probiotics for cancer patients.

1.6.6. LAB in the Management of Glycemic Control

Diabetes remains an overwhelming health problem worldwide despite advancements in healthcare management. Probiotic supplements do not cause clinically significant decreases in Hemoglobin A1c (HbA1c) levels in people with type 2 diabetes, but they do cause marginally clinically significant reductions in fasting glucose and fasting insulin levels. Multi-strain and high-dose probiotics have had a larger positive impact on glucose homeostasis compared to single-strain and low-dose probiotics. Probiotic therapy may also be more successful in people who are older and have a high baseline Body Mass Index (BMI) [127]. The supplementation of 10^8 CFU of *L. casei* 01 for 8 weeks significantly reduced the serum fetuin-A level, fasting blood sugar, insulin concentration, and insulin resistance [128]. In contrast, the supplementation of probiotic yogurt containing *Lactobacillus acidophilus* and *Bifidobacterium lactis* showed no significant effect on fasting blood glucose, whilst there was a reduction in hemoglobin A1c compared to the placebo group [129]. This discrepancy might be due to the period of study, the dosage, and the use of LAB. Further study is needed with longer interventions to better conclude the effectiveness of LAB on blood glucose control. Table 2 has summaries the functional properties of LAB in the management of nutrition.

Table 2. Functional properties of LAB in the management of nutrition.

Therapeutic Effects	Lactic Acid Bacteria (LAB) Strain	Remarks	References
Lactose intolerance	*Lactobacillus acidophilus*,	Method: Supplementation *L. acidophilus* Results: Abdominal symptom (LAB < control)	[96]
	B. animalis, *Lactobacillus plantarum*	Methods: Supplementation of *B. animalis* Results: Abdominal symptoms (no significant difference)	[84]

Table 2. *Cont.*

Therapeutic Effects	Lactic Acid Bacteria (LAB) Strain	Remarks	References
Lactose intolerance	*Lactobacillus plantarum, P. acidilactici*	Method: Supplementation of *Lactobacillus plantarum* and *P. acidilactici* among lactose intolerance patients Results: Total symptom score of lactose intolerance (LAB < control)	[93]
Gastrointestinal problem: diarrhea	*Baccilus cereus*	Method: Supplementation of 20 mL/day *Baccilus cereus* or soluble fiber (control) among patients with diarrhea on enteral feeding Results: Ceasing the diarrhea incident (no significant difference), duration to stop diarrhea (*B. Cerius* group < control)	[105]
	Lactobacillus rhamnosus, Lactobacillus acidophilus, Bifidobacterium lactis, Bifidobacterium longum, Bifidobacterium bifidum	Method: Synbiotics supplementation among diarrhea-dominant IBS for 8 weeks Results: After intervention, feeling of incomplete bowel movements, flatulence, pain, stool pressure, and diarrheal stools (synbiotics group < control)	[104]
Immunomodulatory effect	*Bifidobacterium breve, Lactobacillus casei*	Method: 3 g supplementation of synbiotics (*Bifidobacterium breve* and *Lactobacillus casei*) within 3 days after admission Results: Enteritis and penumonia incidence lowered in synbiotics group compared to control	[118]
	Lactobacillus and Streptococcus lactis	Method: Lactic acid bacteria capsule among depression and diabetes patient Result: Reduction of self-rating anxiety scale, IL-2 and TNF-α, fasting plasma (LAB > control), and increment of CD^{+4} (LAB > control) Adverse effect LAB < control)	[108]
	Lactobacillus casei, Lactobacillus acidophilus, Lactobacillus rhamnosus, Lactobacillus bulgaricus, Bifidobacterium breve, Bifidobacterium longum, Streptococcus thermophiles	Methods: Supplementation of 1 capsule/12 h among VAP multi-trauma patients Results: VAP (intervention group < control)	[130]
	Lactobacillus rhamnosus	Method: Supplementation of 2×10^9 Colony Forming Units (CFU) of *Lactobacillus rhamnosus GG* on a twice daily basis among ventilated medical ICU patients Results: VAP (no significant difference between LAB and the control)	[131]

Table 2. *Cont.*

Therapeutic Effects	Lactic Acid Bacteria (LAB) Strain	Remarks	References
Immunomodulatory effect	*L rhamnosus GG*	Method: Enteral *L rhamnosus GG* twice daily among patients on ventilation Results: VAP incidence (no significant difference between both the intervention group and the control)	[132]
Hepatoprotective effect	*Bifidobacterium animalis*	Method: Supplementation of 300 g synbiotics yogurt (B. animalis and inulin) or conventional (control) among NAFLD patients Results: Grades of NAFLD (synbiotics group < control), reduction in serum concentration of alanine aminotransferase, aspartate aminotransferase, alkaline phosphatase, and γ-glutamyltransferase (synbiotics group > control)	[121]
	Lactobacillus, Bifidobacterium	Methods: Supplementation of probiotics sachet or placebo for 6 months among NAFLD patients Results: No significant difference in LiverFAST analysis (steatosis, fibrosis, and inflammation scores), alanine aminotransferase	[122]
Treatment of cancer	*Bifidobacterium, lactobacillus*	Method: Gastric cancer patient receiving fiber-free nutrition formula (FF group), fiber-enriched nutrition formula (FE group), and fiber- and probiotic-enriched nutrition formula (FEp group) Results: The FEP group had the lowest number of diarrhea and intestinal disorders. No significant difference in the lymphocyte count, albumin, prealbumin, and transferrin levels	[125]
	Bifidobacteria, Lactobacillus	Method: Supplementation of probiotics + glucose solution or glucose solution (control) among colorectal cancer patients undergoing radical resection Results: Increase in intestinal micro-ecological environment and strengthening of the intestinal mucosal barrier function (glucose solution + probiotic group > glucose group), duration of early recovery of inflammatory response (glucose solution + probiotic group > glucose group)	[120]

Table 2. *Cont.*

Therapeutic Effects	Lactic Acid Bacteria (LAB) Strain	Remarks	References
Glycemic control	*Lactobacillus casei*	Method: 10^8 CFU of *L. casei* supplementation for 8 weeks among type 2 diabetes mellitus Result: Serum fetuin-A level, fasting blood sugar, insulin concentration, and insulin resistance significantly decrease among *L. casei* supplementation compared to the control	[128]
	Lactobacillus acidophilus, *Bifidobacterium lactis*	Methods: 200 g/d yogurt containing probiotic 4.65×10^6 CFU/g or placebo group received 200 g/d conventional yogurt Results: No significant different in fasting plasma glucose (FPG), hemoglobin A1c (HbA1c)	[129]

1.7. Challenge of Lactic Acid Bacteria as a Food Nutrient

While lactic acid bacteria have shown promise in the food industry, their use as food nutrients in clinical nutrition poses additional challenges. Not all clinical trials have shown improvement in the health of individuals receiving probiotic medication, and very few have indicated that probiotic strains may be the causal agents of opportunistic illnesses. These very uncommon illnesses are mostly seen in higher-risk categories, such as immunocompromised people. In immunocompromised individuals, there is a chance that specific LAB strains might increase their risk of infection or sepsis, among other safety issues. It is also important to ensure the LAB used as dietary nutrients do not compromise the efficiency or absorption of other drugs [133]. Future studies should emphasize the drug-nutrient relationship in the creation and delivery of the therapeutic effects of LAB. More investigations into the probiotic-drug and probiotic-gut microbiota interactions are required in the near future because the precise mechanisms are still partially understood [134]. Additionally, LABs are known to have immune-modulating effects on the host, making them a prospective therapeutic and preventative choice for a number of illnesses, including inflammatory disease. Understanding the genus and species of the probiotics is crucial to attaining the desired effects on the host, since probiotic effects vary depending on the dose, circumstance, and strain [135]. LAB usage can be considered generally safe for healthy individuals. There is an urgent need for further evidence on adverse events, particularly in immunocompromised hosts and vulnerable groups in both the short and long term [136].

2. Conclusions

Understanding the characteristics of LAB and their application in the management of nutrition is important for ensuring optimal health outcomes. These nonpathogenic bacteria are useful in technological and industrial processes. LAB is characterized as a phylogenetic and functionally diverse taxonomic order of bacteria. By modulating the gut microbiota, LAB supports better digestion, increase nutritional absorption, improve antimicrobial properties, and boost immunological function. The safety profiles of various LAB as a function of various genera, species, and strains, as well as their applicability in a variety of people or populations at risk, have gained substantial interest. The use of LAB to provide health benefits to the host requires the specification of the dosage regimens and the duration of use as recommended by the manufacturer of each individual strain or product based on scientific evidence, and as permitted in the country of sale, as per the Joint FAO/WHO (2002) guidelines on LAB. The minimal daily dose required for any

LAB-containing product to bestow a particular health benefit or advantage should also be specified. Clear proof of this goal should come from in vitro, animal, or human clinical investigations, if feasible.

Author Contributions: Conceptualization, B.N.A.H., N.J.X. and S.N.H.O.; resources, B.N.A.H., N.J.X. and S.N.H.O.; writing—original draft preparation, B.N.A.H., N.J.X. and S.N.H.O.; writing—review and editing, B.N.A.H., N.J.X. and S.N.H.O.; visualization, B.N.A.H., N.J.X. and S.N.H.O. All authors have read and agreed to the published version of the manuscript.

Funding: This research received no external funding.

Data Availability Statement: Not applicable.

Conflicts of Interest: The authors declare no conflict of interest.

References

1. Sørensen, H.M.; Rochfort, K.D.; Maye, S.; MacLeod, G.; Brabazon, D.; Loscher, C.; Freeland, B. Exopolysaccharides of Lactic Acid Bacteria: Production, Purification and Health Benefits towards Functional Food. *Nutrients* **2022**, *14*, 2938. [CrossRef] [PubMed]
2. Hasler, C.M.; Brown, A.C. Position of the American Dietetic Association: Functional Foods. *J. Am. Diet. Assoc.* **2009**, *109*, 735–746. [PubMed]
3. EFSA. *Call for Technical and Toxicological Data on Pullulan (E 1204) Authorised as a Food Additive in the EU*; EFSA: Parma, Italy, 2017.
4. Kruger, M.C.; Ha, P.C.; Todd, J.M.; Schollum, L.M.; Ma, J.; Qin, G.; Lau, E. High-Calcium, Vitamin D Fortified Milk Is Effective in Improving Bone Turnover Markers and Vitamin D Status in Healthy Postmenopausal Chinese Women. *Eur. J. Clin. Nutr.* **2012**, *66*, 856–861. [CrossRef] [PubMed]
5. Rezac, S.; Kok, C.R.; Heermann, M.; Hutkins, R. Fermented foods as a dietary source of live organisms. *Front. Microbiol.* **2018**, *9*, 1785. [CrossRef] [PubMed]
6. Mathur, H.; Beresford, T.P.; Cotter, P.D. Health benefits of lactic acid bacteria (LAB) fermentates. *Nutrients* **2020**, *12*, 1679. [CrossRef]
7. Alsuwaylihi, A.S.; McCullough, F. The safety and efficacy of probiotic supplementation for critically ill adult patients: A systematic review and meta-analysis. *Nutr. Rev.* **2023**, *81*, 322–332. [CrossRef]
8. Linares, D.M.; Gómez, C.; Renes, E.; Fresno, J.M.; Tornadijo, M.E.; Ross, R.P.; Stanton, C. Lactic acid bacteria and bifidobacteria with potential to design natural biofunctional health-promoting dairy foods. *Front. Microbiol.* **2017**, *8*, 846. [CrossRef]
9. Bourrie, B.C.T.; Willing, B.P.; Cotter, P.D. The microbiota and health promoting characteristics of the fermented beverage kefir. *Front. Microbiol.* **2016**, *7*, 647. [CrossRef]
10. Savaiano, D.A.; Hutkins, R.W. Yogurt, cultured fermented milk, and health: A systematic review. *Nutr. Rev.* **2021**, *79*, 599–614. [CrossRef]
11. Ağagündüz, D.; Yılmaz, B.; Şahin, T.Ö.; Güneşliol, B.E.; Ayten, Ş.; Russo, P.; Spano, G.; Rocha, J.M.; Bartkiene, E.; Özogul, F. Dairy Lactic Acid Bacteria and Their Potential Function in Dietetics: The Food–Gut-Health Axis. *Foods* **2021**, *10*, 3099. [CrossRef]
12. Shehata, M.G.; El Sohaimy, S.A.; El-Sahn, M.A.; Youssef, M.M. Screening of isolated potential probiotic lactic acid bacteria for cholesterol lowering property and bile salt hydrolase activity. *Ann. Agric. Sci.* **2016**, *61*, 65–75. [CrossRef]
13. Yu, A.O.; Leveau, J.H.J.; Marco, M.L. Abundance, diversity and plant-specific adaptations of plant-associated lactic acid bacteria. *Environ. Microbiol. Rep.* **2020**, *12*, 16–29. [CrossRef] [PubMed]
14. Mulaw, G.; Sisay Tessema, T.; Muleta, D.; Tesfaye, A. In vitro evaluation of probiotic properties of lactic acid bacteria isolated from some traditionally fermented Ethiopian food products. *Int. J. Microbiol.* **2019**, *2019*, 7179514–7179521. [CrossRef] [PubMed]
15. Davoren, M.J.; Liu, J.; Castellanos, J.; Rodríguez-Malavé, N.I.; Schiestl, R.H. A novel probiotic, *Lactobacillus johnsonii* 456, resists acid and can persist in the human gut beyond the initial ingestion period. *Gut Microbes* **2019**, *10*, 458–480. [CrossRef]
16. Ramos, C.L.; Thorsen, L.; Schwan, R.F.; Jespersen, L. Strain-specific probiotics properties of *Lactobacillus fermentum*, *Lactobacillus plantarum* and *Lactobacillus brevis* isolates from Brazilian food products. *Food Microbiol.* **2013**, *36*, 22–29. [CrossRef]
17. King, S.; Tancredi, D.; Lenoir-Wijnkoop, I.; Gould, K.; Vann, H.; Connors, G.; Sanders, M.; Linder, J.; Shane, A.; Merenstein, D. Does probiotic consumption reduce antibiotic utilization for common acute infections? A systematic review and meta-analysis. *Eur. J. Public Health* **2019**, *29*, 494–499. [CrossRef]
18. Yasmin, I.; Saeed, M.; Khan, W.A.; Khaliq, A.; Chughtai, M.F.J.; Iqbal, R.; Tehseen, S.; Naz, S.; Liaqat, A.; Mehmood, T.; et al. In Vitro Probiotic Potential and Safety Evaluation (Hemolytic, Cytotoxic Activity) of *Bifidobacterium* Strains Isolated from Raw Camel Milk. *Microorganisms* **2020**, *8*, 354. [CrossRef]
19. Li, M.; Wang, Y.; Cui, H.; Li, Y.; Sun, Y.; Qiu, H.-J. Characterization of Lactic Acid Bacteria Isolated From the Gastrointestinal Tract of a Wild Boar as Potential Probiotics. *Front. Vet. Sci.* **2020**, *7*, 49. [CrossRef]
20. Khedid, K.; Faid, M.; Mokhtari, A.; Soulaymani, A.; Zinedine, A. Characterization of lactic acid bacteria isolated from the one humped camel milk produced in Morocco. *Microbiol. Res.* **2009**, *164*, 81–91. [CrossRef]
21. Khan, U.; Selamoglu, Z. Use of Enzymes in Dairy Industry: A Review of Current Progress. *Arch. Razi Inst.* **2020**, *75*, 131–136.

22. Reuben, R.C.; Roy, P.C.; Sarkar, S.L.; Alam, A.R.U.; Jahid, I.K. Characterization and evaluation of lactic acid bacteria from indigenous raw milk for potential probiotic properties. *J. Dairy Sci.* **2020**, *103*, 1223–1237. [CrossRef] [PubMed]
23. Marcial-Coba, M.S.; Knøchel, S.; Nielsen, D.S. Low-moisture food matrices as probiotic carriers. *FEMS Microbiol. Lett.* **2019**, *366*, fnz006. [CrossRef] [PubMed]
24. Gao, J.; Li, X.; Zhang, G.; Sadiq, F.A.; Simal-Gandara, J.; Xiao, J.; Sang, Y. Probiotics in the dairy industry—Advances and opportunities. *Compr. Rev. Food Sci. Food Saf.* **2021**, *20*, 3937–3982. [CrossRef]
25. Campana, R.; van Hemert, S.; Baffone, W. Strain-specific probiotic properties of lactic acid bacteria and their interference with human intestinal pathogens invasion. *Gut Pathog.* **2017**, *9*, 12. [CrossRef]
26. Mokoena, M.P. Lactic acid bacteria and their Bacteriocins: Classification, biosynthesis and applications against Uropathogens: A mini-review. *Molecules* **2017**, *22*, 1255. [CrossRef] [PubMed]
27. Osborne, J. Advances in microbiological quality control. In *Managing Wine Quality*; Woodhead Publishing: Sawston, UK, 2022; pp. 207–241.
28. Hossain, Z. Bacteria: Streptococcus. In *Encyclopedia of Food Safety*; Academic Press: Cambridge, MA, USA, 2014; pp. 535–545.
29. Cai, Y.; Yang, J.; Pang, H.; Kitahara, M. *Lactococcus fujiensis* sp. Nov., a lactic acid bacterium isolated from vegetable matter. *Int. J. Syst. Evol. Microbiol.* **2011**, *61*, 1590–1594. [CrossRef]
30. Onyeaka, H.N.; Nwabor, O.F. Lactic acid bacteria and bacteriocins as biopreservatives. In *Food Preservation and Safety of Natural Products*; Academic Press: Cambridge, MA, USA, 2022; pp. 147–162.
31. Kajihara, T.; Nakamura, S.; Iwanaga, N.; Oshima, K.; Takazono, T.; Miyazaki, T.; Izumikawa, K.; Yanagihara, K.; Kohno, N.; Kohno, S. Clinical characteristics and risk factors of enterococcal infections in Nagasaki, Japan: A retrospective study. *BMC Infect. Dis.* **2015**, *15*, 426. [CrossRef]
32. Ezechukwu, I.; Singal, M.; Igbinosa, O. Aerococcus Viridans: Case report, microbiology, and literature review. *Am. J. Case Rep.* **2019**, *20*, 697–700. [CrossRef] [PubMed]
33. Lorenzo, J.M.; Munekata, P.E.; Dominguez, R.; Pateiro, M.; Saraiva, J.A.; Franco, D. Main groups of microorganisms of relevance for food safety and stability. In *Innovative Technologies for Food Preservation*; Academic Press: Cambridge, MA, USA, 2018; pp. 53–107.
34. Bartowsky, E.J. Wines. Malolactic Fermentation. In *Encyclopedia of Food Microbiology*, 2nd ed.; Batt, C.A., Tortorello, M.L., Eds.; Academic Press: Oxford, UK, 2014; pp. 800–804.
35. Fusco, V.; Quero, G.M.; Cho, G.; Kabisch, J.; Meske, D.; Neve, H.; Bockelmann, W.; Franz, C.M. The genus Weissella: Taxonomy, ecology and biotechnological potential. *Front. Microbiol.* **2015**, *6*, 155. [CrossRef]
36. Endo, A.; Maeno, S.; Tanizawa, Y.; Kneifel, W.; Arita, M.; Dicks, L.; Salminen, S. Fructophilic lactic acid bacteria, a unique group of fructose-fermenting microbes. *Appl. Environ. Microbiol.* **2018**, *84*, e01290-18. [CrossRef]
37. Guindo, C.O.; Morsli, M.; Bellali, S.; Drancourt, M.; Grine, G. A *Tetragenococcus halophilus* human gut isolate. *Curr. Res. Microb. Sci.* **2022**, *3*, 100112. [CrossRef]
38. Fu, S.; Xia, W.; Wang, Q.; Rahman, M.M.; Hao, J.; Ye, S.; Liu, Y.; Li, R. Genomic characterization and pathogenicity analysis of the probiotic *Vagococcus lutrae* strain VL-18 causing severe skin lesions in warm-blooded animals. *Aquaculture* **2020**, *523*, 735166. [CrossRef]
39. Şanlier, N.; Gökcen, B.B.; Sezgin, A.C. Health benefits of fermented foods. *Crit. Rev. Food Sci. Nutr.* **2019**, *59*, 506–527. [CrossRef]
40. Tamang, J.P.; Shin, D.; Jung, S.; Chae, S. Functional properties of microorganisms in fermented foods. *Front. Microbiol.* **2016**, *7*, 578. [CrossRef]
41. Touret, T.; Oliveira, M.; Semedo-Lemsaddek, T. Putative probiotic lactic acid bacteria isolated from sauerkraut fermentations. *PLoS ONE* **2018**, *13*, e0203501. [CrossRef] [PubMed]
42. Shah, A.M.; Tarfeen, N.; Mohamed, H.; Song, Y. Fermented foods: Their health-promoting components and potential effects on gut microbiota. *Fermentation* **2023**, *9*, 118. [CrossRef]
43. Behera, S.S.; Ray, R.C.; Zdolec, N. *Lactobacillus plantarum* with functional properties: An approach to increase safety and shelf-life of fermented foods. *BioMed Res. Int.* **2018**, *2018*, 9361614. [CrossRef] [PubMed]
44. Abbasi, A.; Rad, A.H.; Ghasempour, Z.; Sabahi, S.; Kafil, H.S.; Hasannezhad, P.; Rahbar Saadat, Y.; Shahbazi, N. The biological activities of postbiotics in gastrointestinal disorders. *Crit. Rev. Food Sci. Nutr.* **2022**, *62*, 5983–6004. [CrossRef]
45. Rad, A.H.; Abbasi, A.; Kafil, H.S.; Ganbarov, K. Potential Pharmaceutical and Food Applications of Postbiotics: A Review. *Curr. Pharm. Biotechnol.* **2020**, *21*, 1576–1587. [CrossRef]
46. Sahadeva, R.P.K.; Leong, S.F.; Chua, K.H.; Tan, C.H.; Chan, H.Y.; Tong, E.V.; Wong, S.Y.W.; Chan, H.K. Survival of commercial probiotic strains to pH and bile. *Int. Food Res. J.* **2011**, *18*, 1515–1522.
47. Thorakkattu, P.; Khanashyam, A.C.; Shah, K.; Babu, K.S.; Mundanat, A.S.; Deliephan, A.; Deokar, G.S.; Santivarangkna, C.; Nirmal, N.P. Postbiotics: Current trends in food and pharmaceutical industry. *Foods* **2022**, *11*, 3094. [CrossRef] [PubMed]
48. Moradi, M.; Kousheh, S.A.; Almasi, H.; Alizadeh, A.; Guimarães, J.T.; Yılmaz, N.; Lotfi, A. Postbiotics produced by lactic acid bacteria: The next frontier in food safety. *Compr. Rev. Food Sci. Food Saf.* **2020**, *19*, 3390–3415. [CrossRef]
49. Wang, Y.; Wu, J.; Lv, M.; Shao, Z.; Hungwe, M.; Wang, J.; Bai, X.; Xie, J.; Wang, Y.; Geng, W. Metabolism characteristics of lactic acid bacteria and the expanding applications in food industry. *Front. Bioeng. Biotechnol.* **2021**, *9*, 612285. [CrossRef] [PubMed]
50. Nout, M.J.R. Food Technologies: Fermentation. In *Encyclopedia of Food Safety*; Foods, Materials, Technologies and Risks; Holt, S., Phadnis, R., Eds.; Academic Press: Cambridge, MA, USA, 2014; Volume 3, pp. 168–177.

51. Sharma, R.; Bagrodia, D.; Pandey, M.; Sanodiya, B.S.; Thakur, G.S. Efficacy and potential of lactic acid bacteria modulating human health. *Int. J. Pharma Bio Sci.* **2012**, *3*, 935–948.
52. Chen, C.; Zhao, S.; Hao, G.; Yu, H.; Tian, H.; Zhao, G. Role of lactic acid bacteria on the yogurt flavor: A review. *Int. J. Food Prop.* **2017**, *20*, S316–S330. [CrossRef]
53. Bintsis, T. Lactic acid bacteria as starter cultures: An update in their metabolism and genetics. *AIMS Microbiol.* **2018**, *4*, 665–684. [CrossRef]
54. Eiteman, M.A.; Ramalingam, S. Microbial production of lactic acid. *Biotechnol. Lett.* **2015**, *37*, 955–972. [CrossRef]
55. Castillo Martinez, F.A.; Balciunas, E.M.; Salgado, J.M.; Domínguez González, J.M.; Converti, A.; Oliveira, R.P. Lactic acid properties, applications and production: A review. *Trends Food Sci. Technol.* **2013**, *30*, 70–83. [CrossRef]
56. Xu, Q.; Zang, Y.; Zhou, J.; Liu, P.; Li, X.; Yong, Q.; Ouyang, J. Highly efficient production of D-lactic acid from chicory-derived inulin by lactobacillus bulgaricus. *Bioprocess Biosyst. Eng.* **2016**, *39*, 1749–1757. [CrossRef]
57. Punia Bangar, S.; Suri, S.; Trif, M.; Ozogul, F. Organic acids production from lactic acid bacteria: A preservation approach. *Food Biosci.* **2022**, *46*, 101615. [CrossRef]
58. Hernández-González, J.C.; Martínez-Tapia, A.; Lazcano-Hernández, G.; García-Pérez, B.E.; Castrejón-Jiménez, N.S. Bacteriocins from lactic acid bacteria. A powerful alternative as antimicrobials, probiotics, and Immunomodulators in veterinary medicine. *Animals* **2021**, *11*, 979. [CrossRef] [PubMed]
59. Todorov, S.D. Bacteriocins from lactobacillus plantarum production, genetic organization and mode of action: Produção, organização genética E modo de ação. *Braz. J. Microb.* **2009**, *40*, 209–221. [CrossRef] [PubMed]
60. Kumariya, R.; Garsa, A.; Rajput, Y.; Sood, S.; Akhtar, N.; Patel, S. Bacteriocins: Classification, synthesis, mechanism of action and resistance development in food spoilage causing bacteria. *Microbial. Pathog.* **2019**, *128*, 171–177. [CrossRef] [PubMed]
61. Shin, J.; Gwak, J.; Kamarajan, P.; Fenno, J.; Rickard, A.; Kapila, Y. Biomedical applications of nisin. *J. Appl. Microbiol.* **2016**, *120*, 1449–1465. [CrossRef]
62. Mauriello, G. Control of microbial activity using antimicrobial packaging. In *Antimicrobial Food Packaging*; Academic Press: San Diego, CA, USA, 2016; pp. 141–152.
63. Negash, A.W.; Tsehai, B.A. Current applications of Bacteriocin. *Int. J. Microbiol.* **2020**, *2020*, 4374891. [CrossRef]
64. Kalschne, D.L.; Geitenes, S.; Veit, M.R.; Sarmento, C.M.; Colla, E. Growth inhibition of lactic acid bacteria in ham by nisin: A model approach. *Meat Sci.* **2014**, *98*, 744–752. [CrossRef]
65. Trabelsi, I.; Ben Slima, S.; Ktari, N.; Triki, M.; Abdehedi, R.; Abaza, W.; Moussa, H.; Abdeslam, A.; Ben Salah, R. Incorporation of probiotic strain in raw minced beef meat: Study of textural modification, lipid and protein oxidation and color parameters during refrigerated storage. *Meat Sci.* **2019**, *154*, 29–36. [CrossRef]
66. Wu, Z.; Wu, J.; Cao, P.; Jin, Y.; Pan, D.; Zeng, X.; Guo, Y. Characterization of probiotic bacteria involved in fermented milk processing enriched with folic acid. *J. Dairy Sci.* **2017**, *100*, 4223–4229. [CrossRef]
67. Capozzi, V.; Russo, P.; Dueñas, M.T.; López, P.; Spano, G. Lactic acid bacteria producing B-group vitamins: A great potential for functional cereals products. *Appl. Microbiol. Biotechnol.* **2012**, *96*, 1383–1394. [CrossRef]
68. Kariluoto, S.; Aittamaa, M.; Korhola, M.; Salovaara, H.; Vahteristo, L.; Piironen, V. Effects of yeasts and bacteria on the levels of folates in rye sourdoughs. *Int. J. Food Microbiol.* **2006**, *106*, 137–143. [CrossRef]
69. Greppi, A.; Hemery, Y.; Berrazaga, I.; Almaksour, Z.; Humblot, C. Ability of lactobacilli isolated from traditional cereal-based fermented food to produce folate in culture media under different growth conditions. *LWT* **2017**, *86*, 277–284. [CrossRef]
70. Khalili, M.; Rad, A.; Khosroushahi, A.; Khosravi, H.; Jafarzadeh, S. Application of probiotics in folate bio-fortification of yoghurt. *Probiotics Antimicrob.* **2020**, *12*, 756–763. [CrossRef]
71. Averianova, L.A.; Balabanova, L.A.; Son, O.M.; Podvolotskaya, A.B.; Tekutyeva, L.A. Production of vitamin B2 (Riboflavin) by microorganisms: An overview. *Front. Bioeng. Biotechnol.* **2020**, *8*, 570828. [CrossRef] [PubMed]
72. Melini, F.; Melini, V.; Luziatelli, F.; Ficca, A.G.; Ruzzi, M. Health-promoting components in fermented foods: An up-to-Date systematic review. *Nutrients* **2019**, *11*, 1189. [CrossRef] [PubMed]
73. Sanalibaba, P.; Cakmak, G.A. Exopolysaccharides production by lactic acid bacteria. *Appl. Microbiol.* **2016**, *2*, 1000115. [CrossRef]
74. Singh, P.; Saini, P. Food and health potentials of Exopolysaccharides derived from lactobacilli. *Microbiol. Res. J. Int.* **2017**, *22*, 1–14. [CrossRef]
75. Liu, C.; Lu, J.; Lu, L.; Liu, Y.; Wang, F.; Xiao, M. Isolation, structural characterization and immunological activity of an exopolysaccharide produced by *Bacillus licheniformis* 8-37-0-1. *Biores. Technol.* **2010**, *101*, 5528–5533. [CrossRef]
76. Flemming, H. EPS—Then and now. *Microorganisms* **2016**, *4*, 41. [CrossRef]
77. Majee, S.; Avlani, D.; Gopa, R.; Biswas, G. Rheological behaviour and pharmaceutical applications of bacterial exopolysaccharides. *J. App. Pharm. Sci.* **2017**, *7*, 224–232.
78. Angelin, J.; Kavitha, M. Exopolysaccharides from probiotic bacteria and their health potential. *Int. J. Biol. Macromol.* **2020**, *162*, 853–865. [CrossRef]
79. Baruah, R.; Das, D. Heteropolysaccharides from lactic acid bacteria: Current trends and applications. *J. Prob. Health* **2016**, *4*, 141. [CrossRef]
80. Korcz, E.; Varga, L. Exopolysaccharides from lactic acid bacteria: Techno-functional application in the food industry. *Trends Food Sci. Technol.* **2021**, *110*, 375–384. [CrossRef]

81. Prete, R.; Alam, M.K.; Perpetuini, G.; Perla, C.; Pittia, P.; Corsetti, A. Lactic acid bacteria Exopolysaccharides producers: A sustainable tool for functional foods. *Foods* **2021**, *10*, 1653. [CrossRef] [PubMed]
82. De Vuyst, L. Heteropolysaccharides from lactic acid bacteria. *FEMS Microbiol. Rev.* **1999**, *23*, 153–177. [CrossRef] [PubMed]
83. Han, X.; Yang, Z.; Jing, X.; Yu, P.; Zhang, Y.; Yi, H.; Zhang, L. Improvement of the texture of yogurt by use of Exopolysaccharide producing lactic acid bacteria. *BioMed Res. Int.* **2016**, *2016*, 7945675. [CrossRef]
84. Nehal, F.; Sahnoun, M.; Smaoui, S.; Jaouadi, B.; Bejar, S.; Mohammed, S. Characterization, high production and antimicrobial activity of exopolysaccharides from *Lactococcus lactis* F-mou. *Microb. Pathog.* **2019**, *132*, 10–19. [CrossRef]
85. Alizadeh Behbahani, B.; Jooyandeh, H.; Falah, F.; Vasiee, A. Gamma-aminobutyric acid production by *Lactobacillus brevis* A3: Optimization of production, antioxidant potential, cell toxicity, and antimicrobial activity. *Food Sci. Nutr.* **2020**, *8*, 5330–5339. [CrossRef]
86. Cui, Y.; Miao, K.; Niyaphorn, S.; Qu, X. Production of gamma-aminobutyric acid from lactic acid bacteria: A systematic review. *Int. J. Mol. Sci.* **2020**, *21*, 995. [CrossRef]
87. Lyu, C.; Liu, L.; Huang, J.; Zhao, W.; Hu, S.; Mei, L.; Yao, S. Biosynthesis of γ-aminobutyrate by engineered lactobacillus brevis cells immobilized in gellan gum gel beads. *J. Biosci. Bioeng.* **2019**, *128*, 123–128. [CrossRef]
88. Sahab, N.R.; Subroto, E.; Balia, R.L.; Utama, G.L. γ-aminobutyric acid found in fermented foods and beverages: Current trends. *Heliyon* **2020**, *6*, e05526. [CrossRef] [PubMed]
89. Yunes, R.; Poluektova, E.; Dyachkova, M.; Klimina, K.; Kovtun, A.; Averina, O.; Orlova, V.; Danilenko, V. GABA production and structure of gadB/gadC genes in *Lactobacillus* and *Bifidobacterium* strains from human microbiota. *Anaerobe* **2016**, *42*, 197–204. [CrossRef]
90. Coolbear, T.; Weimer, B.; Wilkinson, M.G. Lactic acid bacteria | lactic acid bacteria in flavor development. Encyclopedia. *Dairy Sci.* **2011**, *73*, 160–165.
91. Smit, G.; Smit, B.A.; Engels, W.J. Flavour formation by lactic acid bacteria and biochemical flavour profiling of cheese products. *FEMS Microbiol. Rev.* **2005**, *29*, 591–610. [CrossRef]
92. Misselwitz, B.; Butter, M.; Verbeke, K.; Fox, M.R. Update on lactose malabsorption and intolerance: Pathogenesis, diagnosis and clinical management. *Gut* **2019**, *68*, 2080–2091. [CrossRef]
93. Cano-Contreras, A.D.; Minero Alfaro, I.J.; Medina López, V.M.; Amieva Balmori, M.; Remes Troche, J.M.; Espadaler Mazo, J.; Perez Lopez, N. Efficacy of i3.1 probiotic on improvement of lactose intolerance symptoms. *J. Clin. Gastroenterol.* **2020**, *56*, 141–147. [CrossRef] [PubMed]
94. Sen, M. Role of probiotics in health and disease—A review. *Int. J. Adv. Life Sci. Res.* **2019**, *2*, 1–11.
95. Gyawali, R.; Oyeniran, A.; Zimmerman, T.; Aljaloud, S.O.; Krastanov, A.; Ibrahim, S.A. A comparative study of extraction techniques for maximum recovery of β-galactosidase from the yogurt bacterium *Lactobacillus delbrueckii* ssp. bulgaricus. *J. Dairy Res.* **2020**, *87*, 123–126. [CrossRef] [PubMed]
96. Pakdaman, M.N.; Udani, J.K.; Molina, J.P.; Shahani, M. The effects of the DDS-1 strain of lactobacillus on symptomatic relief for lactose intolerance-a randomized, double-blind, placebo-controlled, crossover clinical trial. *Nutr. J.* **2015**, *15*, 56. [CrossRef]
97. Roškar, I.; Švigelj, K.; Štempelj, M.; Volfand, J.; Štabuc, B.; Malovrh, Š.; Rogelj, I. Effects of a probiotic product containing *Bifidobacterium animalis* subsp. *animalis* IM386 and *Lactobacillus plantarum* MP2026 in lactose intolerant individuals: Randomized, placebo-controlled clinical trial. *J. Funct. Foods* **2017**, *35*, 1–8. [CrossRef]
98. Ahn, S.I.; Kim, M.S.; Park, D.G.; Han, B.K.; Kim, Y.J. Effects of probiotics administration on lactose intolerance in adulthood: A meta-analysis. *J. Dairy Sci.* **2023**, *106*, 4489–4501. [CrossRef]
99. Zhou, P.; Bian, W.; Li, Y.; Chang, L.; Wang, J.; Feng, X. Prevalence of malnutrition based on GLIM criteria for completeness of diagnosis and future risk of malnutrition based on current malnutrition diagnosis: Systematic review and meta-analysis. *Front. Nutr.* **2023**, *10*, 1174945.
100. Wang, J.; Ke, H.; Liu, K.X.; Qu, J.M. Effects of exogenous probiotics on the gut microbiota and clinical outcomes in critically ill patients: A randomized controlled trial. *Ann. Palliat. Med.* **2021**, *10*, 1180–1190. [CrossRef]
101. Ranjha, M.M.A.N.; Shafique, B.; Batool, M.; Kowalczewski, P.L.; Shehzad, Q.; Usman, M.; Manzoor, M.F.; Zahra, S.M.; Yacub, A.; Aadil, R.M. Nutritional and health potential of probiotics: A review. *Appl. Sci.* **2021**, *11*, 11204. [CrossRef]
102. Sanjukta, M.; Swastik, A. A brief overview on probiotics: The health friendly microbes. *Biomed Pharmacol. J.* **2021**, *14*, 1869–1880.
103. Liu, X.; Zhang, Y.; Chu, J. Effect of probiotics on the nutritional status of severe stroke patients with nasal feeding that receive enteral nutrition: A protocol for systematic review and meta-analysis of randomized controlled trials. *Medicine* **2021**, *100*, e25657. [CrossRef] [PubMed]
104. Skrzydło-Radomańska, B.; Prozorow-Król, B.; Cichoż-Lach, H.; Majsiak, E.; Bierła, J.B.; Kosikowski, W.; Szczerbinski, M.; Gantzel, J.; Cukrowska, B. The effectiveness of synbiotic preparation containing lactobacillus and bifidobacterium probiotic strains and short chain fructooligosaccharides in patients with diarrhea predominant irritable bowel syndrome—A randomized double-blind, placebo-controlled study. *Nutrients* **2020**, *12*, 1999.
105. de Castro Soares, G.G.; Marinho, C.H.; Pitol, R.; Andretta, C.; Oliveira, E.; Martins, C.; Riella, M.C. Sporulated Bacillus as alternative treatment for diarrhea of hospitalized adult patients under enteral nutrition: A pilot randomized controlled study. *Clin. Nutr. ESPEN* **2017**, *22*, 13–18. [CrossRef]

106. Lee, Z.Y.; Lew, C.C.H.; Ortiz-Reyes, A.; Patel, J.J.; Wong, Y.J.; Loh, C.T.I.; MArtindale, R.G.; Heyland, D.K. Benefits and harm of probiotics and synbiotics in adult critically ill patients. A systematic review and meta-analysis of randomized controlled trials with trial sequential analysis. *Clin. Nutr.* **2023**, *42*, 519–531. [CrossRef] [PubMed]
107. Azad, M.; Kalam, A.; Sarker, M.; Wan, D. Immunomodulatory effects of probiotics on cytokine profiles. *BioMed Res. Int.* **2018**, *2018*, 8063647. [CrossRef]
108. Wei, F.; Zhou, L.; Wang, Q.; Zheng, G.; Su, S. Effect of Compound Lactic Acid Bacteria Capsules on the Small Intestinal Bacterial Overgrowth in Patients with Depression and Diabetes: A Blinded Randomized Controlled Clinical Trial. *Dis. Markers* **2022**, *2022*, 6721695. [CrossRef]
109. Plaza-Diaz, J.; Ruiz-Ojeda, F.J.; Gil-Campos, M.; Gil, A. Mechanisms of action of probiotics. *Adv. Nutr.* **2019**, *10* (Suppl. 1), S49–S66. [CrossRef]
110. Cazorla, S.I.; Maldonado-Galdeano, C.; Weill, R.; De Paula, J.; Perdigón, G.D. Oral administration of probiotics increases paneth cells and intestinal antimicrobial activity. *Front. Microbiol.* **2018**, *9*, 736. [CrossRef]
111. Zmora, N.; Suez, J.; Elinav, E. You are what you eat: Diet, health and the gut microbiota. *Nat. Rev. Gastroenterol. Hepatol.* **2019**, *16*, 35–56. [CrossRef]
112. Batra, P.; Soni, K.D.; Mathur, P. Efficacy of probiotics in the prevention of VAP in critically ill ICU patients: An updated systematic review and meta-analysis of randomized control trials. *J. Intensive Care* **2020**, *8*, 81. [CrossRef] [PubMed]
113. Mahmoodpoor, A.; Hamishehkar, H.; Asghari, R.; Abri, R.; Shadvar, K.; Sanaie, S. Effect of a probiotic preparation on ventilator-associated pneumonia in critically ill patients admitted to the intensive care unit: A prospective double-blind randomized controlled trial. *Nutr. Clin. Pract.* **2019**, *34*, 156–162. [CrossRef]
114. Li, H.; Xie, Z.; Zhang, Y.; Liu, Y.; Niu, A.; Liu, Y.; Zhang, L.; Guan, L. Rosa rugosa polysaccharide attenuates alcoholic liver disease in mice through the gut-liver axis. *Food Biosci.* **2021**, *44*, 101385. [CrossRef]
115. BAPEN. BAPEN Publishes Results of Biggest Malnutrition Survey Ever Undertaken (Scotland). Available online: https://www.bapen.org.uk/media-centre/press-releases/377-bapen-publishes-results-of-biggest-malnutrition-survey-ever-undertaken-scotland (accessed on 14 December 2020).
116. Zaher, S. Nutrition and the gut microbiome during critical illness: A new insight of nutritional therapy. *Saudi J. Gastroenterol. Off. J. Saudi Gastroenterol. Assoc.* **2020**, *26*, 290. [CrossRef] [PubMed]
117. Kamel, N.A.; Soliman, M.M.; Abo-Zeid, M.A.; Shaaban, M.I. Effect of AntiInflammatory and Antimicrobial Cosupplementations on Sepsis Prevention in Critically Ill Trauma Patients at High Risk for Sepsis. *Front. Pharmacol.* **2021**, *12*, 792741. [CrossRef]
118. Shimizu, K.; Ogura, H.; Kabata, D.; Shintani, A.; Tasaki, O.; Ojima, M.; Ikeda, M.; Shimazu, T. Association of prophylactic synbiotics with reduction in diarrhea and pneumonia in mechanically ventilated critically ill patients: A propensity score analysis. *J. Infect. Chemother.* **2018**, *24*, 795–801. [CrossRef]
119. Sun, X.; Shi, J.; Kong, L.; Shen, Q.; Zeng, X.; Wu, Z.; Guo, Y.; Pan, D. Recent insights into the hepatoprotective effects of lactic acid bacteria in alcoholic liver disease. *Trends Food Sci. Technol.* **2022**, *125*, 91–99. [CrossRef]
120. Bajaj, J.S. Alcohol, liver disease and the gut microbiota. *Nat. Rev. Gastroenterol. Hepatol.* **2019**, *16*, 235–246. [CrossRef]
121. Bakhshimoghaddam, F.; Shateri, K.; Sina, M.; Hashemian, M.; Alizadeh, M. Daily consumption of synbiotic yogurt decreases liver steatosis in patients with nonalcoholic fatty liver disease: A randomized controlled clinical trial. *J. Nutr.* **2018**, *148*, 1276–1284. [CrossRef]
122. Mohamad Nor, M.H.; Ayob, N.; Mokhtar, N.M.; Raja Ali, R.A.; Tan, G.C.; Wong, Z.; Shafiee, N.H.; Wong, Y.P.; Mustangin, M.; Nawawi, K.N.M. The effect of probiotics (MCP® BCMC® strains) on hepatic steatosis, small intestinal mucosal immune function, and intestinal barrier in patients with non-alcoholic fatty liver disease. *Nutrients* **2021**, *13*, 3192. [CrossRef]
123. Caggianiello, G.; Kleerebezem, M.; Spano, G. Exopolysaccharides produced by lactic acid bacteria: From health-promoting benefits to stress tolerance mechanisms. *Appl. Microbiol. Biotechnol.* **2016**, *100*, 3877–3886. [CrossRef] [PubMed]
124. Górska, A.; Przystupski, D.; Niemczura, M.J.; Kulbacka, J. Probiotic bacteria: A promising tool in cancer prevention and therapy. *Curr. Microbiol.* **2019**, *76*, 939–949. [CrossRef]
125. Zhao, R.; Wang, Y.; Huang, Y.; Cui, Y.; Xia, L.; Rao, Z.; Rao, Z.; ZHou, Y.; Wu, X. Effects of fiber and probiotics on diarrhea associated with enteral nutrition in gastric cancer patients: A prospective randomized and controlled trial. *Medicine* **2017**, *96*, e8418. [CrossRef]
126. Kaźmierczak-Siedlecka, K.; Skonieczna-Żydecka, K.; Hupp, T.; Duchnowska, R.; Marek-Trzonkowska, N.; Połom, K. Next-generation probiotics–do they open new therapeutic strategies for cancer patients? *Gut Microbes* **2022**, *14*, 2035659. [CrossRef] [PubMed]
127. Zhang, C.; Wang, C.; Li, S.; Yu, L.; Tian, F.; Zhao, J.; Zhang, H.; Chen, W.; Zhai, Q. Meta-analysis of randomized controlled trials of the effects of probiotics on type 2 diabetes in adults. *Clin. Nutr.* **2022**, *41*, 365–373. [CrossRef] [PubMed]
128. Khalili, L.; Alipour, B.; Asghari Jafarabadi, M.; Hassanalilou, T.; Mesgari Abbasi, M.; Faraji, I. Probiotic assisted weight management as a main factor for glycemic control in patients with type 2 diabetes: A randomized controlled trial. *Diabetol. Metab. Syndr.* **2019**, *11*, 1–9. [CrossRef]
129. Mirjalili, M.; Sharif, A.S.; Sangouni, A.A.; Emtiazi, H.; Mozaffari-Khosravi, H. Effect of probiotic yogurt consumption on glycemic control and lipid profile in patients with type 2 diabetes mellitus: A randomized controlled trial. *Clin. Nutr. ESPEN* **2023**, *54*, 144–149. [CrossRef] [PubMed]

130. Nazari, B.; Amani, L.; Ghaderi, L.; Khanbabayi Gol, M. Effects of probiotics on prevalence of ventilator-associated pneumonia in multitrauma patients hospitalized in neurosurgical intensive care unit: A randomized clinical trial. *Trauma Mon.* **2020**, *25*, 262–268.
131. Anandaraj, A.M.; Pichamuthu, K.K.; Hansdak, S.G.; Samuel, P.; Irodi, A.; Valsa, S.; Peter, J.V. A Randomised Controlled Trial of Lactobacillus in the Prevention of Ventilator Associated Pneumonia. *J. Clin. Diagn. Res.* **2019**, *13*, 1–4. [CrossRef]
132. Johnstone, J.; Meade, M.; Lauzier, F.; Marshall, J.; Duan, E.; Dionne, J.; Arabi, Y.; Heels-Ansdell, D.; Thabane, L.; Lamarche, D.; et al. Effect of probiotics on incident ventilator-associated pneumonia in critically ill patients: A randomized clinical trial. *JAMA* **2021**, *326*, 1024–1033. [CrossRef] [PubMed]
133. Katkowska, M.; Garbacz, K.; Kusiak, A. Probiotics: Should all patients take them? *Microorganisms* **2021**, *9*, 2620. [CrossRef] [PubMed]
134. Purdel, C.; Ungurianu, A.; Adam-Dima, I.; Margină, D. Exploring the potential impact of probiotic use on drug metabolism and efficacy. *Biomed. Pharmacother.* **2023**, *161*, 114468. [CrossRef]
135. Tegegne, B.A.; Kebede, B. Probiotics, their prophylactic and therapeutic applications in human health development: A review of the literature. *Heliyon* **2022**, *8*, e09725. [CrossRef]
136. Merenstein, D.; Pot, B.; Leyer, G.; Ouwehand, A.C.; Preidis, G.A.; Elkins, C.A.; Hill, C.; Lewis, Z.T.; Shane, A.L.; Zmora, N.; et al. Emerging issues in probiotic safety: 2023 perspectives. *Gut Microbes* **2023**, *15*, 2185034. [CrossRef]

MDPI AG
Grosspeteranlage 5
4052 Basel
Switzerland
Tel.: +41 61 683 77 34

Foods Editorial Office
E-mail: foods@mdpi.com
www.mdpi.com/journal/foods

www.ingramcontent.com/pod-product-compliance
Lightning Source LLC
Chambersburg PA
CBHW061514210926
51850CB00008B/1910
9783725815739